Emerging Communication Technologies Based on Wireless Sensor Networks

Current Research and Future Applications

Emerging Communication Technologies Based on Wireless Sensor Networks

Current Research and Future Applications

Edited by
Mubashir Husain Rehmani • Al-Sakib Khan Pathan

CRC Press
Taylor & Francis Group
Boca Raton   London   New York

CRC Press is an imprint of the
Taylor & Francis Group, an **informa** business

CRC Press
Taylor & Francis Group
6000 Broken Sound Parkway NW, Suite 300
Boca Raton, FL 33487-2742

© 2016 by Taylor & Francis Group, LLC
CRC Press is an imprint of Taylor & Francis Group, an Informa business

No claim to original U.S. Government works

Printed on acid-free paper
Version Date: 20160211

International Standard Book Number-13: 978-1-4987-2485-2 (Hardback)

Visit the Taylor & Francis Web site at
http://www.taylorandfrancis.com

and the CRC Press Web site at
http://www.crcpress.com

Contents

SECTION IV WSNs PROTOCOLS AND ALGORITHMS

Preface

Overview

With the advancement of technology and miniaturization of electronic devices, applications of wireless sensor networks (WSNs) can be seen in diverse areas in our lives. In fact, these WSNs have gained a lot of attention from both the research community and industry, making them easily available on the market. An extensive amount of research, easy availability, and cheap cost make them useful in various types of futuristic applications as well. Given today's trend, WSNs are increasingly becoming an essential component for future communication technologies. For instance, whenever we talk about the Internet of Things (IoT), smart cities, or cyber physical systems (CPS), we can find the role of WSNs in these technologies. However, with these diverse applications and underlying communication architectures, new research challenges have also arisen.

This book is all about wireless sensor networks. The main objective is to present different types of emerging communication technologies based on WSNs. It also describes how wireless sensor networks can be integrated with other communication technologies. Despite the availability of several previously published books on WSNs, there is a clear need for a book that contains critical information about WSNs and their use in emerging communication technologies. Additionally, many of the new techniques, including cognitive radio sensor networks, wireless nanosensor networks, and modern applications, should be covered as well. In fact, WSNs have very wide applications, now ranging from wireless body area networks to the use of wireless sensor networks in the Internet of Things. Recently, we have seen other WSN-based emerging applications in smart homes, smart cities, and satellite communications as well.

Organization of the Book

The book is composed of fourteen selected chapters, divided into four sections.

The first section contains two chapters that focus on an overview of wireless sensor networks. In Chapter 1, the authors present background on the fundamentals of such networks, basic units, and all the relevant information. Then, the principal working method of WSNs is discussed. The authors then discuss different types of WSNs. After that, the characteristics of different types of emerging technologies based on WSNs are discussed.

In Chapter 2, renewable energy sources, battery replenishment strategies, and application-specific energy challenges of WSNs are reviewed. The chapter starts with a brief discussion on the concepts of power consumption and saving modes. Then different types of energy harvesting techniques are presented. The authors also introduce wireless power transfer and its relevant techniques. Finally, application-specific problems of WSNs are identified.

The second section of the book contains three chapters. This section is dedicated to the issues related to wireless body area networks. Chapter 3 discusses wearable wireless sensor networks and their applications, standards, and research trends. In Chapter 4, the authors discuss routing schemes devised for wireless body area networks. Finally, Chapter 5 discusses thermal-aware routing protocols for wireless body area networks.

In the third section, we grouped the chapters that focus on different emerging communication technologies based on WSNs. Among the five chapters in this section, Chapter 6 is about electromagnetic wireless nanosensor networks. Chapter 7 discusses the use of wireless sensor networks in the IoT, while Chapter 8 focuses on the management of wireless sensor networks through satellite networks. Use of wireless sensor networks in smart homes is discussed in Chapter 9. Finally, use of cognitive radio technology in conjunction with wireless sensor networks is covered in Chapter 10.

Four chapters are grouped within the fourth and the last section of this book. The topics generally related to typical wireless sensor networks are covered here. Chapter 11 is about energy-efficient data collection in wireless sensor networks. Chapter 12 is about key distribution mechanisms in wireless sensor networks. Chapter 13 discusses distributed data gathering algorithms for mobile wireless sensor networks, and finally, Chapter 14 presents a novel mobility scheme for WSNs that supports IPv6.

Target Audience of This Book

This book can be used as a textbook for graduate students taking courses related to wireless sensor networks and advanced topics in wireless networks. It offers knowledge about wireless sensor networks and their applications with the aim of imparting this critical and usable knowledge to the readers.

List of Key Terms

6LoWPAN
Cloud computing
Cognitive radio
Cognitive radio sensor networks (CRSNs)
Collaborative beamforming
Cyber physical systems
Delay-tolerant networks
Direct-access architecture
Full function device
Gateway-based access architecture
Internet of Things (IoT)
IP-WSN
IP-WSN mobility
M2M systems
MIPv6
PMIPv6
Reduced function device
Satellite-based WSNs
Sensor nodes
Smart grid
SPMIPv6
Wireless sensor networks (WSNs)
ZigBee

Acknowledgments

We would like to express sincere thanks to Allah Subhanahu Wa-ta'ala (the Almighty, be he glorified in the highest), that by his grace and bounty, we were able to work on this book. Last but not least, we would like to acknowledge the continuous support of our families as well.

Mubashir Husain Rehmani
COMSATS Institute of Information Technology, Wah Cantt., Pakistan

Al-Sakib Khan Pathan
Islamic University in Madinah, Kingdom of Saudi Arabia

Editors

 **Mubashir Husain Rehmani** is currently an assistant professor in the Department of Electrical Engineering, COMSATS Institute of Information Technology (CIIT), Wah Cantt., Pakistan. He earned his PhD degree under the supervision of Dr. Aline Carneiro Viana and Prof. Serge Fdida from LIP6, Université of Pierre et Marie Curie (Sorbonne Universités), Paris, France, in 2011; MS degree in networks and telecommunications from L2S, Laboratory of Signals and Systems, Supelec, and University of Paris-Sud 11, Paris, France, in 2008; and BEng degree in computer systems engineering from Mehran University of Engineering and Technology, Jamshoro, Pakistan, in 2004. He was a postdoctoral fellow at the LIGM, Université Paris Est, Marne La Vallée, France, in 2012. In May 2013, the Higher Education Commission (HEC), government of Pakistan, selected him to be an HEC-approved supervisor.

He is currently an editor of *IEEE Communications Surveys and Tutorials* and associate editor of *IEEE Communications Magazine, Access, Computers, and Electrical Engineering* (*CAEE*), the *Journal of Network and Computer Applications* (*JNCA*), *Ad Hoc Sensor Wireless Networks* (*AHSWN*), *Wireless Networks*, and the *Journal of Communications and Networks* (*JCN*). He has also served as a guest editor for *Ad Hoc Networks, Future Generation Computer Systems Journal, Access, Pervasive and Mobile Computing,* and *Computers and Electrical Engineering.* Dr. Rehmani served on the Technical Program Committees (TPCs) for the IEEE VTC Spring 2016, IEEE ICC 2015, IEEE WoWMoM 2014, IEEE ICC 2014, ACM CoNEXT Student Workshop 2013, IEEE ICC 2013, and IEEE IWCMC 2013 conferences. He is a member of the Institute of Electrical and Electronics Engineers (IEEE) and IEEE Communications Society.

Dr. Rehmani is the author of *Cognitive Radio Sensor Networks: Applications, Architectures, and Challenges* (IGI Global, 2014). He has published 15 journal papers, 14 book chapters, and more than 8 international conference papers. He is the founding member of the IEEE Special Interest Group (SIG) on Green and Sustainable Networking and Computing with Cognition and Cooperation.

He earned an MS leading to a PhD scholarship from the Higher Education Commission of Pakistan. He was awarded CIIT's Research Productivity Award in 2014. In 2013 and 2014, he earned an appreciation letter from the director of CIIT for his dedicated services. His main research interests include cognitive radio ad hoc networks, the smart grid, the cognitive radio–based smart grid, wireless energy transfer, flying ad hoc networks, wireless sensor networks, and mobile ad hoc networks.

 **Al-Sakib Khan Pathan** earned his PhD degree (MS leading to PhD) in computer engineering in 2009 from Kyung Hee University, Seoul, South Korea. He earned his BSc degree in computer science and information technology from the Islamic University of Technology (IUT), Dhaka, Bangladesh, in 2003. He is currently relocating to his new position as an assistant professor in the Faculty of Computer and Information Systems at Islamic University in Madinah, Madinah al-Munawwarah, Kingdom of Saudi Arabia. From August 2010 to July 2015, he served as an assistant professor in the Computer Science Department at the International Islamic University Malaysia (IIUM), Selangor, Malaysia. Before that, until June 2010, he was an assistant professor in the Computer Science and Engineering Department at BRAC University, Dhaka, Bangladesh, and also worked as a researcher at Networking Lab, Kyung Hee University, from September 2005 to August 2009. His research interests include wireless sensor networks, network security, and e-services technologies. Currently, he is working on some multidisciplinary issues.

He is a recipient of several awards and best paper awards and has more than 170 publications in these areas. He has served as a chair, organizing committee member, and Technical Program Committee (TPC) member of numerous international conferences and workshops, including GLOBECOM, ICC, GreenCom, AINA, WCNC, HPCS, ICA3PP, IWCMC, VTC, and HPCC. He was awarded the IEEE Outstanding Leadership Award and Certificate of Appreciation for his role in the IEEE GreenCom'13 conference. He is currently serving in various editorial positions, including as an associate technical editor of *IEEE Communications Magazine,* editor of *Ad Hoc and Sensor Wireless Networks* and the *International Journal of Sensor Networks,* area editor of the *International Journal of Communication Networks and Information Security,* associate editor of the *International Journal of Computational Science and Engineering,* and guest editor of many special issues of top-ranked journals. He has also edited or authored 14 published books. One of his books has been included twice in Intel Corporation's Recommended Reading List for Developers, in the second half of 2013 and first half of 2014; three books were included in IEEE Communications Society's (IEEE ComSoc) Best Readings in Communications and Information Systems Security, 2013; two other books were indexed with all the chapters in Elsevier's acclaimed abstract and citation database, Scopus, in February 2015; and a seventh book was translated to the simplified Chinese language from the English version. Also, two of his journal papers and one conference paper were included under different categories in IEEE ComSoc's Best Readings Topics on Communications and Information Systems Security, 2013. He also serves as a referee of numerous renowned journals. He has earned awards for his reviewing activities, such as one of the most active reviewers of the *International Arab Journal of Information Technology* (IAJIT), two times, in 2012 and 2014, and recognized reviewer status of *Computers and Electrical Engineering* (March 2014) and *Ad Hoc Networks* (April 2014).

As part of his academic duties, he has supervised two PhD students to completion. He is a senior member of the Institute of Electrical and Electronics Engineers (IEEE) and several IEEE technical committees.

Contributors

 **Ubaid Abbasi** earned his PhD from CNRS-LaBRI, University of Bordeaux 1, Bordeaux, France. His main research interests are video streaming in peer-to-peer networks, overlay organization, and the Internet of Things.

 **Adesoji A. Adesina** is a chemical engineer with more than 30 years of experience in industry and academia. He earned a PhD (1986) from the University of Waterloo, Waterloo, Canada, and has published extensively in catalysis and reaction engineering, with a focus on energy and environmental technologies. He is presently the founder and CEO of Atodatech LLC, a private engineering consulting company in Pasadena, California, and for many years, he was a professor of chemical engineering at the University of New South Wales, Sydney, Australia, where he established a first-class research facility in catalytic and multiphase reaction engineering. He is a chartered engineer and fellow of the Institution of Chemical Engineers–UK.

 **Ozgur Baris Akan** earned his PhD degree in electrical and computer engineering from the Broadband and Wireless Networking Laboratory, School of Electrical and Computer Engineering, Georgia Institute of Technology, in 2004. He is currently a full professor with the Department of Electrical and Electronics Engineering, Koc University, Istanbul, Turkey, the director of the Next-Generation and Wireless Communications Laboratory, and an IEEE fellow (class of 2016). His current research interests include wireless, nanoscale, and molecular communications, and information theory. He is an associate editor of the *IEEE Transactions on Communications*, the *IEEE Transactions on Vehicular Technology*, the *International Journal of Communication Systems* (Wiley), the *Nano Communication Networks Journal* (Elsevier), and the *European Transactions on Technology*.

 **Fayaz Akhtar** recently completed his MS degree in electrical engineering from COMSATS Institute of Information Technology, Pakistan. He is currently in the process of enrolling in a PhD program. His research interests include the Internet of Things, cognitive radio networks, energy harvesting for MEMs, and D2D communications.

Muhammad Mahtab Alam has been a research scientist since January 2014 at the Qatar Mobility Innovations Center, Doha. He is working in the National Priority Research Program of the Qatar Foundation on wearable WSNs for rescue and critical applications. In 2013, he was an assistant professor in the Swedish College of Engineering and Technology, Rahimyar Khan, Pakistan. From 2009 to 2012, he worked on his PhD at the INRIA Research Center and IRISA Laboratory, Rennes, France. From 2007 to 2009, he was a research engineer at the Center for Software Defined Radio (CSDR) at Aalborg University, Aalborg, Denmark. During the past 8 years, he has contributed to a number of European projects, such as GEODES, GRECO, and MS-SDR. He is a member of the Institute of Electrical and Electronics Engineers (IEEE) and the Institution of Engineering and Technology (IET)–UK. His research interests are in the fields of self-organized and self-adaptive wireless sensor and body area networks, in particular, energy-efficient communication protocols and algorithms, radio-link and mobility modeling, digital signal processing, algorithm architecture optimizations, and software-defined radios.

Zuneera Aziz is a telecommunication engineer and is currently a lecturer in the Department of Telecommunication Engineering, Mehran University of Engineering and Technology (MUET), Jamshoro, Pakistan. She earned her master's in communication systems and networks from MUET. Aziz had the privilege of being an exchange student at the University of Malaga, Spain, during the final semester of her bachelor studies, under the banner of the Erasmus Mundus Mobility of Life Program. She is currently performing research in the domain of wireless sensor networks (WSNs) and specifically on wireless body area networks (WBANs). She also has expertise in the field of microwave engineering. Her research interests include ensuring reliability and security in WSNs and WBANs. She has several research papers and a journal paper to her credit. She has been a reviewer and subreviewer in several conferences, including IMTIC '15, ICCCN '14, and IMTIC '13.

Rakesh Kumar Bansal earned his bachelor's and master's of engineering from Thapar University (then Thapar Institute of Engineering and Technology—Deemed University), Patiala, India. Later, he earned his PhD (engineering) from Punjabi University, Punjab, India. He started his professional career in technical education with the now Thapar University in 1986 and is presently professor at Giani Zail Singh—Punjab Technical University, Bathinda (Punjab), India. His areas of interest include wireless sensor networks, fault-tolerant real-time scheduling in parallel systems, and instrumentation applications. He has 45 publications at the international level.

Savina Bansal earned her bachelor of science and later bachelor of engineering from Punjab University, Chandigarh, India. She completed her master's at Thapar University (then Thapar Institute of Engineering and Technology—Deemed University), Punjab, India, and PhD at the Indian Institute of Technology Roorkee, Roorkee, India, in parallel and distributed computing. She started her professional career with the Punjab state government enterprise PUNWIRE in 1988 and later shifted to technical education at (now) Thapar University, Patiala in Punjab, India, and is currently professor at Giani Zail Singh—Punjab Technical University, Bathinda (Punjab). She specializes in wireless sensor networks, energy efficiency and multiprocessor scheduling, fault-tolerant real-time computing, grid computing, and irregular interconnection networks. She has 65 publications at the international level. She is a fellow of the Institution of Engineers (India) and Institution of Electronics and Telecommunication Engineers and a lifetime member of the Computer Society of India (CSI) and Advanced Computing & Communication Society (ACCS) Indian Society for Technical Education (ISTE).

Nafeesa Bohra earned her MEng from Pakistan and gained experience of more than 5 years as a research associate at the University of Passau (UP), Passau, Germany. At UP, she was involved in the EU projects RESUMENet, AUTO I, and LEO-MESH Nets. She is currently working as an assistant professor in the Department of Telecommunication Engineering, Mehran University of Engineering and Technology, Jamshoro, Pakistan. She has served as a reviewer and Technical Program Committee (TPC) member for national and international conferences. She has published more than 20 scientific articles in international refereed journals and proceedings of international conferences and workshops. Her research interests include distributed network monitoring for wired and wireless networks, wireless body area networks, and peer-to-peer networks.

Oktay Cetinkaya earned his BSc degrees in electrical engineering and electronics and communication engineering from Yildiz Technical University, Istanbul, Turkey, in 2013 and 2014, respectively. He is currently a research assistant at Next-Generation and Wireless Communications Laboratory and is pursuing his PhD degree at Koc University. His current research interests include energy efficiency and harvesting operations in smart environments with advanced management, monitoring, and controlling systems.

Chun Tung Chou is an associate professor at the School of Computer Science and Engineering, University of New South Wales, Sydney, Australia. He earned his BA in engineering science from the University of Oxford, Oxford, UK, and his PhD in control engineering from the University of Cambridge, Cambridge, UK. He has published more than 150 articles on various topics, including systems and control, wireless networks, and communications. His current research interests are molecular communication, nanoscale communication, compressive sensing, and embedded networks.

Socrates Costicoglou earned a degree in computer engineering and informatics from the University of Patras, Patras, Greece, and an MSc degree in computer engineering from Syracuse University, Syracuse, New York. He joined Space Hellas, Athens, Greece, in 1998, and since then, he has actively participated in and coordinated several R&D projects related to satellite technologies and applications. He is currently the director of the IT, Applications, and R&D Division of Space Hellas. He is a member of the executive board of the Hellenic Space Technologies and Applications Cluster (si-Cluster) in Greece, as well as being a member of the Technical Chamber of Greece since 1991.

Emad Felemban earned his master's and PhD from The Ohio State University, Columbus, and earned his BSc from the King Fahd University of Petroleum and Minerals (KFUPM) in Dhahran, Saudi Arabia, with first honors. Currently, he is working as an assistant professor at the College of Computer and Information System in Umm Al-Qura University, Mecca, Saudi Arabia. He specializes in wireless networks and, more specifically, wireless sensor networks. His research interests include the algorithms and architectural design of sensor networks, performance modeling of wireless systems, cyber physical systems, and smart city environments.

Georgios Gardikis earned his diploma (2000) and PhD (2004) in electrical and computer engineering from the National Technical University of Athens, Athens, Greece. His research interests are in the fields of data networking and multimedia delivery systems, specializing in network management, monitoring, and virtualization technologies. He currently works as an R&D project manager at Space Hellas, Athens, Greece, supervising a number of national and European research projects. He is a certified Project Management Professional (PMP), senior member of the Institute of Electrical and Electronics Engineers (IEEE) and Communications and Broadcast Technology Societies, member of the Project Management Institute, and member of the Technical Chamber of Greece since 2001. He is the author or coauthor of more than 50 publications in international journals and refereed conferences and workshops.

Elyes Ben Hamida is currently a machine-to-machine and Internet-of-Things product manager and R&D expert at the Qatar Mobility Innovations Center (QMIC) in Doha. He has more than 9 years of R&D experience in the fields of wireless networking, telecommunications, and security. He has been involved in several international and industrial R&D projects, and he had the chance to lead the design and specification of several products currently on the market. He is an inventor of four patents and has published more than 40 publications. He is currently leading two funded Qatar National Research Fund (QNRF) projects on wearable wireless sensor networks (NPRP 6-1508-2-616) and security in intelligent transport systems (NPRP 7-1113-1-199). He earned a PhD (2009) and an MS (2006) degree from INSA, Lyon, France, and a DiplIng (2005) degree from INSAT, Tunis, Tunisia, all in the fields of computer networks and telecommunications.

Mahbub Hassan is a full professor in the School of Computer Science and Engineering, University of New South Wales, Sydney, Australia. He is a distinguished lecturer at the Institute of Electrical and Electronics Engineers (IEEE) (COMSOC) for 2013–2016. He worked as a visiting professor at Osaka University, Osaka, Japan; University of Nantes, Nantes, France; and National Information and Communications Technology Australia (NICTA), Sydney. He is currently an editor of *IEEE Communications Surveys and Tutorial* and has previously served as a guest editor for *IEEE Network* and associate technical editor for *IEEE Communications Magazine*. He has coauthored 3 books, 1 U.S. patent, and more than 150 refereed articles. Professor Hassan has earned a PhD from Monash University, Melbourne, Australia, and an MSc from the University of Victoria, Victoria, Canada, both in computer science. His current research interests include mobile networks, nanoscale wireless sensor networks, self-powered and energy harvesting wireless networks, Internet of Things, and wearable computing.

Md. Motaharul Islam earned his BS in computer science and information technology from the Islamic University of Technology, Dhaka, Bangladesh, in 2002. He earned his master of business administration in management information systems in 2008, and in 2013, he completed his PhD in computer engineering from the Innovative Cloud and Security (ICNS) Laboratory of Kyung Hee University, Seoul, South Korea. He has worked on the faculty at the Institute of Scientific Instrumentation, University Grants Commission (UGC), Dhaka, Bangladesh, since 2006. Currently, he is an assistant professor in the Department of Computer Science and Engineering at the Islamic University of Technology. His research areas are the smart Internet of Things, network virtualization, and IP-WSN.

Muhammad Islam is currently pursuing his MS in the Computer Sciences Department at COMSATS Institute of Information Technology, Wah Cantt., Pakistan. He earned his BS from the same department in 2015. His research interests include cognitive radio networks, wireless sensor networks, and cloud computing.

Syed Qaisar Jalil has been a BS student in the Computer Sciences Department (telecommunication and networking) at COMSATS Institute of Information Technology, Wah Cantt., Pakistan, since 2011. He is interested in research related to the fields of communication and networking, specifically cloud computing, wireless sensor networks, cognitive radio sensor networks, distributed computing, and mobile ad hoc networks.

Faiz Haider Khan is currently a master's degree student at Technische Universität München, Germany. He earned his BS in electrical (telecommunication) engineering from COMSATS Institute of Information Technology, Wah Cantt., Pakistan. His current research interests include wireless sensor networks, cognitive radio networks, and free space optics.

 **Zeeshan Ali Khan** holds the position of assistant professor in the College of Computer Science and Information Technology at the University of Dammam, Dammam, Kingdom of Saudi Arabia. He earned his PhD degree in electronics engineering from the University of Nice-Sophia Antipolis, Nice, France, in 2011. His research interests include low-power design techniques for wireless sensor networks and embedded systems.

 **Tariq Jamil Saifullah Khanzada** has been working as an associate professor in the Department of Computer System Engineering, Mehran University of Engineering and Technology (UET), Jamshoro, Pakistan, since 2012, and previously, was an assistant professor and lecturer at Mehran UET, in 2004 and 2001, respectively. Prior to joining Mehran UET, he earned his first-class first position in computer diploma, worked as a private network administrator, was a faculty member at Petroman, and had established the Muhammad Institute of Science and Technology, Mirpur Khas, Pakistan, as a pioneer head of faculty. During his doctoral studies, Dr. Khanzada presented his research work in more than a dozen countries, including Germany, France, the United States, Canada, the Netherlands, Italy, and the Czech Republic. He worked for the ViERforES project and Inlite project, carried out under the German Government Research Ministry (DFG) and European Aeronautical Defense System (EADS), Germany, from 2008 to 2010. Furthermore, he conducted lectures and exercises for the international Master of Electrical Electronics Industry Training (EEIT) course at Otto-von-Guericke University, Magdeburg, Germany, during the same period. Dr. Khanzada is a member of the Board of Studies, Board of Faculty, Examination Factotum, Departmental Management Review Committee, Departmental Syllabus Design Committee, and Departmental Thesis Review Committee, and an approved PhD supervisor of the Higher Education Commission (HEC) in Pakistan. Dr. Khanzada has served as a Technical Program Committee (TPC) member, session chair, or keynote speaker at the IMTIC '15, NCMCS '15, FIT 2014, INMIC 2014, MCCT '14, FIT '13, NCIA '13, IMTIC '13, IMTIC '12, NCIT '13, UIG '13, M-DOC '12, and NeX 11 national conferences. He has more than 30 international conferences and journal publications and has coauthored a book on computers. He is a member of the Institute of Electrical and Electronics Engineers (IEEE), a reviewer of *IEEE Transactions on Communication*, and a member of the Pakistan Engineering Council. His research interests include indoor positioning systems, orthogonal frequency division multiplexing (OFDM), single carrier transmissions, and wireless communications.

 **Abdelmajid Khelil** earned his MSc in electrical engineering in 2000 and PhD in computer science in 2007, both from the University of Stuttgart, Stuttgart, Germany. Currently, he is senior consultant at Bosch Software Innovations, Waiblingen, Germany, in the field of the Internet of Things (IoT) and smart cities. Before that, he was the leader of the Internet of Vehicles (IoV) research area at the Huawei European Research Center in Munich, Germany. From 2006 to 2012, he was a leader of two research teams at the Technical University of Darmstadt, Darmstadt, Germany, in the Dependable, Embedded Systems and Software (DEEDS) Group: Dependable WSN Team and the P2P Security and Critical Infrastructure Protection Team. His main research interests include the areas of IoT, machine-to-machine network, IoV, dependable mobile data management, wireless

sensor networks, participatory sensing, body sensor networks for healthcare, mobile transaction processing, peer-to-peer networks, and critical infrastructure protection.

Charilaos Kourogiorgas earned his diploma from the School of Electrical and Computer Engineering at the National Technical University of Athens (NTUA), Athens, Greece. Since December 2011, he has worked toward his PhD at NTUA's School of Electrical and Computer Engineering, on the topic of channel modeling for next-generation high-data-rate wireless terrestrial and satellite communication systems. His research interests include channel modeling for low- (L- and S-bands) and high- (above the Ku-band) frequency bands and the performance assessment of wireless terrestrial and satellite communication systems. He is a coauthor of 14 journal papers and more than 30 conference papers. In 2014, he was awarded the Young Scientist Award from the International Union of Radio Science (URSI). In December 2014, he visited the Jozef Stefan Institute, Ljubljana, Slovenia, as a researcher for a short-term scientific mission granted by the European Association on Antennas and Propagation.

Natarajan Meghanathan is currently an associate professor of computer science at Jackson State University, Jackson, Mississippi. He graduated with a PhD in computer science from the University of Texas at Dallas in 2005. He has published more than 140 peer-reviewed articles in several international journals and conference proceedings. His research has been funded through the U.S. National Science Foundation, Air Force Research Lab, NASA, and the Army Research Lab. He serves as the editor in chief of three international journals and is an active member on the editorial boards of more than 10 journals, as well as on the organizing and technical committees of several international conferences. His research interests are in the areas of wireless ad hoc networks, sensor networks, network science and graph theory, software security, and computational biology.

Muhammad Mostafa Monowar currently works as an assistant professor in the Department of Information Technology, Faculty of Computing and Information Technology, King Abdulaziz University, Jeddah, Kingdom of Saudi Arabia. He earned his PhD in computer engineering in 2011 from Kyung Hee University, Seoul, South Korea. He earned his BSc degree in computer science and information technology from the Islamic University of Technology (IUT), Gazipur, Bangladesh, in 2003. His research interests include wireless networks, especially ad hoc, sensor, and mesh networks, including routing protocols, medium access control mechanisms, internet protocol and transport layer issues, cross-layer design, and quality of service provisioning. He has served as an editor for a number of books published by the Taylor & Francis Group. He has also worked as a guest editor for several special issues of top-ranked journals. He has served as a program committee member for several international conferences and workshops, such as IADIS, DNC, and ICCCS, and as a program vice chair of IEEE HPCC 2013.

Athanasios D. Panagopoulos earned his diploma degree in electrical and computer engineering (summa cum laude) and doctor of engineering degree from the National Technical University of Athens (NTUA), Athens, Greece, in July 1997 and April 2002, respectively. From January 2005 to May 2008, he was head of the Satellite Division of the Hellenic Authority for Information and Communication Security and Privacy. From May 2008 to May 2013, he was a lecturer at NTUA's School of Electrical and Computer Engineering and is now an assistant professor. He has published more than 125 papers in international journals and transactions and more than 160 papers for conference proceedings. He has also published more than 25 chapters in international books. He is chairman of the IEEE Greek Communication Chapter. He is also associate editor of *IEEE Transactions on Antennas and Propagation* and *IEEE Communication Letters*.

Georgios T. Pitsiladis earned his diploma in electrical and computer engineering (ECE) from the National Technical University of Athens (NTUA), Athens, Greece, in September 2006. In December 2012, he earned his DrIng degree from NTUA. Since 2007, he has been an associate researcher and project engineer with the Mobile Radio Communications Laboratory, ECE, NTUA, participating in various industry and research-oriented projects. His research interests include connectivity in wireless multihop networks, fading and high-frequency propagation models, vehicular networks, and stochastic geometry. He has published several papers in international journals and conference proceedings, as well as book chapters. He also serves as a reviewer for several international journals and conference proceedings. Finally, he has been a member of the Greek Technical Chamber (TEE) since 2007.

Marios I. Poulakis earned his diploma in electrical and computer engineering (ECE) from the National Technical University of Athens (NTUA), Athens Greece, and his MSc degree in economics and administration of telecommunication networks from the National and Kapodistrian University of Athens, Greece, in July 2006 and December 2008, respectively. In May 2014, he earned his DrIng degree from NTUA. Since 2007, he has been an associate researcher and project engineer with the Mobile Radio Communications Laboratory, ECE, NTUA, participating in various industry and research-oriented projects. From August 2009 to January 2011, he was a radio frequency engineer with the Non-Ionizing Radiation Office of the Greek Atomic Energy Commission. His research interests include wireless and satellite communications, as well as cognitive radio networks, with an emphasis on optimization mechanisms for quality of service–driven scheduling and resource management. He has published more than 20 papers in international journals and conference proceedings, as well as book chapters.

Umair Mujtaba Qureshi is a lecturer in the Department of Telecommunication Engineering, Mehran University of Engineering and Technology (MUET), Jamshoro, Pakistan. He completed his BE in telecommunications from MUET in 2011 and his ME in communication systems and networks in 2014. He had the honor of having an Erasmus Mundus Mobility of Life Program scholarship at the Universidad de Malaga, Malaga, Spain, in 2010, where he did research projects on voice over IP (VoIP) and medical image processing. He also has several research papers and one journal paper to his credit and has produced many research papers. He was the youngest National Information and Communications Technology (ICT) award winner in 2012. He has been a reviewer and subreviewer at several conferences, including IMTIC '15, ICCCN '14, and IMTIC '13.

Mubashir Husain Rehmani earned his BEng degree in computer systems engineering from Mehran University of Engineering and Technology, Jamshoro, Pakistan, in 2004; MS degree from the University of Paris XI, Paris, France, in 2008; and PhD degree from the Pierre and Marie Curie University, Paris, France, in 2011. He is currently an assistant professor at the COMSATS Institute of Information Technology, Wah Cantt., Pakistan. He was a postdoctoral fellow at the University of Paris Est, Paris, France, in 2012. His current research interests include cognitive radio ad hoc networks, the smart grid, wireless sensor networks, and mobile ad hoc networks. Dr. Rehmani served on the Technical Program Committees for the IEEE ICC 2015, IEEE WoWMoM 2014, IEEE ICC 2014, ACM CoNEXT Student Workshop 2013, IEEE ICC 2013, and IEEE IWCMC 2013 conferences. He is currently an associate editor of *IEEE Communications Magazine, Computers and Electrical Engineering*, and *Access*. He is serving as a guest editor of *Pervasive and Mobile Computing, Ad Hoc Networks*, and *Computers and Electrical Engineering*.

Muhammad Farrukh Shahid has worked as a lecturer in the Department of Electrical Engineering, COMSATS Institute of Information Technology, Lahore, Pakistan, since 2013. He completed his master's of engineering with distinction in major subjects from Mehran University of Engineering and Technology (UET), Jamshoro, Pakistan, in 2013. He secured his first position in bachelor of telecommunication engineering from Mehran UET in 2010. His honors and awards include receiving a silver medal from Mehran UET upon securing the first position in bachelor of telecommunication engineering in 2010; being an invited guest speaker at the IEEE Seminar on Software Defined Radio at Mehran UET; receiving certificates for giving research paper presentations at Bahria University, Karachi, Pakistan, and the IEEEP 26th Seminar at Karachi; receiving certificates for attending seminars on digital filter design and MATLAB® learning at Mehran UET; being a 4-year Merit Scholar recipient for the bachelor of telecommunication engineer at Mehran UET; attending as a judge the IEEE quiz competition held at COMSATS, Lahore, in 2013; receiving a certificate for attending a workshop on lab view learning at COMSATS, Lahore, in 2014; and being selected for an expert panel discussion on space technologies at Space Conference 2014 in Islamabad, Pakistan, organized by Space and Upper Atmosphere Research Commission (SUPARCO) and IST. He is an official member of the LabVIEW Academy and also a researcher in the Antenna and RADAR Research Group at COMSATS, Lahore. He has been associated with professional activities. The following include his training and workshops: he organized semester projects for antennas and Digital

Signal Processing (DSP) in the Electrical Engineering Department at COMSATS, Lahore, in May 2015; conducted a 1-week faculty workshop on LabVIEW at COMSATS, Lahore, in January 2015; conducted a 1-day seminar on LabVIEW and industry integration with the collaboration of the National Instruments (NI) LabVIEW Academy in Islamabad; conducted a workshop on emerging wireless networks at COMSATS, Lahore, in May 2014; conducted a workshop on voice over IP at COMSATS, Lahore, in October 2013; and conducted a seminar on software-defined radio technology at Mehran UET in 2010. He is the author of more than 6 international conference and two international journal papers.

Faisal Karim Shaikh earned his MEng in Pakistan and PhD from the Technical University of Darmstadt, Darmstadt, Germany. He is currently doing postdoctoral work at the University of Umm Al-Qura, Mecca, Saudi Arabia, Kingdom of Saudi Arabia. He is an associate professor at the Mehran University of Engineering and Technology, Jamshoro, Pakistan. He has served as a Technical Program Committee chair and member for several national and international conferences. He has published more than 40 scientific articles in international refereed journals and proceedings of conferences and workshops. His research interests include dependable wireless sensor networks, mobile ad hoc networks, and vehicular and body area networks.

Sukhwinder Sharma earned his BTech degree with honors from Kurukshetra University, Kurukshetra, India, in 2005 and later his MTech degree in computer engineering from Punjabi University, Patiala, India. He is currently a PhD research scholar in computer engineering at Punjab Technical University, Jalandhar, India. Currently, he works as an assistant professor at Baba Banda Singh Bahadur Engineering College, Fatehgarh Sahib, India. He has been teaching undergraduate and postgraduate computer engineering students for the last 10 years. His current areas of interest are wireless sensor networks and computer networks. He has contributed more than 10 research papers to various international journals.

Stavroula Vassaki earned her diploma degree in electrical and computer engineering (ECE) in September 2006 from the National Technical University of Athens (NTUA), Athens, Greece, and in December 2008, she earned her MSc degree in economics and the administration of telecommunication networks from the National and Kapodistrian University of Athens, Athens, Greece. In September 2013, she earned her DrIng degree from NTUA. Since 2006, she has been an associate researcher and project engineer with the Mobile Radio Communications Laboratory, ECE, NTUA, participating in various research-oriented projects. From August 2009 to January 2011, she was a radio frequency engineer with the Non-Ionizing Radiation Office of the Greek Atomic Energy Commission. Her research interests focus on wireless and satellite communications networks and include resource management mechanisms, optimization theory, and game theory. She has published more than 20 papers in international journals and conference proceedings, as well as book chapters.

 **Eisa Zarepour** is a postdoctoral research associate in the Networked Systems and Security Group, School of Computer Science and Engineering, University of New South Wales (UNSW), Sydney, Australia. He earned his bachelor's degree in computer engineering from the University of Razi, Kermanshah, Iran, and master's degree in software engineering from Sharif University of Technology, Tehran, Iran, in 2003 and 2006, respectively. He also earned his PhD degree in electromagnetic wireless nanoscale sensor networks from UNSW in 2015. While pursuing his PhD, he published 15 peer-reviewed papers and received five awards. His current research interest is mainly in the area of nanoscale communication, specializing in designing efficient communication protocols for wireless nanoscale sensor networks, mobile data communication, and privacy.

OVERVIEW OF WIRELESS SENSOR NETWORKS

Chapter 1

Role of Wireless Sensor Networks in Emerging Communication Technologies: A Review

Muhammad Islam, Syed Qaisar Jalil,
and Mubashir Husain Rehmani

Contents

Abstract

A wireless sensor network (WSN) is a deployment of several small-sized chips called nodes, equipped with sensors to monitor physical or environmental conditions, such as temperature, sound, and pressure, and then cooperatively send back their data through the network to a central location for processing. WSNs are emerging rapidly and have a vast field of research. WSNs are cheap, smaller in size, and have intelligent nodes; due to these characteristics, a number of emerging technologies are based on them. These emerging technologies include the Internet of Things (IoT), cognitive radio sensor networks, cloud computing, cyber physical systems, smart grid, and vehicular ad hoc sensor networks. To the best of our knowledge, no work has been done in the literature so far to combine all these emerging technologies in a single source. In this chapter, our goal is to combine WSN-based emerging technologies as a single source so that reviewers can form opinions regarding these technologies.

1.1 Introduction

The concept of wireless sensor networks (WSNs) is the result of development in both microelectronics chip technology (i.e., size, cost, and technology) and information and communication technology (ICT). A lot of research work has been done in the field of WSNs in the last couple of years. WSNs are an emerging technology that is being used to collect information from various environmental phenomena. It has a number of diversified application domains, including home, office, automation and control, transportation, health care, environmental monitoring, security and surveillance, and vehicle monitoring and detection. It consists of small devices called nodes that have sensing or detecting and wireless communicating capabilities. The sensors are deployed in a region for a specific task, where they collect the data and send it to a central location known as a sink node.

The data to be collected can be temperature, humidity, pressure, acoustic, or seismic in nature. A sensor can sense a single phenomenon, or many sensors can be fitted together for sensing multiple phenomena. Nodes can be deployed in two ways: structured and unstructured. In structured deployment, the nodes are deployed with a preplanned policy, while in unstructured deployment, the nodes are deployed randomly. Consequently, the nodes in unstructured deployment should have self-managing and connecting capabilities. A node usually has a radio interface, power source (battery), processor, and memory as its necessary components.

Many new technologies are based on WSNs, including cognitive radio sensor networks (CRSNs), Internet of Things (IoT), cloud computing, wireless sensor and actuator networks (WSANs), the smart grid, vehicular ad hoc networks (VANETs), wireless nanosensor networks (WNSNs), machine-to-machine (M2M) communication, cyber physical systems (CPSs), underwater sensor networks (USNs), delay-tolerant networks (DTNs), and disaster response networks (DRNs), as shown in Figure 1.1. Many of these technologies have been brought into existence, and fast research is going on for other technologies to be brought into existence. The design, energy, and protocol requirements for wireless networks change from one emerging technology to another. For example, the structure of nodes, network structure, and communication medium for WSNs above the surface of the earth (terrestrial WSNs) will be different from those deployed for underwater monitoring. In the near feature, these emerging technologies will become part of our daily life. Many researchers have exposed the different problems of these emerging technologies and propose different solutions.

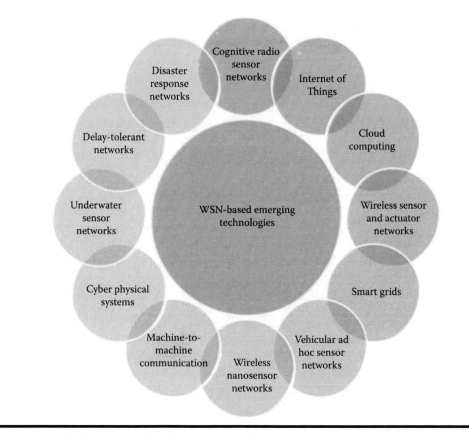

Figure 1.1   Wireless sensor-based emerging technologies.

VANET is one of the emerging technologies of WSNs in which vehicles are equipped with sensors to make an ad hoc network and share sensed information of the surroundings with other vehicles connected in an ad hoc mode. Information sharing among vehicles makes it easier for drivers to drive safely. Also, it decreases the chances of accidents and other tragedies. Smart grid is also a promising emerging technology of WSNs. To meet the power quality and power availability demands of the 21st century, researchers have proposed the smart grid, which can be implemented using smart and intelligent sensors connected through WSNs. The techniques used in energy generation, power transmission, and electricity distribution were designed and implemented almost a century ago, which means they are outdated and cause power losses and other issues. A smart grid will solve these issues and make power distribution and transmission more efficient.

To the best of our knowledge, no such work has been done for combining the emerging technologies of WSNs into a single source. However, in this chapter, we introduce the emerging technologies of WSNs and merge them on a single platform where one can get details regarding them. The main contributions of the chapter are as follows:

- We introduce the emerging technologies based on WSNs.
- We discuss the role of WSNs in these emerging technologies.
- This chapter can serve as a single point of reference for the major references.

This chapter is organized as follows: In Section 1.2, a brief overview of wireless sensor networks is discussed. Emerging WSN-based technologies are discussed in Section 1.3. Sections 1.4 through 1.15 discuss cognitive radio sensor networks, the Internet of Things, cloud computing, wireless sensor and actuator networks, the smart grid, vehicular ad hoc sensor networks, wireless nanosensor networks, machine-to-machine communication, cyber physical systems, underwater sensor networks, delay-tolerant networks, and disaster response networks, respectively. Finally, Section 1.16 concludes the chapter.

1.2 Brief Overview of WSNs

Wireless sensor networks are one of the most interesting and vigorously growing research areas in recent decades. WSNs comprise many (hundreds to thousands) micro-sized, cheap chips that are powered with low-cost batteries and interconnected by wireless media. These chips are called nodes, which could be of many types, including acoustic, radar, low-sampling-rate magnetic, thermal, and visual [1]. These WSNs are deployed for the achievement of a specific goal or task, such as ZebraNet, which is used for zebra monitoring. Today, a lot of applications, including environmental control, mine information, the smart grid and smart cities, wild animal tracking, vehicle tracking, disaster relief, home security, underwater investigations, war applications, airplane surveillance [2], and body sensor networks [3,4], exploit WSNs. Many of these applications have been realized, while for others, researchers from around the world are developing protocols and algorithms to realize them. WSNs can be classified as terrestrial, underground, and underwater.

1.2.1 Terrestrial Wireless Sensor Networks

WSNs deployed above the surface of the earth for monitoring or sensing a phenomenon are known as terrestrial wireless sensor networks (TWSNs) [5]. They are used for surface exploration, environmental monitoring, security and surveillance, and industry monitoring [6]. The wireless media connecting the nodes use radio frequency (RF) for communication, which enables TWSNs to communicate over long ranges. Energy efficiency is crucial for TWSNs; therefore, the system should be energy efficient.

1.2.2 Underwater Wireless Sensor Networks

Underwater wireless sensor networks (UWSNs) are deployed in water for oceanographic data collection, pollution monitoring, underwater navigation, and underwater wildlife monitoring [7]. For example, the UWSN deployed for underwater pollution monitoring consists of wireless nodes capable of sensing the proportion of different elements and compounds in water, while wildlife monitoring sensor nodes use cameras. In comparison with TWSNs, here the attenuation is more pronounced, due to it preferring an acoustic medium instead of RF [8]. Apart from attenuation, UWSNs have more challenges than TWSNs, such as nonreachable limited power, less bandwidth, and long propagation delay.

1.2.3 Underground Wireless Sensor Networks

Underground wireless sensor networks (UGWSNs) are buried in the ground for monitoring underground conditions; this includes wireless nodes placed in caves and mines for various types of data collection [9,10]. UWSNs can be deployed for monitoring soil quality, water level, and underground mineral information, as well as for agricultural purposes. Radio frequency can be used as a wireless medium in the underground environment, but it faces strong attenuation due to electromagnetic waves [9]. The authors of [10] point out that seismic waves and magnetic induction (MI) can be better for communication in underground environments. The fading of signals in UWSNs is more pronounced and unavoidable.

From the perspective of node deployment, WSNs can be of two types: ad hoc WSNs and preplanned WSNs. Ad hoc WSNs are self-organized and without infrastructure, therefore making them more attractive for most applications, such as disaster conditions, where infrastructure networks fail. The nodes are randomly deployed, usually by a vehicle in the region of interest (ROI). On the other hand, preplanned WSNs are structured in nature, where nodes are placed in predefined areas, for example, those discussed in [7–10]. Another classification can be made from the perspective of the mobility of nodes. One type is static WSNs, in which the nodes are fixed and nonmoving; for example, the Navy can use fixed UWSNs for security and surveillance of boundaries. The other type is mobile WSNs; they have moving and self-propelled nodes. An example of a mobile WSN is a wireless sensor and actuator network, where the nodes are attached to a vehicle (robot) moving according to a control decision [11,12].

1.3 Emerging WSN-Based Technologies

WSNs and ad hoc networks are interesting areas that provide a strong base for many emerging technologies to be realized in the near future. In WSNs, the small nodes have some basic capabilities, namely, sensing, computation, memory, acting as a small power source (battery), and communicating through a radio interface. These capabilities have resulted in the opening of new research trends in various domains, including, but not limited to, the following: environmental monitoring, physical system monitoring, surveillance and security for military purposes, home monitoring, agriculture, health care, smart transportation, industrial monitoring, machine-to-machine communication, oceanographic monitoring, wild tracking, mine monitoring, and rescue operations. The intelligent nodes, ease of communication, low power, and low cost are some of the characteristics that make WSN a base for all modern emerging technologies given in Figure 1.1.

WSNs can be classified into two categories based on their infrastructure—ad hoc and preplanned—as discussed in Section 1.1. The working of a WSN without infrastructure makes it really attractive for extreme situations such as flood, earthquake, and disaster situations, where the infrastructure network or static network fails to offer connectivity. The WSN characteristics are application specific, and they will change from application to application for best results. For example, terrestrial WSNs use radio frequency for communication, but UWSNs prefer using an acoustic medium for communication due to the high resistance of water to RF and the fading effect. In Table 1.1, emerging technologies are listed, with a basic introduction given for each. All technologies in the table are based on WSNs and have some common properties; for example, wireless nodes, wireless link connection, data gathering capability, and sharing collected data with other nodes or base stations.

Today many technologies have become an essential part of human life, and most work is done by machines in industries and factories to meet the needs of the growing population and improve industrialization. No doubt this makes life standardized and comfortable, but along with these advantages, doing work this way is proving to be a serious cause of damage and loss in the form of expulsion and fire in factories in recent years. Therefore, modern ICT is

Table 1.1 Quick Reference Table for WSN-Based Emerging Technology

WSN-Based Emerging Technology	References	Discussed Point
Cognitive radio sensor networks (CRSNs)	[13]	Design principles, network architecture, potential application areas
	[14]	Cognitive radio functions, conceptual CR-WSN design, issues in realizing CR-WSN
	[15]	CRSN applications and challenges
Internet of Things (IoT)	[1,16–19]	IoT as ubiquitous communication of numerous devices
	[20]	Effective data collection through satellite-routed sensing system
	[21]	Coverage in IoT
Cloud computing	[22,23]	Cloud computing introduction and models, cloud monitoring
Wireless sensor and actuator networks (WSANs)	[24]	Introduction, models, applications
Smart grids	[25]	Smart grid
	[26]	Introduction, demand management
	[27,28]	Cyber security for smart grids
	[29]	Communication architectures
Vehicular ad hoc networks (VANETs)	[30,31]	Overview of the field (VANETs), architecture, security, and applications
	[32]	Cooperative target tracking in VANETs
	[33]	Routing in VANETs
Wireless nanosensor networks (WNSNs)	[34]	View on electromagnetic wireless nanosensor networks
Machine-to-machine (M2M) communication	[35]	Machine-to-machine communication
	[36]	M2M service platforms
Cyber physical systems (CPSs)	[37,38]	Understanding of cyber physical systems, features, research progress
Underwater wireless sensor networks (UWSNs)	[39]	Introduction, applications, challenges
Delay-tolerant networks (DTNs)	[40]	Introduction, properties
Disaster response networks (DRNs)	[41]	Human detection system in disaster situations

becoming an essential part of all industry, business, education, and exploration domains. WSNs are making it possible, proving to be the best candidate to put intelligence in machines, keep computations, sense defects, and inform the authorities before any unwanted situations occur. The WSNs' low energy, low cost, self-network configuration, and ease of communications are capabilities that have proven they can be deployed anywhere and everywhere machines and devices are found.

WSN infrastructure changes from one emerging technology to another and is used in each emerging technology at different levels. We can divide its role into two categories: minor scale and major scale. For example, in smart grids, WSNs have a major role, as is they are used at all three levels of power production, power transmission, and power distribution, or the consumer end. In the IoT, WSNs have a minor role because they only sense physical environmental factors and send the sensed data to the core network. In CRSNs, WSNs only sense the spectrum and have minor utilization. Table 1.2 summarizes the purposes or utilizations and divides them into major or minor roles in each emerging technology.

1.4 Cognitive Radio Sensor Networks

The wireless sensor network consists of wireless sensor nodes that can collect data and share it with other nodes and pass it to a main location called a gateway. The wireless sensor networks are one of the emerging technologies of the recent decades. A lot of work has been done in this field, while there's a lot more stuff to do. The WSN can be deployed for environmental monitoring, VANETs, smart grids, indoor telemedicine, home monitoring, and animal monitoring. The main advantage of WSN is that it provides low-cost services in a smart way. The conventional WSNs mostly operate in fixed and static unlicensed bands (industrial, medical, scientific), but the rapid growth of low-cost wireless applications in unlicensed bands has given birth to spectrum scarcity among those bands, which necessitates limiting the growth of new technologies. According to the Federal Communications Commission (FCC), the current static spectrum policy exhibits low-spectrum utilization; as a result, about 70% of the licensed spectrum remains unused. These are called white spaces.

Recently, a technology called cognitive radio (CR), in which the nodes intelligently sense and adapt according to their environment, has been proposed to address the issue of spectrum scarcity. The emergence of cognitive radio into conventional WSNs has created a new paradigm of networks called CRSNs. In general, CRSNs can be defined as distributed networks of wireless sensor nodes equipped with CR technology that collect readings and share them in a multihop manner with each other according to application-specific requirements [13]. On the other hand, CR-based wireless sensor nodes can operate in licensed bands as secondary users (SUs), while not interfering with the primary, or licensed, users (PUs) [42], which will result in the best and most efficient use of the spectrum [43]. Several functions have been proposed for cognitive radio, known as the cognitive cycle (CC), including spectrum sensing, spectrum decision, spectrum sharing, and spectrum handoff [13,14,44].

CRSNs have opened a new area of applications that includes, but is not limited to, environmental monitoring [45], smart grids, indoor telemedicine, home monitoring, and animal monitoring. CRSNs can be deployed everywhere for WSNs for efficient use of the spectrum, with other advantages, such as dynamic spectrum access, opportunistic use of channels for bursty traffic, reduction of power consumption, deployment of multiple concurrent WSNs, and communications under different spectrum conditions [13]. CRSNs can be deployed for

Table 1.2   Role of WSNs in Emerging Communication Technologies

Technology Name	Purpose/Utilization	WSN Utilization	Role of WSN
Cognitive radio sensor networks (CRSNs)	To overcome spectrum scarcity by using dynamic spectrum access [13]	To sense spectrum	Minor
Internet of Things (IoT)	To connect environment to Internet	Provide convergence [46]	Minor
Cloud computing	To provide massive data and information [47]	Edge of the cloud computing	Minor
Wireless sensor and actuator networks (WSANs)	Constitute core of the WSANs	WSNs themselves and connection between WSNs and actors	Major
Smart grids	To fulfill the needs of the 21st century and minimize the failures [48]	Generation, transmission, and consumption	Major
Vehicular ad hoc networks (VANETs)	To drive with safety and providing free accident information [32]	Vehicle to vehicle and infrastructure to vehicle	Major
Wireless nanosensor networks (WNSNs)	To enable wireless communication at nanoscale	Core of WNSNs	Major
Machine-to-machine (M2M) communication	To connect machines with the cloud through wireless media	Infrastructure	Minor
Cyber physical systems (CPSs)	Component of cyber physical system enabling CPS to provide timely input	Virtual machine senses and controls physical phenomena	Minor
Underwater sensor networks (USNs)	To enable effective collection of oceanographic data	Core networks	Major
Delay-tolerant networks (DTNs)	To enable pervasive and interplanetary communication	Collection of environmental data	Major
Disaster response networks (DRNs)	To enable communication in extreme situations [41]	Core of the network	Major

the detection of military and public security applications such as (1) chemical, biological, radiological, and nuclear attacks; (2) command control; (3) battlefield surveillance; (4) battle damage information; (5) intelligent assistance; and (6) targeting [15]. In health care, CRSNs can be deployed for providing accurate information, without any delay, with the sensors fitted on the patient's body. The sensors collect data and pass it to the care providers sitting in a nearby room, reducing errors and making it easier for the care providers, as well as doctors and specialists.

Some indoor applications include emergency events, intelligent buildings, factory automation, home monitoring systems, and personal entertainment. These applications exploit CRSNs for reliable and accurate data transfer [13]. Delay-sensitive data applications require real-time efficient communications that cannot be achieved with conventional WSNs. CRSNs can be deployed for the retrieval of videos, images, audio streaming, and delay-sensitive data from an event region [49]. By using the channels opportunistically, cognitive radio sensor networks provide reliable and smooth data transfer in real-time applications. Real-time surveillance, such as biodiversity mapping, habitat monitoring, environmental monitoring, disaster relief operations, tunnel and bridge monitoring, and irrigation and vehicle tracking, requires less access time and minimum delay, which is only possible with CRSNs.

1.5 Internet of Things

Technology is changing our daily life very quickly by opening its emerging areas in recent years. The most dominant is wireless technology, having smart capabilities and cost-effective characteristics. Today some countries, such as the United States and Korea, have deployed WSNs for public services, while others are planning the deployment of VANETs and smart cities. The nodes in WSNs can sense and collect any type of data (audio, video, sound, temperature) and pass it to a central location for processing and controlling purposes. The central point may be further connected to the Internet for sharing data with other locations, therefore making the way for each and every device to communicate with other devices or objects [1,46,50]. This capability makes it possible to include everything on the Internet, ranging from mobile phones to environmental control.

The WSNs and their emerging technologies have opened opportunities for future technologies, referred to as the IoT. IoT can be generally defined as millions of heterogeneous interconnected things (machines or devices) through their wireless terminals. It is not limited to the connection of mobile phones, personal computers, and other smart devices at any time, but every physical thing or object can be connected to the Internet [17–19]. One such concept is smart cities, which assist citizens in roaming and emergency situations, thus making life more standardized, secure, and comfortable [51,52]. The IoT circle includes enormous applications, such as homeland security, environmental control, smart cities, smart grid, smart metering, smart roads, noise control and maps in bar areas, security and emergency, and other physical systems [50].

Connecting such enormous objects from different environments in a single network so that the collected data can be analyzed and the system makes effective and correct decisions is a challenging task. Two big problems are (1) the bandwidth, energy, and delivery time from the WSN perspective and (2) the lack of standardization (between heterogeneous devices) from the communication perspective. One solution is to integrate mobile ad hoc networks (MANET) and WSNs, proposed in [53], for smart cities. Because homeland security, surveillance, and environmental control require minimum delay, MANETs are proving to be better for this scenario, as they have less latency, no need for infrastructure, and self-configurable networks.

The concurrent collection of data from numerous devices and communications between them is a challenging task for network technologies such as ZigBee, Wi-Fi, near-field communication (NFC), and Bluetooth. Therefore, the authors of [20] proposed a satellite-routed sensor system (SRSS) technology for data collection from such numerous and heterogeneous devices. The satellite system is able to collect data concurrently from numerous devices and terminals [54]. Also, they argue that their proposed system will work in inadequate situations such as disasters [55,56]. The satellite will collect data from sensor terminals concurrently and send it to a ground station for controlling and managing the system to act.

A lot of work is going on to prepare smart systems to be a part of IoT. For example, [57] discusses a cloud or sky-based camera image system for the smart grid. The camera will catch images for analysis to predict future energy use. The authors of [46] discuss M2M communication for road traffic management systems in IoT. With these applications, IoT faces serious challenges, such as coverage problems from the WSN perspective [58–61] and others from the network heterogeneity perspective.

1.6 Cloud Computing

The advent of the computer was followed by networking between any two systems far away from each other for sharing resources, information, and data. WSNs and distributed computing are emerging technologies from the last 10 years. The growing networking between devices and applications resulted in massive data and information, which require resources on a large scale. For organizations, it has become difficult to provide such large-scale storage and networking resources. For this, the modern technology in literature is cloud computing, where the resources are provided by a third party, thus reducing cost at the consumer level [62].

Cloud computing is a sharing of computing resources, such as memory or processing, where a program or application runs on a connected server or servers, rather than local servers or personal devices, to handle applications. Cloud computing has many advantages over the traditional client–server model, such as high-performance computing, reduction of capital costs, less maintenance, continuous availability, scalability, and more. Cloud computing provides a virtual private network (VPN) for consumers. With cloud computing, organizations can use network resources at low cost, which increases their efficiency. Cloud computing can be divided into four primary models for different business scenarios and requirements [22]:

1. Private cloud: Operated within one organization and managed by the organization itself or a third party. It does not offer high cost efficiency, but it provides security.
2. Public cloud: Operated and owned by third parties. It reduces capital expenditure and operational costs.
3. Community cloud: Created by multiple organizations and managed either by one or more of those organizations or by any third party. It reduces operational IT cost.
4. Hybrid cloud: Combination of private and public cloud. It provides security as well as reduces operational cost.

A WSN has a large number of nodes, connected together, over a large geographic area for monitoring different environmental conditions. Sensor nodes are inexpensive and can be deployed over a large geographic area, but on the other hand, sensor nodes have very limited resources, such as energy, bandwidth, computational speed, and storage. With the increasing demand of WSNs in various domains and different applications, sensor nodes require special attention to limited resources. The sensed data requires management on a large scale, which is challenging for organizations. In many WSN applications, complex computations, large storage, and sometimes security are also required. Sensor nodes have limited battery power. They cannot perform complex computations for weather forecasting. Also, they have limited storage, so they cannot store large amounts of data for transport monitoring, irrigation, disaster forecasting, and volcano monitoring systems. Security is also one of the major concerns in sensor nodes for military applications. Cloud computing will provide easy management and cost reduction to address these sensor node challenges [62].

Some mobile sink–based wireless sensor networks through cloud computing are used for the military, weather forecasting, transport monitoring, irrigation, disaster forecasting, volcano monitoring, glacier monitoring, forest fire detection, pipenets, pipeline monitoring, mine safety monitoring, target detection and tracking, and health monitoring. But still there are issues and challenges in WSNs with cloud computing, such as data storage on the cloud, speed of mobile sink, multiple mobile sinks, coordination between multiple mobile sinks, security, multiple clouds, and mobile sink traversal time.

1.7 Wireless Sensor and Actuator Networks

Wireless sensor networks are an emerging research area that has many applications, ranging from civil to military purposes in the last decade [63–66]. For example, a WSN deployed in a smart city can sense heat, the pH of water for agricultural purposes, humidity in air, radiation, roadside information, traffic information, and preearthquake information and forward the data to a central location. No doubt, the WSN has a large application domain, but most WSNs are static. To really interact with the physical environment, WSNs can be extended with actors. Actors are vehicles, for example, robots, electrical motors, or humans, that process the control signal received from a station or nodes into physical activity or action [24].

Coverage in these WSNs is a challenging task that needs further improvement for better results, because WSN nodes are usually static, and in node failure situations, the coverage problem is created. The WSNs, together with actors, form WSANs, known as next-generation WSNs. WSANs have proven to address the coverage challenge in WSNs [11,12]. WSANs change their environment, unlike WSNs, which cannot change it. In WSANs, the sensors are fitted on an autonomous mobile device (robot, unmanned vehicle), which can change its location, which is controlled by certain policies. The deployment region is usually called the ROI. Initially, the device can be deployed randomly in the ROI [67,68]; later on, in the case of sensing holes (created due to node failure), the robot can carry sensors to that area, and through this self-deployment, it covers all ROIs [69–71]. In [72–77], similar deployments have been proposed. The robot will reload a sensor from a central point and drop it at a sensing hole [75]. But reloading a sensor each time is not cost-effective; therefore, it has been proposed that the robot pick spare nodes already deployed in the ROI and place them in emerging sensing holes [72–77]. In [12], the authors discussed the algorithms for self-relocation of sensors. A coverage repair system for sensors, based on a robot-assisted relocation system, has been proposed in [78].

WSANs can be deployed for battlefield surveillance; microclimate control in buildings; nuclear, biological, and chemical attack detection [1]; home monitoring [79]; and environmental monitoring. Two networks work together in WSANs: the sensor network, which senses an event, and the actor network, which performs actions accordingly. For example, in a fire, the sensors can detect humans, and the actor nodes (directed by sensor nodes) perform evacuation and rescue operations in the specified region. Coordination in WSANs is necessary for achieving the best results [80] because better coordination will result in identifying the effected region. In addition, coordination of actor to actor and sensor to actor can avoid collision between actors and prevent the network from being flooded by actors. On the battlefield, the sensor node position should change to enhance coverage of the network using autonomous vehicles. In homes, WSANs can assist the elderly and mentally disabled people [81].

On the other hand, WSANs can be used in planetary exploration projects where the vehicle will be landed on the planet. Then the actors loaded with sensors can be dropped to form a network and collect meaningful data by maintaining coordination between them. The collected information will be sent to the central position, for example, landing vehicle, which can process and send it a far distance. WSANs have extended the domain of WSNs, but they have some challenges from both a WSN perspective and a sensor-to-actor perspective. The coordination between sensors and actors should be reliable so that correct detection is possible.

1.8 Smart Grid

Electricity is a necessary part of our lives these days. Home appliances, air conditioning, electric kettles, fans—every device uses electricity for its operation. The electricity network, which consists of generation, transmission, and distribution, is mostly old-fashioned. With vigorous development in other areas of life, this old system of electricity has become insufficient. In addition, especially in North America, it has been reported to have caused huge financial losses. These and some other power-related serious issues demand the involvement of ICT in power grids, which will not only solve these power-related issues but also will add efficiency to the power grid. WSNs are proving to be the best candidate to lessen these losses due to their smart sensing, calculating, and communicating capabilities at low costs. WSNs and emerging technologies provide opportunities for management applications in smart grids [82].

The integration of WSN into power grids, called smart grids, will be the future power grids. WSNs can provide efficient services in power generation, transmission, and distribution. Traditional power generation mostly relies on the wired network. Consequently, the network covers a specific region, leaving other regions uncovered. The reason is the limitations of wired networks in harsh places. WSNs, on the other hand, can be deployed anywhere without any infrastructure. The power generation can be controlled and managed in a smart way, and without the human intervention. WSN nodes can sense temperature, light, and the mechanical effect of the machinery of the generation system and send the data to the managing and control team, which enables the team to handle situations before unwanted problems occur. This can avoid system outages, which results in an increase of the overall performance of power generation.

Power transmission and distribution can be monitored by WSNs in the smart grid, which is very helpful for efficiently dealing with the demand and supply of electricity. Also, it is a challenging task due to the huge geographic area of WSNs. Four key WSN application areas in the smart grid are theft, sagging conductors, insulation breakdown, and fault detection and location [83]. With these four applications of WSN in the smart grid, it is easy to manage the power transmission and distribution network efficiently compared to the old electricity grid.

A huge number of applications of WSN require infrastructure that should be reusable, flexible, and manageable, so that every application of WSN can use it concurrently. For this, researchers have presented the concept of a WSN infrastructure as a WSN cloud that provides services to multiple application and data collection systems that adhere to the cloud computing paradigm [84]. In this infrastructure, every end user or system can use a specific set of services. Using this infrastructure, we can reduce cost, and it is easy to manage, rather than having individual infrastructures for specific purposes. A huge deployment of WSNs with this infrastructure leads toward the concept of a smart city.

Demand management is an important issue in the smart grid. The smart grid is the integration of ICT into the electricity grids to monitor and regulate energy generation and demand [26], so smart meters and sensors are required for the implementation of the smart grid. These interconnected smart meters and sensors, from power generation to distribution up to the household level, are used to find the peak hours of electricity in specific areas so that power generation and distribution can be achieved accordingly. It reduces cost as well as makes it easy to manage the demand of electricity for specific areas.

Security is one of the main challenging tasks in the smart grid and smart home. Reliable and efficient communication is required between humans and devices. In home area networks, secure real time monitoring and control to devices can be achieved by the secure access gateway (SAG) in [85]. In the architecture and design of a SAG for home area networks, security at the network and physical layer is enhanced.

WSN can be deployed in a wide geographic area with low cost, making it more promising than other solutions. On the other hand, in the smart grid in a wide geographic area, with the conventional battery lifetime in WSN, it is difficult to recharge. Especially in harsh environments, it is not feasible to use conventional WSN batteries. For this problem, researchers use RF–based wireless energy transfer and have proposed the sustainable wireless rechargeable sensor network (SuReSense), which charges multiple sensors from several landmark locations [82].

1.9 Vehicular Ad Hoc and Sensor Networks

The transportation system has become more important for humans in recent years due to the vigorously growing number of vehicles and road systems. Transportation is a part of human life these days, demanding advancement and use of technology in it, like in other systems (electricity smart grids, smart cities, and smart dams). It is reported that today, many of human deaths are due to accidents and machines. Traffic is one of U.S. society's greatest time wasters. In large cities, the traffic problems are becoming critical for the government and residents. Advancement in wireless sensor networks has resulted in new trends in many application areas, including transportation systems. This transportation system will be known as an intelligent transportation system (ITS).

The roadside infrastructure and communicating cars collectively make a vehicular ad hoc sensor network. Localization and tracking are the key things for making an ITS [86,87]. The vehicle position is very important and is required for geographic routing [87]. Vehicle-to-vehicle (V2V) and infrastructure-to-vehicle (I2V) communications opened new trends of applications, including cooperative collision avoidance, cooperative lane changing, safe and comfortable driving, blind spot detection, warning systems, ambulance approach, roadside information, pedestrian and animal crossings, traffic congestion information, parking lot information, and road defects at a distance. Cooperative target tracking [32] is a novel challenge for vehicular ad hoc sensor networks, as it differs from traditional mobile ad hoc networks; it has a changing and complex environment that consists of many things. The vehicle surroundings have a large spatiotemporal-dependent property.

VANET consists of two types of nodes: vehicular nodes (VNs) and roadside infrastructure (RSI) nodes. The cars are equipped with wireless sensors so that they can establish communications with each other as well with the roadside infrastructure nodes. Most companies are now showing a trend to deploy wireless nodes in theirs car [88]. In accident situations, WSNs (VNs and RSI) will alert the nearby station, thus making it easy and efficient for stations to find the location and manage injuries in a timely manner. The growing number of vehicles makes it difficult for drivers to drive safely in such a dense environment. With vehicle-to-vehicle and vehicle-to-infrastructure communication, they can drive safely by knowing the surroundings of the vehicle from each side.

The vehicle dynamic environment is very difficult to consider and requires many sensors. In a dense environment, a vehicle may receive data from many nodes, which makes it difficult to consider. In situations such as accidents, strict time constraints are required. The authors of [89,90] disused a variety of motion models having different complexities and parameters. The vehicle ad hoc sensor network has a number of challenges, including filtering problems, data decimation from many sensors, and model problems of motion selection [32]. Correct information calculation sharing is required during cooperation between vehicles to make correct decisions. The packet delay and incorrect sharing of information may result in serious problems.

1.10 Wireless Nanosensor Networks

A wireless nanosensor network (WNSN) is a network of nanoscale devices that take advantage of the unique properties of novel nanomaterials and nanoparticles to detect new types of events at the nanoscale and perform specific functionalities, such as computing, data storing, sensing, and actuation. Nanotechnology has applications in the medical field, such as studying drug effects in the body on a nanoscale, and security and surveillance against chemical and biological attacks on battlefields. The delivery of a drug at the right time [91] results in efficient care of the patient. Advanced nanodevices can be created by integrating several of these nanocomponents in a single entity [92]. Communication at the nanoscale among nanoscale devices is difficult to achieve, but two promising approaches, presented in [93], are electromagnetic wireless communication and molecular communication.

Nanosensors operate on very high frequency (hundreds of terahertz), so simple miniaturization of classical antennas is not a solution; to overcome this limitation, graphene-based nanoantennas and nanotransceivers have been proposed [94–97]. Energy storage is another important issue in nanosensors, as nanobatteries can hold a limited amount of energy, and manually recharging or replacing them is not feasible. To overcome this limitation, novel nanoscale energy harvesting systems have been proposed [98–100].

The classical medium access control (MAC) protocol cannot be used for WNSN because of the peculiarities of the terahertz band [101], temporal energy fluctuations of nanosensors due to the behavior of power nanogenerators [102], and its requirement for ultralow complexity protocols due to its limited processing capabilities [34], so an energy- and spectrum-aware MAC protocol to achieve perpetual WNSNs has been proposed in [92]. But still, WNSNs have lot of unsolved issues and challenges. WNSNs have a number of applications, including advanced health monitoring systems, surveillance networks for chemical and biological attack prevention, environmental monitoring, and military applications.

1.11 M2M Communication

Machines have proven to change human lives, especially the wireless sensor nodes, cellular networks, and CPSs of recent decades [50,103,104]. Machines are taking the place of humans in every field and performing in a smart and efficient way compared with them. The communicating autonomous wireless nodes have created a new way of wireless communication and interconnection [105,106]. This new way of communication is M2M communication. M2M communication, along with IoT, has diversified applications; some of them and their corresponding applications are given in [30,107–111]. The number of communicating machines can be in the trillions, making communication a great challenge in the realization of the system.

The enormous number of machines and their heterogeneous properties are a challenging problem for proposing its architecture. The proposed cloud-based communication architecture consists of cloud infrastructure and a connected ocean (extremely large number) of machines [35]. The current wireless technologies, such as Wi-Fi (IEEE 802.11b), Bluetooth (IEEE 802.15.1), and ZigBee (IEEE 802.15.4), don't suffice for M2M communication; as a result, a comprehensive connection among machines distributed over a large area is required [35]. The cellular networks may be the best candidate because of their autonomous connecting capabilities, coverage, and mobility property.

There are three architectural models for M2M communication: Third Generation Partnership Project (3GPP), IEEE 802.16p, and the one proposed by the European Telecommunication Standard Institute (ETSI) [112]. According to authors of [112], 3GPP and IEEE 802.16p resemble each other in functionality; that is, both define cellular reference architectural models. Unlike the above two reference architectures, ETSI devised a service-oriented architectural model for M2M communication. The authors of [112] proposed yet another model called Long-Term Evolution—Advanced (LTE-A) architecture [113], which combines both 3GPP and ETSI architectures. In addition, they argue that their proposed architecture causes efficient communication between a large number of machines. Security is a big issue for M2M communication [114], which is a challenging problem for researchers. In addition, some other challenges are discussed in [35].

1.12 Cyber Physical Systems

MANET and WSN were the fastest-growing areas of research in the past two decades and continue to be. A reasonable amount of work has been done in these two areas up to now, and lot more future work is still unexplored. MANET is generally the extension of infrastructure networks [115], while WSN is deployed for the collection of data from the environment and passing the sensed data to a central location. The cyber physical system is promising for mergence of the physical world into a virtual world, making strong interactions between the two integrating elements. A lot of CPS applications, including in areas such as social networking, navigation, gaming, intelligent transportation, manufacturing, biosignaling information, health care, robotics, and military purposes, have been discussed in the literature.

A CPS consists of sensors (static or dynamic) from multiple WSNs, together with actuator networks directed by a central decision-making system. CPS involves data flow between the virtual systems and physical systems or real-time systems based on intelligence. For example, in a CPS deployed for climate control, the sensed data can be sent to a central location for investigation. The sensed data stimulate the computational elements through a network, resulting in activation of the controlling system. The CPS has some similarities with the MANET and WSN, but it differs in other aspects [116]. CPS has to face many sensor networks of different levels, their data, and coordination between those networks instead of just being confined to a single WSN. CPS is a cross-domain collection of WSNs [117], each having different functionality and security threats, resulting in the flow of heterogeneous information.

CPS has a lot of applications developed in different areas; some of them have been discussed in [118]. The actuator and robot networks (discussed in detail in Section 1.7) have also emerged in recent years. For example, the climate control system will sense the environment and send the sensed data to the intelligent decision-making section, which will direct the actuators to maintain good environmental factors [80,119,120]. The authors of [116] have discussed CPS applications and platforms in detail, including areas such as health care, navigation and rescue, intelligent transportation systems, and social networking and gaming. Apart from all the applications of CPS, it has challenges related to networking, QoS issues, and interference issues due to its cross-domain nature. The data in CPS is not confined to local devices, which exposes it to the outside world, resulting in serious security and privacy challenges.

1.13 Underwater Sensor Networks

More than 70% of the Earth's surface is covered with water, but we still have very limited underwater wireless technologies due to large propagation delay, low communication bandwidth, node mobility, high error rate, and a harsh underwater environment. Underwater wireless sensor network is one of the enabling technologies for the development of future ocean observation systems and wireless sensor networks. Currently, we have seismic imaging tasks for offshore oil fields carried out by ship, but it is costly, so it can only be carried out once every 2–3 years. On the other hand, an underwater wireless sensor network would have very low cost and can be deployed permanently.

The system architecture of an underwater wireless sensor network typically has four types of nodes [121]. First, large numbers of sensor nodes are deployed on the sea floor, at the lowest layer, where they collect data through attached sensors and communicate with other nodes through short-range acoustic modems. Second, control nodes are placed at the top layer for connection to the Internet. They are powered and can be onshore or offshore with large storage capacity. The third type is supernodes, which have access to high-speed networks. They can efficiently relay data to base stations. The fourth type is robotic submersibles, which interact with systems via acoustic communications.

Existing solutions of wireless sensor networks for the surface cannot be applied directly to underwater wireless sensor networks. With the system architecture in [121] ensuring all safety measures, we still have risks involved, such as a harsh underwater environment, fishing trawlers, underwater life, or failure of waterproofing. So, redundancy is required at basic deployment. Also, battery power consumption is another important issue. Applications of underwater wireless sensor networks include but are not limited to information of oil fields inside the oceans, seismic monitoring, equipment monitoring and leak detection, support for swarms of underwater robots, pollution control, climate recording, prediction of natural disturbances, search and survey missions, and study of marine life.

1.14 Delay-Tolerant WSNs

The peer-to-peer network is demanding attention from researchers due to its glorious distributed connection features, network availability, quality of service (QoS), cost efficiency, and other attractive features. It is used for content sharing, video streaming, and other applications. The most flourishing of these are WSNs. WSN emerging areas are underwater systems, environmental judging and monitoring, underground investigation systems, military missions and rescue, and medical systems. For example, MANET is used for video streaming, home monitoring, data networks, military rescue, and information sharing. Another class of networks, called DTNs, is also very common. DTN refers to networks that exhibit long communication delays, such as space communications and interplanetary communications. These characteristics make traditional network protocols insufficient for DTNs.

Intermittently connected wireless sensor network is another attractive class of networks. It combines the features of both WSNs and DTNs, such as limited range, cost efficiency, small storage, low energy, and small computation capability, forming Intermittently Connected Delay-Tolerant Wireless Sensor Networks (ICDT-WSNs). It has some unique characteristics, such as the absence of end-to-end connectivity, the absence of fixed routes from source to destination, non-symmetric data, variable delay, mobility, and high chances of errors in data, making it different from traditional networks. Its application areas include human rescue, environmental monitoring, and wild tracking. Underwater WSNs, underground WSNs, and mobile WSNs are some of the networks that use this concept. An example of mobile WSNs is the ZebraNet project, in which sensors are carried by animals in a large wildlife area. In all these networks, the path from source to destination is unstable and does not always exist, which results in a big problem in protocol design.

Delay-tolerant wireless sensor and actuator networks consist of static sensors for environmental monitoring and mobile actors for data collection. The dynamic environment makes it very difficult for mobile actors to communicate, resulting in data dissemination—a challenging problem. The authors of [122] presented a protocol named Energy-Efficient Peer-to-Peer Message Dissemination (EMD), which meets the requirements of a delay-tolerant WSAN and solves the data dissemination problem.

1.15 Disaster Response Networks

The advances in technology are making human life more comfortable and secure by opening new areas of research, especially in wireless sensor networks. This is being done in two main ways: (1) adding smart facilities to human life, usually planned tasks, and (2) securing life in instant and unexpected situations, such as flooding, earthquakes, and terrorist attacks. In these disaster scenarios, machines are proving better than humans in rescuing and protecting human life and avoiding huge infrastructure damage. The connection between machines and humans is important, but in these scenarios, the traditional wired network fails, as floods or earthquakes destroy the infrastructure in the affected area. Wireless emerging technologies such as wireless sensor and actuator networks (WSANs) are proving to be the best candidates in this scenario.

A disaster response network is a collection of wireless nodes and autonomous machines (static or dynamic) connected with each other that can sense an activity or an object and respond to a central location. The authors of [123,124] discuss applications based on WSANs for humanitarian rescue and relief after disasters. WSANs enable the rescue team to communicate with each other during disaster situations, which is impossible with infrastructure-based networks. First responders can report medical data, resulting in less causalities. In extreme cases, such as in volcano hazard monitoring, atomic reactor disasters, battlefields, and chemical attacks, where human contact is very difficult, WSANs can be used to provide valuable information [125].

Currently, robot networking is attracting attention from researchers in different applications. In [41], the authors present a system where multiple mobile robots cooperatively map the disaster region and the sensors collect the data, which is sent to a central location for controlling and managing, using the same system as in [126–128]. The sensors sense human existence and a send signal, which directs the nearest robot to that place. Also, they argue that their proposed system is better than a human-assisted system because of its autonomous deployment, intelligence, and flexibility. The robots will map an unknown region, known as multirobot simultaneous location and mapping (SLAM) [129,130]. Robots can be used in very critical situations such as planetary exploration and reconnaissance, adding flexibility to the work of exploration. The robots can cover large areas by providing information from the target area [131,132]. Using the sensor nodes (mobile detectors) for disaster situations, they can be placed randomly or deterministically depending on the application environment [133]. The authors of [128] explain a system for WSN restoration, which acts like a postdisaster information gathering system.

The robot WSANs have many other applications; for example, sensors are fitted onto a device called MARVIN presented in the AWARE project, which uses unmanned aerial vehicles with ground WSANs for cooperative tracking and surveillance [134,135]. In [136], the authors proposed a mobile robot–assisted sensor network for unmanned monitoring and exploration purposes. The authors of [137] proposed a WSN construction system by using a teleport robot, while [77] proposes a system for the replacement of failed wireless nodes using robots. The reliance on WSN makes it more exposed to attackers and hackers, making the security issue more pronounced in such networks [138].

1.16 Conclusion

Wireless sensor networks have several potential applications, ranging from home security to smart cities and future technology, the IoT. The two important features of WSNs—smart and easy service and low-cost service—make them a promising candidate for enormous applications. As a result, a large number of emerging technologies are based on WSNs, including cyber physical systems, the Internet of Things (IoT), VANETs, cloud computing, M2M communication, DTNs, CRSNs, underwater sensor networks, disaster response networks, wireless sensor and actuator networks, nanosensor networks, and smart grids. This chapter introduced each emerging technology in detail, enabling readers to take a quick survey of WSN-based emerging technologies.

References

1. I. F. Akyildiz, W. Su, Y. Sankarasubramaniam, E. Cayirci, Wireless sensor networks: A survey, *Computer Networks* 38 (2002) 393–422.
2. K. Bur, P. Omiyi, Y. Yang, Wireless sensor and actuator networks: Enabling the nervous system of the active aircraft, *IEEE Communications Magazine* 48 (2010) 118–125.
3. M. Domingo, A context-aware service architecture for the integration of body sensor networks and social networks through the IP multimedia subsystem, *IEEE Communications Magazine* 49 (1) (2011) 102–108.
4. M. Quwaider, J. Rao, S. Biswas, Body-posture-based dynamic link power control in wearable sensor networks, *IEEE Communications Magazine* 48 (2010) 134–142.
5. J. Yick, B. Mukherjee, D. Ghosal, Wireless sensor network survey, *Computer Networks* 52 (2008) 2292–2330.
6. V. Gungor, G. Hancke, Industrial wireless sensor networks: Challenges, design principles, and technical approaches, *IEEE Transactions on Industrial Electronics* 56 (10) (2009) 4258–4265.

7. I. F. Akyildiz, D. Pompili, T. Melodia, Underwater acoustic sensor networks: Research challenges, *Ad Hoc Networks* 3 (2005) 257–279.
8. J. Heidemann, W. Ye, J. Wills, A. Syed, Y. Li, Research challenges and applications for underwater sensor networking, in *Wireless Communications and Networking Conference 2006 (WCNC 2006)*, Las Vegas, NV, 2006, vol. 1, pp. 228–235.
9. X. Li, A. Nayak, D. Simplot-Ryl, I. Stojmenovic, Sensor placement in sensor and actuator networks, in *Wireless Sensor and Actuator Networks: Algorithms and Protocols for Scalable Coordination and Data Communication*, ed. A. Nayak, I. Stojmenovic, Wiley, Hoboken, NJ, 2010, pp. 263–294.
10. I. F. Akyildiz, E. P. Stuntebeck, Wireless underground sensor networks: Research challenges, *Ad Hoc Networks* 4 (2006) 669–686.
11. X. Li, R. Falcon, A. Nayak, I. Stojmenovic, Servicing wireless sensor networks by mobile robots, *IEEE Communications Magazine* 50 (7) (2012) 147–154.
12. X. Li, A. Nayak, D. Simplot-Ryl, I. Stojmenovic, Sensor placement in sensor and actuator networks, in *Wireless Sensor and Actuator Networks: Algorithms and Protocols for Scalable Coordination and Data Communication*, ed. A. Nayak, I. Stojmenovic, Wiley, Hoboken, NJ, 2010, p. 300.
13. O. Akan, O. Karli, O. Ergul, Cognitive radio sensor networks, *IEEE Network* 23 (4) (2009) 34–40.
14. K.-L. Yau, P. Komisarczuk, P. Teal, Cognitive radio-based wireless sensor networks: Conceptual design and open issues, in *Proceeding of the IEEE 34th Conference on Local Computer Networks 2009 (LCN 2009)*, 2009, pp. 955–962.
15. G. P. Joshi, S. Y. Nam, S. W. Kim, Radio wireless sensor networks: Applications, challenges, and research trends, *Sensors* 13 (2013) 11196–11228.
16. J. Pan, Y. Hou, L. Cai, Y. Shi, X. Shen, Topology control for wireless sensor networks, in *9th ACM Conference on Mobile Computing and Networking (MOBICOM '03)*, San Diego, CA, 2003, pp. 286–299.
17. L. Xu, L. Rongxing, L. Xiaohui, S. Xuemin, C. Jiming, L. Xiaodong, Smart community: An Internet of Things application, *IEEE Communications Magazine* 49 (2011) 68–75.
18. L. Atzori, A. Iera, G. Morabito, SIoT: Giving a social structure to the Internet of Things, *IEEE Communications Letters* 15 (11) (2011) 1193–1195.
19. M. Zorzi, A. Gluhak, S. Lange, A. Bassi, From today's intranet of things to a future Internet of Things: A wireless- and mobility-related view, *IEEE Wireless Communications* 17 (6) (2010) 44–51.
20. K. Yuichi, N. Hiroki, M. F. Zubair, K. Nei, Effective data collection via satellite-routed sensor system (SRSS) to realize global-scaled Internet of Things, *Sensors Journal* 13 (10) (2013) 3645–3654.
21. L. Liang, Z. Xi, M. Huadong, Percolation theory-based exposure-path prevention for wireless sensor networks coverage in Internet of Things, *Sensors Journal* 13 (10) (2013) 3625–3636.
22. S. K. Dash, S. Mohapatra, P. K. Pattnaik, A survey on application of wireless sensor network using cloud computing, *International Journal of Computer Science and Engineering Technologies* 1 (4) (2010) 50–55.
23. G. Aceto, A. Botta, W. de Donato, A. Pescap, Cloud monitoring: A survey, *Computer Networks* 57 (9) (2013) 2093–2115.
24. H. Liu, A. Nayak, I. Stojmenovic, Applications, Models, Problems, and Solution Strategies, John Wiley & Sons, Hoboken, NJ, 2010, pp. 1–32.
25. X. Fang, S. Misra, G. Xue, D. Yang, Smart grid #x2014; the new and improved power grid: A survey, *IEEE Communications Surveys Tutorials* 14 (4) (2012) 944–980.
26. Z. Zhu, S. Lambotharan, W. H. Chin, Z. Fan, Overview of demand management in smart grid and enabling wireless communication technologies, *IEEE Wireless Communications* 19 (3) (2012) 48–56.
27. Y. Yan, Y. Qian, H. Sharif, D. Tipper, A survey on cyber security for smart grid communications, *IEEE Communications Surveys Tutorials* 14 (4) (2012) 998–1010.
28. W. Wang, Z. Lu, Cyber security in the smart grid: Survey and challenges, *Computer Networks* 57 (5) (2013) 1344–1371.
29. W. Wang, Y. Xu, M. Khanna, A survey on the communication architectures in smart grid, *Computer Networks* 55 (15) (2011) 3604–3629.
30. H. Hartenstein, K. Laberteaux, A tutorial survey on vehicular ad hoc networks, *IEEE Communications Magazine* 46 (6) (2008) 164–171.

31. R. G. Engoulou, M. Bellache, S. Pierre, A. Quintero, VANET security surveys, *Computer Communications* 44 (0) (2014) 1–13.

32. H. S. Ramos, A. Boukerche, R.W. Pazzi, A. C. Frery, A. A. F. Loureiro, Cooperative target tracking in vehicular sensor networks, *IEEE Wireless Communications* 19 (5) (2012) 66–73.

33. F. Li, Y. Wang, Routing in vehicular ad hoc networks: A survey, *IEEE Vehicular Technology Magazine* 2 (2) (2007) 12–22.

34. I. F. Akyildiz, J. M. Jornet, Electromagnetic wireless nanosensor networks, *Nano Communication Networks* 1 (1) (2010) 3–19.

35. K.-C. Chen, S.-Y. Lien, Machine-to-machine communications: Technologies and challenges, *Ad Hoc Networks* 18 (0) (2014) 3–23.

36. J. Kim, J. Lee, J. Kim, J. Yun, M2M service platforms: Survey, issues, and enabling technologies, *IEEE Communications Surveys Tutorials* 16 (1) (2014) 61–76.

37. R. Baheti, H. Gill, Cyber-physical systems, in *The Impact of Control Technology*, ed. T. Samad, A. Annaswamy, IEEE Control Systems Society, New York, 2011, pp. 161–166.

38. J. Shi, J. Wan, H. Yan, H. Suo, A survey of cyber-physical systems, in *Wireless Communications and Signal Processing (WCSP) 2011*, Nanjing, China, 2011, pp. 1–6.

39. A. Davis, H. Chang, Underwater wireless sensor networks, in *Oceans 2012*, Hampton Roads, VA, 2012, pp. 1–5.

40. J. Shen, S. Moh, I. Chung, Routing protocols in delay tolerant networks: A comparative survey, in *23rd International Technical Conference on Circuits/Systems, Computers and Communications (ITC-CSCC 2008)*, Shimonoseki City, Japan, 2008, pp. 6–9.

41. G. Tuna, V. C. Gungor, K. Gulez, An autonomous wireless sensor network deployment system using mobile robots for human existence detection in case of disasters, *Ad Hoc Networks Part A* 13 (0) (2014) 54–68.

42. I. Akyildiz, W.-Y. Lee, M. C. Vuran, S. Mohanty, A survey on spectrum management in cognitive radio networks, *IEEE Communications Magazine* 46 (4) (2008) 40–48.

43. J. Wang, M. Ghosh, K. Challapali, Emerging cognitive radio applications: A survey, *IEEE Communications Magazine* 49 (3) (2011) 74–81.

44. K.-C. Chen, Y.-J. Peng, N. Prasad, Y.-C. Liang, S. Sun, Cognitive radio network architecture: Part I—General structure, in *Proceedings of the 2nd International Conference on Ubiquitous Information Management and Communication*, Suwon, Korea, Republic of Korea, 2008, pp. 114–119.

45. J. Qi, S. Shimamoto, A cognitive mobile sensor network for environment observation, *Telematics and Informatics* 29 (1) (2012) 26–32.

46. L. Foschini, T. Taleb, A. Corradi, D. Bottazzi, M2M-based metropolitan platform for IMS-enabled road traffic management in IOT, *IEEE Communications Magazine* 49 (11) (2011) 50–57.

47. K. Ahmed, M. Gregory, Integrating wireless sensor networks with cloud computing, in *2011 Seventh International Conference on Mobile Ad-Hoc and Sensor Networks (MSN)*, Beijing, China, 2011, pp. 364–366.

48. M. Erol-Kantarci, H. T. Mouftah, Wireless sensor networks for smart grid applications, in *2011 Saudi International Electronics, Communications and Photonics Conference (SIECPC)*, 2011, pp. 1–6.

49. A. O. Bicen, V. C. Gungor, O. B. Akan, Delay-sensitive and multimedia communication in cognitive radio sensor networks, *Ad Hoc Networks* 10 (5) (2012) 816–830.

50. L. Atzori, A. Iera, G. Morabito, The Internet of Things: A survey, *Computer Networks* 54 (15) (2010) 2787–2805.

51. L. Filipponi, A. Vitaletti, G. Landi, V. Memeo, G. Laura, P. Pucci, Smart city: An event driven architecture for monitoring public spaces with heterogeneous sensors, in *2010 Fourth International Conference on Sensor Technologies and Applications (SENSORCOMM)*, Venice, Italy, 2010, pp. 281–286.

52. G. Yovanof, G. Hazapis, An architectural framework and enabling wire-less technologies for digital cities and intelligent urban environments, *Wireless Personal Communications* 49 (3) (2009) 445–463.

53. B. Paolo, C. Giuseppe, C. Antonio, F. Luca, Convergence of MANET and WSN in IOT urban scenarios, *Sensor Journal* 13 (10) (2013) 3558–3567.

54. I. Bisio, M. Marchese, Efficient satellite-based sensor networks for information retrieval, *IEEE Systems Journal* 2 (4) (2008) 464–475.

55. H. Nishiyama, Y. Tada, N. Kato, N. Yoshimura, M. Toyoshima, N. Kadowaki, Toward optimized traffic distribution for efficient network capacity utilization in two-layered satellite networks, *IEEE Transactions on Vehicular Technology* 62 (3) (2013) 1303–1313.

56. M. Shimada, T. Tadono, A. Rosenqvist, Advanced land observing satellite (ALOS) and monitoring global environmental change, *Proceedings of the IEEE* 98 (5) (2010) 780–799.

57. R. Yao, W. Wang, M. Farrokh-Baroughi, H. Wang, Y. Qian, Quality-driven energy-neutralized power and relay selection for smart grid wireless multimedia sensor based IOTs, *IEEE Sensors Journal* 13 (10) (2013) 3637–3644.

58. C.-F. Huang, Y.-C. Tseng, The coverage problem in a wireless sensor network, in *Proceedings of the 2nd ACM International Conference on Wireless Sensor Networks and Applications (WSNA '03)*, New York, 2003, pp. 115–121.

59. S. Kumar, T. H. Lai, J. Balogh, On k-coverage in a mostly sleeping sensor network, in *Proceedings of the 10th Annual International Conference on Mobile Computing and Networking (MobiCom '04)*, New York, 2004, pp. 144–158.

60. P.-J. Wan, C.-W. Yi, Coverage by randomly deployed wireless sensor networks, *IEEE Transactions on Information Theory* 52 (6) (2006) 2658–2669.

61. B. Wang, Coverage problems in sensor networks: A survey, *ACM Computing Surveys* 43 (4) (2011) 32:1–32:53.

62. Y. Jadeja, K. Modi, Cloud computing—Concepts, architecture and challenges, in *2012 International Conference on Computing, Electronics and Electrical Technologies (ICCEET)*, 2012, pp. 877–880.

63. X. Chen, J. Shen, J. Wu, Enhanced delegation forwarding in delay tolerant networks, *International Journal of Parallel, Emergent and Distributed Systems* 26 (5) (2011) 331–345.

64. M. Jorgic, I. Stojmenovic, M. Hauspie, D. Simplot-Ryl, Localized algorithms for detection of critical nodes and links for connectivity in ad hoc networks, in *Proceedings of the Third Annual IFIP Mediterranean Ad Hoc Networking Workshop (Med-Hoc-Net)*, 2004, pp. 360–371.

65. S. Ruhrup, H. Kalosha, A. Nayak, I. Stojmenovic, Message-efficient beaconless georouting with guaranteed delivery in wireless sensor, ad hoc, and actuator networks, *IEEE/ACM Transactions on Networking* 18 (1) (2010) 95–108.

66. I. Stojmenovic, A. Nayak, J. Kuruvila, F. Ovalle-Martinez, E. Villanueva-Pena, Physical layer impact on the design and performance of routing and broadcasting protocols in ad hoc and sensor networks, *Computer Communications* 28 (10) (2005) 1138–1151.

67. R. Mulligan, H. M. Ammari, Coverage in wireless sensor networks: A survey, *Network Protocols and Algorithms* 2 (2) (2010) 27–53.

68. D. Simplot-Ryl, I. Stojmenovic, J. Wu, Energy efficient backbone construction, broadcasting, and area coverage in sensor networks, *Handbook of Sensor Networks* (2005) 343–380.

69. X. Li, H. Frey, N. Santoro, I. Stojmenovic, Strictly localized sensor self-deployment for optimal focused coverage, *IEEE Transactions on Mobile Computing* 10 (11) (2011) 1520–1533.

70. X. Li, N. Santoro, Zoner: A zone-based sensor relocation protocol for mobile sensor networks, in *Proceedings of the 2006 31st IEEE Conference on Local Computer Networks*, Tampa, FL, 2006, pp. 923–930.

71. X. Li, N. Santoro, I. Stojmenovic, Mesh-based sensor relocation for coverage maintenance in mobile sensor networks, in *Ubiquitous Intelligence and Computing*, ed. J. Indulska, J. Ma, L. Yang, T. Ungerer, J. Cao, vol. 4611 of Lecture Notes in Computer Science, Springer, Berlin, 2007, pp. 696–708.

72. M. Batalin, G. Sukhatme, Coverage, exploration and deployment by a mobile robot and communication network, *Telecommunication Systems* 26 (2004) 181–196.

73. C. Chang, C. Chang, Y. Chen, H. R. Chang, Obstacle-resistant deployment algorithms for wireless sensor networks, *IEEE Transactions on Vehicular Technology* 58 (2009) 2925–2941.

74. C. Chang, J. Sheu, Y. Chen, S. Chang, An obstacle-free and power efficient deployment algorithm for wireless sensor networks, *IEEE Transactions on Systems Man and Cybernetics Part A* 39 (2009) 795–806.

75. R. Falcon, X. Li, A. Nayak, Carrier-based focused coverage formation in wireless sensor and robot networks, *IEEE Transactions on Automatic Control* 56 (2011) 2406–2417.

76. G. Fletcher, X. Li, A. Nayak, I. Stojmenovic, Back-tracking based sensor deployment by a robot team, in *Proceedings of IEEE SECON*, Boston, 2010, pp. 385–393.

77. Y. Mei, C. Xian, S. Das, Y. Hu, Y.-H. Lu, Sensor replacement using mobile robots, *Computer Communications* 30 (2007) 2615–2626.

78. X. Li, G. Fletcher, A. Nayak, I. Stojmenovic, Randomized carrier-based sensor relocation in wireless sensor and robot networks, *Ad Hoc Networks* 11 (7) (2013) 1951–1962.

79. E. Petriu, N. D. Georganas, D. Petriu, D. Makrakis, V. Groza, Sensor-based information appliances, *IEEE Instrumentation Measurement Magazine* 3 (4) (2000) 31–35.

80. I. Akyildiz, I. Kasimoglu, Wireless sensor and actor networks: Research challenges, *Ad Hoc Networks* 2 (2004) 351–367.

81. S. Dengler, A. Awad, F. Dressler, Sensor/actuator networks in smart homes for supporting elderly and handicapped people, in *21st International Conference on Advanced Information Networking and Applications Workshops 2007 (AINAW 07)*, Niagara Falls, Ontario, Canada, 2007, vol. 2, pp. 863–868.

82. M. Erol-Kantarci, H. Mouftah, Suresense: Sustainable wireless rechargeable sensor networks for the smart grid, *IEEE Wireless Communications* 19 (3) (2012) 30–36.

83. S. J. Isaac, G. P. Hancke, H. Madhoo, A. Khatri, A survey of wireless sensor network applications from a power utility's distribution perspective, in *AFRICON 2011*, Livingstone, Australia, 2011, pp. 1–5.

84. M. S. Aslam, S. Rea, D. Pesch, Service provisioning for the WSN cloud, in *2012 IEEE 5th International Conference on Cloud Computing (CLOUD)*, Honolulu, HI, 2012, pp. 962–969.

85. T. Li, J. Ren, X. Tang, Secure wireless monitoring and control systems for smart grid and smart home, *IEEE Wireless Communications* 19 (3) (2012) 66–73.

86. A. Boukerche, H. A. Oliveira, E. F. Nakamura, A. A. Loureiro, Vehicular ad hoc networks: A new challenge for localization-based systems, *Computer Communications* 31 (12) (2008) 2838–2849.

87. S. Yousefi, M. S. Mousavi, M. Fathy, Vehicular ad hoc networks (VANETs): Challenges and perspectives, in *2006 6th International Conference on ITS Telecommunications Proceedings*, Chengdu, China, 2006, pp. 761–766.

88. F. Z. J.-J. H. Chih-Yin Lin, Tzong-Chen Wu, New identity-based society oriented signature schemes from pairings on elliptic curves, *Applied Mathematics and Computation* 160 (2005) 245–260.

89. X. R. Li, V. P. Jilkov, Survey of maneuvering target tracking: Part I: Dynamic models, *IEEE Transactions on Aerospace and Electronic Systems* 39 (4) (2003) 1333–1364.

90. R. Schubert, E. Richter, G. Wanielik, Comparison and evaluation of advanced motion models for vehicle tracking, in *2008 11th International Conference on Information Fusion*, Cologne, Germany, 2008, pp. 1–6.

91. S. Parveen, R. Misra, S. Sahoo, Nanoparticles: A boon to drug delivery, therapeutics, diagnostics and imaging, *Nanomedicine: Nanotechnology, Biology and Medicine* 8 (2) (2012) 147–166.

92. P. Wang, J. M. Jornet, M. Abbas Malik, N. Akkari, I. F. Akyildiz, Energy and spectrum aware MAC protocol for perpetual wireless nanosensor networks in the terahertz band, *Ad Hoc Networks* 11 (8) (2013) 2541–2555.

93. N. Agoulmine, K. Kim, S. Kim, T. Rim, J.-S. Lee, M. Meyyappan, Enabling communication and cooperation in bio-nanosensor networks: Toward innovative healthcare solutions, *IEEE Wireless Communications* 19 (5) (2012) 42–51.

94. J. M. Jornet, I. F. Akyildiz, Graphene-based plasmonic nano-antenna for terahertz band communication in nanonetworks, *IEEE Journal on Selected Areas in Communications* 31 (12) (2013) 685–694.

95. M. Tamagnone, J. Gomez-Diaz, J. R. Mosig, J. Perruisseau-Carrier, Reconfigurable terahertz plasmonic antenna concept using a grapheme stack, *Applied Physics Letters* 101 (21) (2012) 214102.

96. T. Otsuji, S. B. Tombet, A. Satou, M. Ryzhii, V. Ryzhii, Terahertz wave generation using graphene: Toward new types of terahertz lasers, *IEEE Journal of Selected Topics in Quantum Electronics* 19 (1) (2013) 8400209–8400209.

97. L. Vicarelli, M. Vitiello, D. Coquillat, A. Lombardo, A. Ferrari, W. Knap, M. Polini, V. Pellegrini, A. Tredicucci, Graphene field-effect transistors as room-temperature terahertz detectors, *Nature Materials* 11 (10) (2012) 865–871.

98. Z. L. Wang, Towards self-powered nanosystems: From nanogenerators to nanopiezotronics, *Advanced Functional Materials* 18 (22) (2008) 3553–3567.
99. L. Gammaitoni, I. Neri, H. Vocca, Nonlinear oscillators for vibration energy harvesting, *Applied Physics Letters* 94 (16) (2009) 164102.
100. F. Cottone, H. Vocca, L. Gammaitoni, Nonlinear energy harvesting, *Physical Review Letters* 102 (8) (2009) 080601.
101. J. M. Jornet, I. F. Akyildiz, Channel modeling and capacity analysis for electromagnetic wireless nanonetworks in the terahertz band, *IEEE Wireless Communications* 10 (10) (2011) 3211–3221.
102. J. M. Jornet, I. F. Akyildiz, Joint energy harvesting and communication analysis for perpetual wireless nanosensor networks in the terahertz band, *IEEE Transactions on Nanotechnology* 11 (3) (2012) 570–580.
103. E. A. Lee, Cyber physical systems: Design challenges, in *2008 11th IEEE International Symposium on Object Oriented Real-Time Distributed Computing (ISORC)*, Orlando, FL, 2008, pp. 363–369.
104. R. R. Rajkumar, I. Lee, L. Sha, J. Stankovic, Cyber-physical systems: The next computing revolution, in *Proceedings of the 47th Design Automation Conference*, 2010, pp. 731–736.
105. G. Wu, S. Talwar, K. Johnsson, N. Himayat, K. D. Johnson, M2M: From mobile to embedded Internet, *IEEE Communications Magazine* 49 (4) (2011) 36–43.
106. S.-Y. Lien, K.-C. Chen, Y. Lin, Toward ubiquitous massive accesses in 3GPP machine-to-machine communications, *IEEE Communications Magazine* 49 (4) (2011) 66–74.
107. Y. Zhang, R. Yu, S. Xie, W. Yao, Y. Xiao, M. Guizani, Home M2M networks: Architectures, standards, and QoS improvement, *IEEE Communications Magazine* 49 (4) (2011) 44–52.
108. X. Li, R. Lu, X. Liang, X. Shen, J. Chen, X. Lin, Smart community: An Internet of Things application, *IEEE Communications Magazine* 49 (11) (2011) 68–75.
109. J. H. Porter, P. C. Hanson, C.-C. Lin, Staying afloat in the sensor data deluge, *Trends in Ecology and Evolution* 27 (2) (2012) 121–129.
110. K.-C. Chen, Machine-to-machine communications for healthcare, *Journal on Computer Science and Engineering* 6 (2) (2012) 119–126.
111. I. Mezei, V. Malbasa, I. Stojmenovic, Robot to robot, *IEEE Robotics and Automation Magazine* 17 (4) (2010) 63–69.
112. A. Lo, Y. Law, M. Jacobsson, A cellular-centric service architecture for machine-to-machine (M2M) communications, *IEEE Wireless Communications* 20 (5) (2013) 143–151.
113. S. Parkvall, A. Furuskar, E. Dahlman, Evolution of LTE toward IMT—Advanced, *IEEE Communications Magazine* 49 (2) (2011) 84–91.
114. P. McDaniel, S. McLaughlin, Security and privacy challenges in the smart grid, *IEEE Security Privacy* 7 (3) (2009) 75–77.
115. S. Basagni, M. Conti, S. Giordano, I. Stojmenovic, *Mobile Ad Hoc Networking*, John Wiley & Sons, Hoboken, NJ, 2004.
116. W. Fang-Jing, K. Yu-Fen, T. Yu-Chee, Review from wireless sensor networks towards cyber physical systems, *Pervasive and Mobile Computing* 7 (2011) 397–413.
117. L. Han, S. Potter, G. Beckett, G. Pringle, S. Welch, S.-H. Koo, G. Wickler, A. Usmani, J. L. Torero, A. Tate, Firegrid: An e-infrastructure for next-generation emergency response support, *Journal of Parallel and Distributed Computing* 70 (11) (2010) 1128–1141.
118. V. L. Narasimhan, A. A. Arvind, K. Bever, Greenhouse asset management using wireless sensor-actor networks, in *International Conference on Mobile Ubiquitous Computing, Systems, Services and Technologies 2007 (UBICOMM '07)*, Papeete, France, 2007, pp. 9–14.
119. T. Melodia, D. Pompili, I. F. Akyildiz, A communication architecture for mobile wireless sensor and actor networks, in *2006 3rd Annual IEEE Communications Society on Sensor and Ad Hoc Communications and Networks 2006 (SECON '06)*, Reston, VA, 2006, vol. 1, pp. 109–118.
120. T. Melodia, D. Pompili, V. Gungor, I. Akyildiz, A distributed coordination framework for wireless sensor and actor networks, in *International Symposium on Mobile Ad Hoc Networking and Computing*, Chicago, 2005, pp. 99–110.
121. J. Heidemann, W. Ye, J. Wills, A. Syed, Y. Li, Research challenges and applications for underwater sensor networking, in *Wireless Communications and Networking Conference 2006 (WCNC 2006)*, Las Vegas, NV, 2006, pp. 228–235.

122. S. He, X. Li, J. Chen, P. Cheng, Y. Sun, D. Simplot-Ryl, EMD: Energy-efficient P2P message dissemination in delay-tolerant wireless sensor and actor networks, *IEEE Journal on Selected Areas in Communications* 31 (9) (2013) 75–84.

123. N. Pogkas, G. Karastergios, C. Antonopoulos, S. Koubias, G. Papadopoulos, An ad-hoc sensor network for disaster relief operations, in *10th IEEE Conference on Emerging Technologies and Factory Automation 2005 (ETFA 2005)*, Catania, Sicily, Italy, 2005, vol. 2, p. 9.

124. E. Cayirci, T. Coplu, Sendrom: Sensor networks for disaster relief operations management, *Wireless Networks* 13 (3) (2007) 409–423.

125. W.-Z. Song, R. Huang, M. Xu, B. Shirazi, R. Lahusen, Design and deployment of sensor network for real-time high-fidelity volcano monitoring, *IEEE Transactions on Parallel and Distributed Systems* 21 (11) (2010) 1658–1674.

126. R. Murphy, J. Kravitz, S. Stover, R. Shoureshi, Mobile robots in mine rescue and recovery, *IEEE Transactions on Robotics and Automation* 16 (2009) 91–103.

127. T. Suzuki, K. Kawabata, Y. Hada, Y. Tobe, Deployment of wireless sensor network using mobile robots to construct an intelligent environment in a multi-robot sensor network, *Advances in Service Robotics* (2008) 315–328.

128. T. Suzuki, R. Sugizaki, K. Kawabata, Y. Hada, Y. Tobe, Autonomous deployment and restoration of sensor network using mobile robots, *International Journal of Advanced Robotic Systems* 7 (2) (2010) 105–114.

129. J. W. Fenwick, P. M. Newman, J. J. Leonard, Cooperative concurrent mapping and localization, in *IEEE International Conference on Robotics and Automation 2002 (ICRA '02)*, 2002, vol. 2, pp. 1810–1817.

130. S. B. Williams, G. Dissanayake, H. Durrant-Whyte, Towards multivehicle simultaneous localisation and mapping, in *Proceedings of Robotics and Automation 2002 (ICRA '02)*, Washington, DC, 2002, vol. 3, pp. 2743–2748.

131. D. Fox, J. Ko, K. Konolige, B. Limketkai, D. Schulz, B. Stewart, Distributed multi-robot exploration and mapping, *Proceedings of the IEEE* 94 (2006) 325–1339.

132. W. Burgard, M. Moors, C. Stachniss, F. Schneider, Coordinated multi-robot exploration, *IEEE Transactions on Robotics* 21 (2005) 376–378.

133. M. Younis, K. Akkaya, Strategies and techniques for node placement in wireless sensor networks: A survey, *Ad Hoc Networks* 6 (2008) 621–655.

134. A. Ollero, K. Kondak, E. Previnaire, F. C. I. Maza, M. Bernard, J. Martinez, P. Marron, K. Herrmann, L. V. Hoesel, J. Lepley, E. de Andres, Integration of aerial robots and wireless sensors and actuator networks: The aware project, in *Proceedings of the IEEE International Conference on Robotics and Automation*, Anchorage, AK, 2010, pp. 1104–1105.

135. A. Ollero, M. Bernard, M. La Civita, L. van Hoesel, P. J. Marron, J. Lepley, E. de Andres, Aware: Platform for autonomous self-deploying and operation of wireless sensor-actuator networks cooperating with unmanned aerial vehicles, in *Safety, Security and Rescue Robotics 2007 (SSRR 2007)*, Rome, Italy, 2007, pp. 1–6.

136. Y. Wang, H. Wu, Delay/fault-tolerant mobile sensor network (DFTMSN): A new paradigm for pervasive information gathering, *IEEE Transactions on Mobile Computing* 6 (9) (2007) 1021–1034.

137. H. Sato, K. Kawabata, T. Yugo, H. Kaetsu, T. Suzuki, Wireless camera nodes deployment by a teleoperated mobile robot for construction of sensor network, in *ICROS-SICE International Joint Conference*, 2009, pp. 3726–3730.

138. I. M. Khalil, A. Khreishah, F. Ahmed, K. Shuaib, Dependable wireless sensor networks for reliable and secure humanitarian relief applications, *Ad Hoc Networks* 13 (2014) 94–106.

Chapter 2

A Review on Renewable Energy Sources, Battery Replenishment Strategies, and Application-Specific Energy Challenges of Wireless Sensor Networks

---

Faiz Haider Khan, Mubashir Husain Rehmani, and Fayaz Akhtar

Contents

Abstract

The emergence of wireless sensor networks (WSNs) has not only paved the way for several life-easing applications, but also played a vital role in technological advancements. The main factor that has always affected the potential of WSNs is energy. Wireless sensor nodes are mainly powered by traditional batteries, which can only provide a limited capacity. As a consequence, sensor nodes cannot remain operational for a long duration or, in some scenarios, cannot even last for a few months. Furthermore, the compact size of a sensor node does not allow for the use of large but long-lasting batteries, and replacement of batteries is also not possible in all circumstances. Several models and approaches for prolonging network lifetime have been proposed in the literature. This chapter consolidates such works into a single source. Different replenishable energy techniques and power-saving algorithms for wireless sensor networks are also discussed in this chapter.

2.1 Introduction

A wireless sensor network (WSN) consists of distributed sensor nodes for monitoring physical and environmental conditions that pass their data through a wireless network [1]. Generally, the nodes are low cost and low power, perform multiple functions, and communicate in short distances. These compact nodes are composed of sensing, data processing, and communicating components.

Unlike traditional sensors, the position of sensor nodes is not to be engineered. This feature allows the random deployment of sensor nodes in inaccessible terrain. Another unique feature of WSNs is the cooperative effort of sensor nodes, that is, onboard processing capability. Nodes use this capability for simple computations and to transmit only the required and partially processed data. Because of these features, WSNs have a wide range of applications in environmental monitoring, health, military, industry, and security. Wildlife tracking, flood detection, irrigation, remote monitoring of patients by a doctor, detection of foreign agents in water and air, monitoring of plants, and assembly lines are a few examples.

WSN has some limitations as well. Sensors nodes are limited in power, computation capacity, and memory. These nodes may not have global identification (ID) because of the large number of overhead and large number of sensors. Moreover, protocols and algorithms proposed for traditional wireless ad hoc networks are not suitable for WSNs because of their unique features and application requirements.

Due to the dense deployment of sensor nodes, neighbor nodes are close to each other. Therefore, multihop communication is expected to consume less power than single-hop communication. Multihop communication can effectively overcome the signal propagation effects that are experienced in long-distance wireless communication. Also, these nodes use broadcast communication, while most ad hoc networks employ point-to-point communication paradigms.

Since sensor nodes have a generally limited irreplaceable power source, when designing a node, low power consumption is set as a priority. As a consequence, the protocols used for communication in a sensor network must also have a priority focus on power consumption. These protocols must have inbuilt mechanisms that give the option to end users to prolong network lifetime as a trade-off for lower throughput and higher delay.

In the literature, many different surveys have been presented that describe energy conservation schemes, recharging mechanisms, and comparisons of energy harvesting systems [2–6]. In [7], the authors first divided the sensor node into its components according to their power consumption. Later in the paper, they described different energy scavenging sources (vibration and thermal) and conversion devices. The authors of [8] broadly classified the energy consumption schemes into three categories: (1) duty cycling, (2) data driven, and (3) mobility-based approaches. Furthermore, they subclassified each category and described each one in detail. In [9], the authors tabulated the sources of energy and described its characteristics (ambient, uncontrollable, and predictable). They presented a comparison of different battery technologies and the application-specific approaches for a variety of different nodes. They discussed which type of battery is required for different nodes according to their application. Furthermore, they also presented energy harvesting architectures, energy harvesting sensor nodes, their applications, and the related challenges. In another survey [10], the authors briefly described the design considerations and tabulated state-of-the-art sensor nodes.

Compared to the aforementioned papers, in this chapter, we discuss the energy harvesting techniques, life estimation of renewable energy sources, battery discharge characteristics, classification of wireless power transmission, and related issues. Theories and applications of on-demand wireless charging are also presented in this survey chapter. Furthermore, we overviewed the experimental analysis of sensor node lifetime already presented in the literature.

The major contributions of this chapter are as follows:

■ We discuss the challenges of and work related to wireless battery charging.
■ We overview proposed energy harvesting techniques and their limitations.
■ We describe wireless power transfer techniques and their classifications.
■ We discuss the experimental analysis of sensor node lifetime.
■ We discuss application-specific problems of WSN.

The rest of the chapter is organized as follows: Section 2.2 discusses the power consumption and saving modes. In Section 2.3, we discuss the challenges and problems in different energy harvesting techniques. Section 2.4 discusses the classification of wireless power transmission and its challenges. We discuss battery discharge characteristics and life estimation of renewable energy sources in Section 2.5. In Section 2.6, we discuss experimental analysis of sensor node lifetime. Section 2.7 discusses the application-specific problems of WSNs. Section 2.8 concludes the survey chapter.

2.2 Power Consumption and Saving Modes

Being a microelectronic device, a wireless sensor node is usually equipped with a limited power source (<0.5 Ah, 1.2 V) [1]. In some critical applications, needs and requirements do not permit the replacement of power resources. Therefore, sensor node lifetime strongly depends on the limited battery capacity. Power consumption and power management are of certain importance due to the fact that the malfunctioning of a few nodes can cause significant topological changes and might require the rerouting of packets and reorganization of the network. In other mobile and ad hoc networks, power consumption is an important design factor, but since power resources can be replaced by the user, it is not the primary consideration. The focus is more on the quality of service (QoS) than on power efficiency. In sensor nodes, since power efficiency directly influences the network lifetime, application-specific protocols can be designed by trading off performance metrics with power efficiency. In a sensor field, nodes are responsible for detecting events, processing data, and transmitting the results. Hence, power consumption can be divided into three domains: sensing, communication, and processing. We discuss each of them in detail in Sections 2.2.1 through 2.2.3.

2.2.1 Communication

In all three domains, sensor nodes use the maximum energy in data communication in both transmission and reception. Usually, the energy cost for both of these is nearly identical for short-range communication with low radiation power (~0 dBm). Figure 2.1 shows power consumption in various modes. In transceiver circuitry, power amplifiers, voltage control oscillators, mixers, frequency synthesizers, and phase-locked loops (PLLs) all consume valuable power. In computing power consumption, it is important to consider not only the active power, but also the start-up power consumption in the transceiver circuit. The high value of start-up time (in the order of hundreds of microseconds) makes the start-up power nonnegligible and can be imputed to the lock time of the PLL. The start-up power consumed by a sensor node begins to prevail the active power consumption as the transmission packet size is reduced. As a consequence, the turning operation becomes inefficient because the large amount of power is consumed by the transceiver in turning on and off each time.

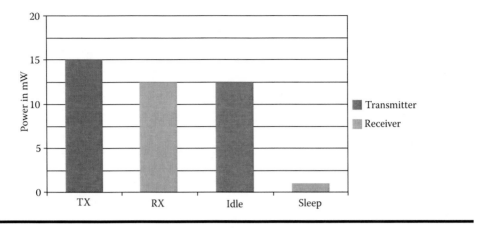

Figure 2.1   Power consumed in different modes.

2.2.2 Data Processing

Energy consumption in data processing is much less than that in data communication, as demonstrated in Figure 2.2. According to the example described in [11], assuming Rayleigh fading and fourth power distance loss, the energy consumption of transmitting 1 KB at a distance of 100 m is approximately equal to executing 3 million instructions by a 100 million instructions per second (MIPS)/W processor. Hence, local data processing is essential for minimizing the power consumption in a sensor network. Therefore, sensor nodes must have inbuilt processing abilities and be capable of communicating with their surroundings, but due to the cost and size restrictions, designers generally choose complementary metal oxide semiconductor (CMOS) technology for the microprocessor. Unfortunately, CMOS has inherent limitations on energy efficiency. A CMOS transistor consumes power every time it is switched, and the switching power is directly related to device capacitance, switching frequency, and square of the voltage swing [1]. Reducing the source voltage means lowering the power consumption in the active state. By reducing the operating frequency, when a microprocessor deals with the time-varying load, it gives a linear decrease in power consumption, but reduces the operating voltage, giving us square (quadratic) gains.

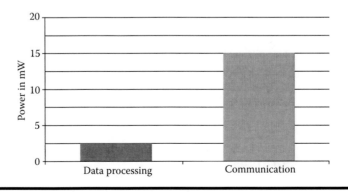

Figure 2.2   Power consumption: CPU vs. transceiver.

On the other side, this reduces the processor peak performance. By recognizing that extreme performance is not always needed, significant energy gains can be obtained. Therefore, operating voltage and frequency are varied dynamically according to the processing requirements. In addition to this, some extra circuitry may be needed for encoding and decoding. For additional circuitry, sensor network algorithms and protocols are designed accordingly to minimize the power expenditures.

2.2.3 Power-Saving Modes of Operation

The operation of power-saving modes must be supported by any type of medium access scheme for the sensor node. For power saving, the most evident mean is to turn the receiver off when it is not needed. Although this method seems to provide significant energy gain and conserve power, due to short packet communication, it consumes more start-up energy. In fact, if we blindly switched off the radio during the idle state to save power, we might utilize more energy than if the radio had been left on. There can be different modes of operation based on the power consumption and latency overhead for a sensor node, which depends on the different states of microprocessor, analog-to-digital (A/D) converter, memory, and transceiver. Various power-saving modes have been proposed and investigated in [12] (Table 2.1).

2.3 Energy Harvesting Techniques

Traditional power sources, such as primary batteries, are energy constrained, and thus in many scenarios fail to meet the lifetime requirements. To prolong operational lifetime, energy harvesting techniques can be employed. In addition to the traditional ones, sensor nodes with energy harvesting capabilities (EH-WSNs) are capable of extracting energy from the surrounding environment and then converting it into usable electrical power [13]. In the traditional scenario of battery-powered nodes, the only way to prolong battery lifetime is to reduce the energy consumption as much as possible. This means that energy-demanding operations such as some security schemes are not possible for energy-limited embedded devices. On the other hand, energy harvesters cunningly draw energy from the environment; for example, in health care, using wearable medical devices, different sources of energy harvesters include indoor light energy, mechanical energy produced by movements, and heat transfer between the human body and the ambient.

2.3.1 CP-ABE

In [13], the authors assess the feasibility of Ciphertext-Policy Attribute-Based Encryption (CP-ABE) in WSNs via an actual implementation in TinyOS for TelosB and Mica2 platforms. This allows determining its energy consumption, memory requirements, and computational complexity, which was guided through the development of specific optimization. Second, the authors designed AGREE, an energy harvesting–aware access control framework for green WSNs. AGREE decreases the energy consumption of CP-ABE schemes by driving the most costly operations to energy harvesting periods.

2.3.2 Clustering Algorithms

Another algorithm proposed in [14] focuses on designing clustering algorithms to maximize network lifetime. A clustered WSN is typically composed of a base station (BS) and a certain number

Table 2.1 Replenishment Strategies Used in Wireless Sensor Networks

Strategy	Replenishment Technique	Simulation/ Implementation	Simulator/Platform
[13]	AGREE, an energy harvesting–aware access control framework, and CPABE	Simulation and experimental implementation	TinyOS for TelosB and Mica2 C and Python for result verification
[14]	Clustering algorithms	Theoretical analysis and simulation	Qualnet for a two-dimensional network
[15]	WRSN based on WISP	Analytical approximation and simulation	Simulation based on the settings of real WISP platforms
[16,17]	Battery recovery effect	Experimentation	TelosB and Imote2
[18]	Mechanical energy harvesting using PZT nanofibers	Experimentation	Piezoelectric nanogenerator based on lead zirconate titanate (PZT) nanofibers
[19]	Piezoelectric vibration-based generator	Simulation and experimentation	PZT-5H with a brass center shim
[20,21]	Thermal energy harvesting for WSN	Experimentally and theoretically	WSN (ABBRF03) [20], ZigBee-based radiator [21]
[22,23]	Wind energy harvesting for WSN	Experimentation	Aeroelastic flutter generator [22], piezoelectric bimorphs [23]
[24,25]	Solar energy harvesting for WSN	Experimentation	PIC 18L452, ChipCon CC1000, temperature and humidity sensor [25], MicaZ, long-range RF, twin batteries [24]
[26–29]	Ambient RF energy harvesting	Experimentation and simulation [26, 28, 29], hardware–software codesign approach [27]	Spiral antenna and HFSS [26], 50 Ω COTS 3 dBi antenna connected at the output of the agilent RF source N5158A [27], microwave generator [28], IEEE 802.15.4/ZigBee and development kit TWE-EK-001 [29]
[30]	AmbiMax: Autonomous energy harvesting platform for multisupply WSN	Experimentation	Eco platform

(Continued)

Table 2.1 (Continued)   Replenishment Strategies Used in Wireless Sensor Networks

Strategy	Replenishment Technique	Simulation/ Implementation	Simulator/Platform
[31]	EDF scheduling for real-time energy harvesting systems	Using modeling	–
[32]	Mode switching scheme using random beamforming for opportunistic energy harvesting	Simulation	–
[33]	Laser-based energy transfer	Experimentation	TelosB and laser u-power beaming
[34–36]	Ultrasonic transcutaneous energy transfer	Experimentation [34,35], simulation [36]	Test tank circuit [34], using flat circular transducers [35]
[37]	Inductive mobile charging	Simulation	–
[38]	Resonance coupling	Experimentation	Strongly coupled magnetic resonances
[39]	SuReSense mobile energy transfer	Simulation	Smart grid scenario

of clusters. Each cluster consists of cluster head (CH) and some non–cluster head (NCH) nodes, as illustrated in Figure 2.3. The CH is responsible for receiving data from NCHs, processing the data, and then forwarding the information to the BS either directly, through other CHs, or through one or multiple relay nodes. Relay nodes are responsible for forwarding the received data from other nodes and may not be responsible for local sensing. In cluster WSNs, transmitting data to a CH nearby rather than to a faraway BS reduces the energy consumption of NCHs. The

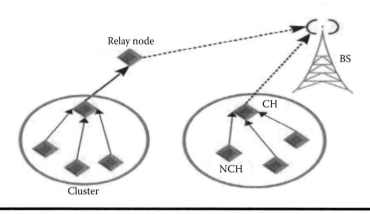

Figure 2.3   Clustering.

life span of CHs is short because they have the burden of the whole cluster for processing and transmitting the data, especially in the absence of relay nodes between CHs and BS. So to prolong the lifetime of clustered WSNs, it is necessary to reduce the energy consumption of CH. As the energy consumption of data transmission largely depends on the communication distance, a good location for CH plays a critical role for prolonging network lifetime. On the other hand, an inappropriate CH location may force CH to communicate with BS over a long distance, thus reducing the lifetime of the network.

Existing work on the lifetime maximizing problem can be classified into two major categories: centralized methods and distributed methods [14]. Centralized methods require knowledge of the sensor locations to achieve global optimization with respect to certain performance metrics. On the other hand, distributed methods make decisions based on local information exchanged between neighboring sensors, achieving a better scalability. Zhang et al. [14] focus on centralized methods, as they can serve as a useful benchmark for the performance of distributed methods. Another important focus of [14] is on energy harvesting techniques. The EH sensors can harvest energy from the environment and convert it into electrical energy with a large number of recharge cycles. But WSNs composed of only EH sensors remain impractical due to their high cost and low number of achievable duty cycles.

2.3.3 WRSN Based on WISP

A promising approach, wireless rechargeable sensor networks (WRSNs), has been introduced in recent years; in WRSNs, battery free nodes scavenge energy from sources present in the environment. These sources include solar, wind, vibrations, and radio frequency (RF) signals. Battery free nodes can provide two major advantages. First, WSNs can operate for a long time before the battery needs to be replaced. Second, for many important applications, sensor nodes can be manufactured in extremely small sizes. In [15], the authors made a wireless identification and sensing platform (WISP) tag of version WISP4.1DL according to Intel schematics. Since WISP tags are gaining attention in the research community and act as an ideal platform for indoor sensing and activity recognition, the authors studied the fundamental design question of how to deploy an RF energy source (readers) in a network to ensure continuous operation of tags; they refer this as an "energy provisioning" problem. This problem is of extreme importance for achieving maintainable system performance while reducing the cost. The authors of [15] further categorized and defined the energy provisioning problem into two forms: point provisioning and path provisioning. Second, they presented a realistic wireless recharge model to record for the harvested energy from a single reader and then extended it to multiple readers. They validated their model by a real hardware experiment. Third, they developed solutions to the point and path provisioning problems. They showed that the number of readers needed for a region can be greatly reduced in the point provisioning problem provided using their proposed deployment, compared to others. In path provisioning, by taking advantage of mobility, a number of readers can be further reduced. Furthermore, for both problems, they derived analytically upper-bound approximation ratios of the proposed solutions to the optimal ones. Their analysis, supported by the simulation results, showed that their deployment method can greatly reduce the number of readers compared to traditional methods.

2.3.4 Battery Recovery Effect

In the literature, different studies in the field of energy optimization consider batteries an ideal energy source; that is, energy can be drained at a constant discharge voltage and stopped and resumed anytime to regain the same voltage. Most commercial batteries provide energy due to a complex internal chemical reaction. These reactions depend on various environmental factors, such as temperature, discharge duration, and discharge current. Among various phenomena, a particularly beneficial one is the battery recovery effect; that is, chemical substances present in the battery, if left idle for a long enough duration, will replenish themselves, hence extending battery lifetime.

The authors of papers [16,17] were motivated by this effect; they first experimentally examined the gain of battery runtime due to the battery recovery effect on commercial sensors. They found that under appropriate duty cycling, such gain can be significant. Their experiments also showed that there exists a saturation threshold after which no consecutive idling period will contribute toward recovery. If we carefully adjust the idle periods of batteries, we can maximize the recovery effect without wasting idling time which may hamper the quality of service. Furthermore, to understand the behavior of the battery recovery effect in the presence of random sensing activities, the authors of [16,17] formulated a Markov chain model and provided an analysis on the gain of battery runtime. Later, they studied the effect of duty cycling and buffering to control the battery recovery effect. In a sensor node, most of the energy is consumed by the RF transceiver operations, compared to processing and sensing.

Therefore, duty cycling is used many times at short intervals to determine the on and off periods of the RF transceiver while retaining the other parts of the sensor module (e.g., sensing and processing units). Hence, proper duty cycling can maximize the battery recovery effect. Lastly, the authors proposed a more energy-efficient duty cycling scheme by adjusting the sleep duration of the RF transceiver, which takes maximum advantage of the battery recovery effect and then analyzes its performance based on the latency of data delivery [16] and duty cycle rates [17].

2.3.5 Mechanical Energy Harvesting Using PZT Nanofibers

Piezoelectric nanowire and nanofiber-based generators have much potential for powering wireless electronics, portable devices, and implantable biosensors via a conversion of mechanical energy into electrical energy [18]. Despite this, the reported piezoelectric voltage constant of semiconductor piezoelectric nanowires is lower than that of lead zirconate titanate (PZT) nanomaterials. Therefore, the voltage constant of piezoelectric nanomaterials, output voltage, and power of the nanogenerator need further improvement for practical applications. Moreover, the performance of the nanogenerator is also dependent on the fabrication method of the semiconductor nanomaterials. It is difficult to grow single-crystal nanowires longer than 50 µm with diameters less than 100 nm [18]. Thus, the short length of nanowires can significantly restrict the output voltage of nanogenerators.

PZT, a ceramic material with high piezoelectric voltage and dielectric constant, is widely used for mechanical-to-electrical energy conversion. Unlike other semiconductor types of piezoelectric materials, for a given volume and the same energy input, it is capable of generating much higher voltages and power outputs. This makes the PZT ideal for nanogenerator or nanobattery applications. In [18], the authors demonstrated a new piezoelectric nanogenerator based on PZT nanofibers with an approximate diameter of 60 nm and length of 500 µm. From this nanogenerator, the peak output voltage was 1.63 V, and the output power was 0.03 µW with a load resistance of 6 MΩ. For preventing damage, PZT nanofibers were embedded in a soft polydimethylsiloxane (PDMS) polymer matrix by the authors.

2.3.6 Piezoelectric Vibration-Based Generator

For powering wireless electronics and sensor network nodes, mechanical vibration has received much attention from various researchers. Most of the wireless electronics run on the onboard battery. As the devices have been miniaturized and the size of network increased by many times, it is impractical to replace the batteries of these devices. Thus, alternative methods are needed for powering wireless devices. Different methods, such as generator-based electromagnetic, electrostatic, and piezoelectric conversion, have already been presented in the literature. Each type of converter has its own benefits and drawbacks according to the application need. Heating, ventilation, and air conditioning (HVAC) ducts, manufacturing and assembly equipment, and household appliances ranging from 0.2 to 10 ms^{-2} in amplitude at frequencies from 60 to 200 Hz are the common vibration sources [4]. The authors of paper [19] focused on the piezoelectric conversion. In [19], authors investigated the modeling and design of a piezoelectric vibration-to-electricity converter; they presented an analytic model of the generator and then further used it as a basis for design optimization. A vibration source of 2.5 ms^{-2} at 120 Hz was chosen by the authors and demonstrated a power output of 375 μW from a 1 cm^3 generator. Moreover, the same generator of 1 cm^3 was also used to power a custom-designed radio of 1.9 GHz.

2.3.7 Thermal Energy Harvesting for WSN

2.3.7.1 Harvesting Energy from Heat Flow at Room Temperature

Electrical energy from heat flow has been harvested for many decades; traditionally, a powerful source for boiling the water was employed that would in turn-drive a generator. Harvesting energy for powering wireless devices using traditional processes is not possible. In [20], the author demonstrated how the temperature difference between the indoors and outdoors can be harvested to generate electricity. To enhance the lifetime of wireless sensor nodes (WSNs), the author had two major objectives. One was to design ultralow power consumption hardware, and the second was to obtain energy from the environment. To fulfill these objectives, he proposed a WSN structure based on CC2520, which was powered by using the thermoelectric energy harvesting technique. He further demonstrated how undesired heat flux between the outdoors and indoors through the windows of building can be harvested to power the WSN. The author concluded [20] that for the elimination of batteries, thermoelectricity is ideal for regions where the outdoor temperature is below zero most of time.

2.3.7.2 Low-Temperature Energy Harvesting System

Alternative to batteries, recent developments in thermoelectric (TE) materials and structures offer unique and attractive power generation capabilities. The production of very low-power, low-cost wireless devices encourages large-scale deployments of WSNs. To overcome the problem of limited lifetime in large-scale deployment of WSNs, the authors of [21] proposed a low-temperature thermal energy harvesting system capable of harvesting heat energy from a temperature gradient and then converting it into electrical energy for powering a ZigBee device. They developed a prototype that was based on three subsystems for the purpose of extracting heat energy from a radiator. The two major benefits of the proposed system were the long lifetime and high efficiency. Their experimental results showed that a maximum of 150 mW could be harvested from the developed prototype, and the system could operate normally even if harvested voltage was as low as 0.45 V.

2.3.8 Wind Energy Harvesting for WSNs

Among several renewable sources, wind energy produces the most energy compared to others, but at the same time, wind generators are too big in size and extremely costly if the size has to be decreased. The authors of [22] designed, constructed, and tested a wind energy harvester (WEH) to determine whether wind energy is viable for powering WSNs. It consisted of a wind generator and power management unit. The wind generator designed by the authors employs aeroelastic flutter to convert wind energy into electrical energy. Wind energy is first converted into mechanical energy using a membrane, and further, the mechanical energy is converted into electrical energy by using an electromagnetic transducer. Two proposed methods in the literature were also studied and discussed by the authors of [22] for the conversion of wind energy to electrical energy. Also, the importance and two methods of designing a power management unit with maximum power point tracking (MPPT) were discussed by the authors. For maximizing the storage energy, the power management unit with MPPT allows it, while the unit without MPPT does not. According to the results, the wind generator was operational at wind speeds of 2–9 m/s and was capable of supplying 70 mW of power. This means that the wind energy harvester had the ability to power WSNs for 8 h when the 0.07 F supercapacitor was fully charged. Moreover, they found that the power management unit had poor power efficiency without MPPT.

In [23], four challenges related to the small-scale electric energy generation from windmills were presented: (1) conversion of the random wind flows into periodic AC mechanical stress, (2) realization of a high enough magnitude of stress (>0.5 N), (3) enhancing the frequency of the stress cycle (>5 Hz), and (4) the cost-effectiveness and light weight of windmills. The authors also presented the analysis of two windmills and discussed their problems and disadvantages. Furthermore, they presented the optimized small-scale piezoelectric windmill. The structure was made up of plastic, had 18 piezoelectric bimorphs, and was highly cost-effective. They tested the windmill at an average speed of 10 mph, and it provided 5 mW of continuous power. Moreover, they found the threshold speed of the windmill, which was 5.4 mph.

2.3.9 Ambient RF Energy Harvesting

Since the 1950s, radio frequency (RF) has been another technique of energy harvesting for powering low-powered electronics and wireless devices in which microwaves are rectified and converted to direct current (DC). Initially, it was proposed for helicopter powering, solar power satellite (SPS), the SHARP system, and recently, the radio frequency identification (RFID) system [26]. The authors of [26] studied two systems for RF energy harvesting and presented the measurement of ambient RF power density. The first system was a broadband system (1–3.5 GHz) without a matching circuit, while the second system was a narrowband system (1.8–1.9 GHz) with a matching circuit. For supplying power to wireless sensor networks, they proposed using commercial RF broadcasting stations, such as Wi-Fi, radar, and Global System for Mobile Communication (GSM). DC power is related to the available RF power source and conversion efficiency by the following equation:

$$P_{dc} = \eta_{RF/DC} \cdot P_{RF}$$

To optimize the DC power, the harvested choice of antenna and frequency is very important. The results in [26] showed that the energy recovered from RF harvesting was not sufficient to power devices but could be used for charging a supercapacitor. In another paper [27], two issues in the development of RF remote powering systems were discussed. The first was related to the limitation on the maximum power with the RF communication standard and spurious emission, while the second was the sensitivity of the RF harvesting circuit. Typically, the RF harvesting circuit consists of a matching network, voltage regulator, voltage rectifier, and power management circuit. Not only is the RF/DC rectifier efficiency important, but also the power dissipation of the matching network and power consumption of the control logic play a critical role in the optimization of the whole RF harvesting module efficiency. In [27], the authors presented an energetically autonomous WSN device designed to enhance safety in the vehicle to connect extra equipment to the main chassis. The developed WSN device incorporates an RF energy harvesting circuit that was optimized by minimizing the power loss of the whole circuit, that is, matching network, RF/DC rectifier, and voltage regulator. They followed a hardware–software codesign approach which combined the ultralow power wireless architecture with smart-task management algorithms and the energy harvesting module. Their developed RF harvesting circuit operated in the 865–868 MHz RFID band. Measurements in [27] showed that RF harvesters can gather up to 50 µW at 3 m from the power source, with efficiencies higher than 30% over a range of 10 dBm.

2.3.9.1 Microwave Power Transmission

An efficient low-cost RF harvesting technique was presented in [28] for the wireless recharging of WSNs. According to the authors, there are three methods for wireless energy transfer—infrared, radio waves, or microwaves—but in the proposed technique, microwaves were used since they provide energy transfer at both low and high power to longer distances. Their developed system had the capability of interfacing with the computer for the purpose of displaying the energy transfer status of sensor nodes. They used pyramidal horn-shaped waveguide antennae in their project for the purpose of carrying electric power. While maintaining the reliability, their developed system cost US$44. The authors of [29] studied how to charge or operate a ZigBee device by intermittent microwave power transmission, also compatible with WSNs. They found that an intermittent microwave power transmitter drove the rectifier circuit at high efficiency. They also affirmed that in high power density with intermittent microwave radiation, the ZigBee device could communicate correctly, while in slight power density with continuous microwave irradiation, it could not communicate. Lastly, in [29] they demonstrated the ZigBee device powered by an intermittent microwave transmitter communicating and working correctly.

2.3.10 AmbiMax: Autonomous Energy Harvesting Platform for Multisupply WSNs

WSNs in many sensing environments can scavenge enough energy from renewable sources to power sensor nodes for an indefinite period. However, there are several issues in designing sensor nodes with energy harvesting capabilities; harvesting efficiency, autonomy of harvesting control, and expandability to multiple sources are a few of them that were under the consideration of the authors [30].

Most wireless systems developed to date lack power conversion efficiency. As a result, due to low harvesting efficiency, the size and cost of the systems are increased. For example, a system with low harvesting efficiency needs a large solar panel to yield the same power level as that from an efficient one, or the system needs large batteries to sustain its operation.

The major problem in energy efficiency was maximum power point tracking or the impedance matching between the source and supply at runtime, where the maximum power point is the point on the I-V curve that maximizes the power output at the given level of light intensity [30].

Different methods for MPPT had originally been proposed for large, complex systems, such as satellites, in the literature, but these algorithms were prohibited for small WSNs. For this reason, researchers proposed running MPPT on the same microcontroller unit (MCU) as a low duty cycle, which performs sensing control and other functions. But this proposed method doesn't work when the MCU is asleep. That's why autonomous MPPT was required.

The last issue in designing sensor nodes with energy harvesting capabilities was expandability to multiple energy sources because this increases the power availability.

To solve all these design issues, the authors of [30] proposed and developed AmbiMax for MPPT and energy storage for WSNs. They described the actual implementation of AmbiMax and further compared the harvesting efficiency with Eco and Mica2 platforms. Their experimental results showed that AmbiMax achieved higher efficiency by performing MPPT on multiple power sources simultaneously and autonomously.

2.3.11 EDF Scheduling for Real-Time Energy Harvesting Systems

In [31], the authors addressed the problem of scheduling that arises in energy harvesting systems with real-time constraints. They addressed the scheduling problem for a single-processor device that executes preemptable time for critical tasks. Their processing model presents three major assumptions: (1) the instantaneous power consumption of any task is no less than the incoming power from the harvesting unit, (2) the energy consumed by any task is not necessarily proportional to its execution time, and (3) their approach to task scheduling is preemptive nonidling. The authors of [31] asked whether the traditional task scheduling algorithm earliest deadline first (EDF) is convenient for energy harvesting systems under nonidling execution settings and showed that EDF has a zero competitive factor but is still optimal for nonidling settings.

2.3.12 Mode Switching Scheme Using Random Beamforming for Opportunistic Energy Harvesting

Harvesting energy from radio signals has gained much attention in recent years and has been successfully implemented in applications such as body sensor networks (BSNs) for medical implants and radio frequency identification (RFID) systems. This technique opens up the path for simultaneous wireless information and power transfer (SWIPT), as radio signals carry both energy and information at the same time. Recently, SWIPT has been investigated by many researchers for various wireless channels, such as the multiantenna channel, the point-to-point additive white Gaussian noise (AWGN) channel, the fading AWGN channel, the multicarrier-based broadcast channel, and the relay channel [32]. The main challenge to achieve maximum wireless energy transfer (WET) and wireless information transfer (WIT) simultaneously is to design efficient receiver architectures. In [32], the authors studied a multiple-input single-output (MISO) multicast SWIPT network with one multiantenna transmitter transmitting common information to single-antenna receivers and at the same harvesting energy at the receiver. The authors [32] then proposed a novel receiver mode switching scheme for SWIPT based on a new application of the conventional random beamforming technique. They assumed that the channel state information (CSI) is only known at the receivers but not available at the transmitter. The utilization of a random beamforming technique in the mode switching scheme at the multiantenna transmitter generates artificial channel variations at each receiver to enable more energy harvesting when the channel power exceeds the given threshold. They investigated the harvested energy and power outage probability, trade-offs in the quasi-static fading channel, and the achievable information rate. They further compared their proposed scheme with periodic receiver mode switching without random transmit beamforming and showed that their proposed scheme was able to achieve better rate–energy trade-offs when the harvested power was sufficiently large. Finally, with the help of simulations, the authors [32] showed that using a single beam for large power harvesting targets gives the best trade-offs between average information and average harvested power.

2.3.13 Ultrasonic Transcutaneous Energy Transfer

In recent years, the use of medical devices has increased significantly, the majority of which are implanted in the body and consume very little power for their operation, typically in the range of hundreds of milliwatts to 10 W [34]. Intrabody electrical sources were presented in the literature for powering implanted devices, such as the one quoted in [34], suggesting intrabody energy harvesting in which biomechanical energy created by the movements of internal organs is converted into electrical energy by using piezoelectric material. The technique presented was able to generate an average power of below 1 mW.

A technique based on the penetration of energy through tissue, called transcutaneous energy transfer (TET), powered those devices that consume an average power of <10 W. Currently, TET devices rely on electromagnetic energy transmission. Electromagnetic TET is capable of transferring energy in the range of watts, but it has some drawbacks, such as low coupling (0.1), which can be the reason for interference with nearby operating devices, such as pacemakers. Despite its drawbacks, electromagnetic TET is still suitable for powering devices that consume large amounts of power. For powering low-power implanted devices (in the range of tens of milliwatts), such as biosensors and glucose indicators for monitoring purposes, ultrasonic transcutaneous energy transfer (UTET) has gained much attention.

In [34], the authors proposed a UTET device capable of transmitting energy up to 40 mm deep via a constant single frequency operating at a 673 kHz acoustic wave. They demonstrated that the proposed UTET has an overall 27% peak power efficiency at 70 mW output power. In their proposed system, first, the electrical energy was converted to a pressure wave, and then it was transmitted transcutaneously. The implanted transducer, which consisted of a PZT plane disc and a thick acoustic matching layer of graphite, converted the received acoustic energy back to electrical energy. On the implanted side, the power rectifier attained 88.5% power transfer efficiency. In summary, the authors [34] discussed design considerations of UTET, such as operating frequency, acoustic layer matching, piezoelectric material selection, and safety issues. The authors of [35] also proposed a UTET device, but unlike [34], it was based on the kerfless transmitter with Gaussian radial distribution. The UTET was operated at 650 kHz with a constant wave and was proposed to power implanted devices up to 50 mm deep. The authors first presented a numerical solution of the Rayleigh double integral and compared it using Comsol multiphysics with a finite-element simulation. They further presented the fabrication of six equal area elements of disc-shaped PZT coated with thin silver layers. Using pig muscle tissue, they experimented and showed a 39.1% peak power transfer efficiency at a power level of 100 mW. Lastly, the authors of [35] presented a power driver for the excitation of a transmitter array having 91.8% efficiency and a rectifier having 89% efficiency for the implanted transducer. The effectiveness of the proposed UTET was further confirmed by the pressure and power transfer measurements within a test tank.

A comparison between ultrasonic and inductive wireless power delivery for miniature biomedical implant devices was presented in [36]. The authors studied, discussed, and investigated several contributions presented in the literature regarding ultrasonic and inductive power delivery and then concluded that none of the researcher groups accounted for high attenuation in human

Table 2.2 Ultrasonic Transmission in Literature

Operating Frequency	Size (Diameter)	Separation between Receiver and Transmitter	Efficiency	Reference
1 MHz	30 mm	7–100 mm	20%	[36]
75 kHz	–	2.5 cm	0.01%	[36]
840 kHz	25–30 mm	5 and 105 mm	21%–35%	[36]
673 kHz	15 mm	5 mm	27%	[36]

tissue, especially at large distances. In human tissue, the attenuation of ultrasound is higher than that in water, that is, 0.9 dB/cm·MHz for soft tissue and 0.002 dB/cm·MHz for water [36]. They built models to compare both power delivery methods in terms of power efficiency. They showed from their simulation results that at a small distance of 1 cm between the source and receiver, the inductive system had 89% efficiency, compared to 39% efficiency for a receiver of 10 mm diameter. On the other hand, at a large distance of 10 cm, the ultrasonic system had better efficiency, that is, 0.2% for ultrasonic and 0.013% for inductive, for a receiver of 10 mm diameter (Table 2.2).

2.4 Wireless Power Transfer

Although a wide range of energy harvesting techniques have been proposed by several researchers for extracting energy from the environment, these techniques are heavily dependent on the environment. Moreover, the size of the energy harvesting device is also a concern, and thus these techniques remain limited in practice. Another promising technology capable of addressing the WSN constraints is wireless power transfer (WPT).

A brief and concise history and recent advances in WPT technology are presented in [40]. The authors examined inductive coupling, electromagnetic radiation, and magnetic resonance coupling. They discussed each one's strengths, weaknesses, and applications and presented the data in a tabular form. In summary, they discussed how various WPT technologies can be employed by WSNs and how their lifetime energy problem can be solved. The following subsections overview some WPT techniques.

2.4.1 Via Strongly Coupled Magnetic Resonance

Nikola Tesla, in the early 20th century, was the first man to devote effort to the transmission of power wirelessly. Due to the excessive use of autonomous electronic devices such as cell phones and laptops in the last decade, wireless power transmission has gained much attention in the research community. The authors of [38] discussed literature regarding WPT, particularly radiative transfer, a feasibility analysis of using resonant objects, a strongly coupled regime of operation, and midrange power transfer. Their main focus was on magnetic resonances. Magnetic resonances are suited to those applications that do not interact with the magnetic field. They used self-resonant coils in a strongly coupled regime and presented an experimental demonstration of transferring nonradiative power at a distance of up to eight times the radial distance of coils. They transferred 60 W power with 40% efficiency at a distance of up to 2 m. Furthermore, they presented a quantitative model that matches the experimental results within 5%. Lastly, they discussed the practical applicability of the system and believed that the efficiency of the proposed scheme and power transfer distances could be improved by silver plating the coils.

2.4.2 Via Mobile Wireless Charging Vehicle

Recent advancement in the field of wireless power transfer offers the potential of removing the performance obstacle of WSNs. The advancement by [38] has given a path to prolong the lifetime of sensor nodes. The authors showed a technique in their work called magnetic resonance coupling through which wireless power transfer is both practical and feasible. They experimentally showed that the energy storage device doesn't need to be in contact with the energy receiving device. They used two magnetic resonance objects having the same resonance frequency to exchange energy efficiently. Furthermore, a wireless power transfer charging device and receiving device don't need to be in the line of sight. The power transfer efficiency, however, decreases with distance. They demonstrated powering a 60 W bulb at a distance of 2m with 40% efficiency. This technique brought clean electrical energy that was efficiently generated to a sensor node periodically and charged its battery wirelessly. It has already been applied in medical sensors and the health care industry.

In [41,42], the authors considered the scenario of a mobile wireless charging vehicle (WCV) periodically traveling inside the sensor network and charging each sensor node's battery wirelessly. The authors introduced the renewable energy cycle and provided necessary and sufficient conditions. These papers proved that the optimal travel path for the WCV is the shortest Hamiltonian cycle. Moreover, the authors also developed a near-optimal solution by a piecewise linear approximation technique and guaranteed its performance.

In another paper [37], the authors designed a system that consists of a mobile wireless power charger, a network of sensor nodes fitted with power receivers, and an energy station responsible for monitoring the energy status of sensor nodes. The designed system is also capable of determining the charging schedule that can be executed by the mobile charger. The authors built a prototype, conducted experiments, and judged the feasibility and performance of the system on a small-scale network. Furthermore, they performed simulations to study and evaluate the performance of the proposed system on a large-scale network. Their results showed that the proposed system using the wireless charging technology effectively increases the battery lifetime of sensor nodes, which in turn prolongs the network lifetime significantly. The authors of [43] noted that the mobile charger optimization problem is (inherently) computationally hard and formulated the wireless recharge problem as the charger dispatch decision problem (CDDP). They proposed distributive and adaptive methods that use only limited network information. Their method can be used together with any underlying routing protocol, unlike other state-of-the-art approaches. Moreover, their protocols are dynamically and distributively adaptable to network diversities; for example, they deal well with heterogeneous node placement. The authors also identified two key issues while proposing alternative strategies for efficient recharging: (1) to what extent each sensor node should be recharged and (2) what good trajectories the mobile charger should follow. Their extensive simulation showed significant performance gain with respect to network lifetime. The authors of [44] proposed a system model that uses mobile nodes to deliver power to deployed sensors in WSNs. In this approach, a mobile node was equipped with charging equipment to inductively recharge sensors. According to [44], by using this approach, WSNs can work perpetually. Moreover, they proposed three charging schemes for inductively charging: region patrol charge (RPC) scheme, region inquire charge (RIC) scheme, and distance- and energy-aware charge (DEC) scheme. They demonstrated the performance of these schemes by simulations. Their simulation results showed that RPC is the simplest with the lowest cost and relatively poor performance. Both RIC and DEC can inductively charge the nodes in time and minimize the energy consumption, while DEC has better performance than RPC, especially in data communication amount.

2.4.3 Via Optical Wireless Recharging Mechanism

Optical wireless communication has distinct advantages over other wireless systems due to its high achievable data rates and low power consumption requirements [45]. Optical wireless recharging can prove to be advantageous in difficult and totally inaccessible environments, such as mines, underwater, and in space, where the replacement of batteries is impossible. Much effort has gone into developing the charging mechanism over short and midrange distances.

In [46], the authors devised a model that optically recharges RF sensor nodes fitted with the recharging facility. They proposed both line of sight (LOS) and quasi-diffuse for recharging purposes. In their analysis, they showed that their proposed model can give better performance and a fairly large increase in network lifetime. In [45], the authors proposed a similar recharge mechanism for juncture nodes that employs free space optics (FSO) using infrared (IR) laser, corner cube retroreflectors (CCRs), and beam-steering lasers. They also proposed a thin-film corner cube retroreflector (TCCR) as a standard to CCR for selective recharging. Furthermore, their proposed recharging mechanism supports both on-demand and scheduled recharging.

2.4.3.1 WSN Power Grid Based on Sensor with Lasers

Wireless energy research removes the requirement for an energy source to be on or near the sensor node. In [33], the authors first discussed the issues of coupling the energy sources with the sensing activities and then presented a practical energy distribution architecture. The proposed architecture allowed decoupling energy supply from sensing activities, enabling the use of abundant energy sources present far away from the sensing environment and solving the problem of unequal energy consumption and restricted lifetime of WSNs. For the purpose of practically enabling decoupling in WSN, they identified four requirements in wireless technology. The authors [33] demonstrated energy transfer for practical decoupling using a laser-based micropower beaming mechanism at a range of 100 m. Furthermore, using an energy distribution protocol, they designed and implemented LAMP architecture for the purpose of managing the energy supply in both mesh and clustered WSN deployments. Lastly, they evaluated and showed that their implementation can support TelosB motes with a 7.4% duty cycle at a distance of more than 100 m, and with the additional cost of $29 per mote, LAMP can perpetually support mesh functionality for up to 40 sensors or 120 nodes in clustered operation.

2.4.4 NDN-Based Real-Time Wireless Recharging

The efficiency and lifetime of a network critically depends on the recharging policy, that is, at what time which nodes need to be recharged and in what pattern. Fundamentally, for the recharging process, sensor nodes periodically report their energy levels, and then an algorithm computes a specific order to recharge the nodes. But this fundamental process doesn't consider important practical issues, which badly limits the applicability in a real environment. In [47], the authors proposed NETWRAP, a Named Data Networking (NDN)–based real-time recharging framework that optimizes the recharging policy under dynamic network conditions. A scalable and efficient energy information collection protocol continuously collects the battery level from all the sensor nodes instead of periodically collecting the energy levels. A mobile vehicle (SenCar) receives this latest energy information and decides the recharging. The recharging of those sensor nodes whose energy levels are below the critical threshold has high priority in order to deal with unpredictable emergencies. The authors borrowed concepts and mechanisms from NDN to design a set of protocols that continuously collect and hand over the information to SenCar, including unpredictable emergencies. Mobile receivers are also supported by this technique, since routing states are constantly updated in intermediate nodes in order to follow the movements of receivers. Getting timely energy information as SenCar changes its location is essential. The authors of [47] also studied the conditions for perpetual operations of the network under the dynamic recharging framework. They derived the probability of an energy-neutral requirement (i.e., energy replenished should be more than or equal to the energy consumed by the node) condition to hold in the network. An administrator can approximate the probability of a network perpetually operating if the network and SenCar parameters are given.

Furthermore, they also discovered that optimal recharging of multiple emergencies is an orienteering problem with knapsack approximation that has high accuracy in typical network environments. Lastly, they compared their real-time framework with static approaches, and with the help of extensive simulations, they demonstrated the effectiveness and efficiency of the proposed framework, which validated the theoretical analysis.

2.4.5 Theories and Applications of On-Demand Wireless Charging

For practical applications of sensor networks in the real world, sensor nodes usually use onboard batteries or supercapacitors for their operation [48]. Due to limited battery lifetime, energy becomes the most precious resource of the system, and its efficient use is of extreme importance. To improve the system's sustainability, recently researchers have been putting their efforts into exploring the concept of mobile energy chargers to replenish the energy supply of individual sensor nodes [49]. The main focus of the existing research is on the offline scenario, in which individual nodes are charged in a periodic and deterministic manner [50]. Yet, due to interaction with the environment, energy consumption analysis demonstrated high diversity. Sensor nodes with integrated energy harvesting modules showed that the amount of harvested energy fluctuates greatly. Also, the uncertainty in the energy demand and supply suggests that the existing periodic and deterministic techniques of recharging may suffer from performance degradation and need further improvement.

After noticing the limitations in the existing techniques, the authors of [51] examined the on-demand energy replenishment in wireless sensor networks and proposed the theoretical foundation for such a process. They investigated how, without prior knowledge of demands that will be raised in the near future, mobile chargers schedule and carry out the charging of individual sensor nodes and what is the achievable performance. For this purpose, they examined nearest-job-next with preemption (NJNP), a simple but efficient discipline for the on-demand charging problem. According to the spatial and temporal properties of sensor nodes, NJNP schedules the charging of individual nodes. Within constant factors of optimal solutions, they proved the performance of NJNP. Their contributions [51] included a mathematical formulation of the on-demand charging problem in WSNs and theoretical establishment of the charging performance with the NJNP discipline. These mathematical results not only showed the performance of NJNP, but also provided the guideline for the design of more advanced charging schemes. Furthermore, the authors of [51] presented four theorems that demonstrate the performance of NJNP within a constant factor with respect to the system throughput and charging latency of individual sensor nodes, under light and high charging demands. Lastly, they demonstrated an example of how their results can guide the system design. In particular, they discussed how to determine the optimal remaining energy level of sensor nodes to send out their charging requests.

2.5 Life Estimation of Renewable Energy Source and Battery Discharge Characteristics

Nowadays, sensor networks are gaining popularity, especially in environmental and patient monitoring applications. Pulse monitoring, blood pressure monitoring, and other data monitoring related to patients are deployed in E-health hospitals [52]. In the case of medical applications of WSNs, additional services are required apart from the traditional ones, that is, reliability, security, and real-time data. These services then need additional energy. To cope with this additional need, a simple model that estimates the energy dissipation of renewable sources for wireless sensor networks was presented in [52]. They used photovoltaic cells for charging batteries and claimed that their proposed model can serve as a cost function for the purpose of providing efficient routing and good quality of service. They concluded that their model allowed them to increase the network lifetime. Due to difficulty and high cost in replacing or recharging drained batteries in a deployed network, prolonging battery life became the principal goal in the design of wireless sensor networks. For the purpose of minimizing energy consumption, this principal goal led to the research and development of various technologies. Achieved battery life is not only the function of the energy consumed by the system, but also the way in which the system drains the battery, and the specific properties of the battery itself. Generally, it is known that making an electronic system more battery-friendly can improve battery performance and hence network lifetime, but there has been very little work done that examines the opportunities for battery-efficient design for wireless sensor networks. Specifically, there is insufficient experimental data available that measures the sensitivity of actual batteries for the design and environmental parameters of WSNs.

In paper [53], authors presented a systematic experimental analysis of battery discharge in WSNs. They studied that how the design and environmental parameters interact with electro chemistry dependent battery characteristics such as thermal effects and recovery, rate capacity and their impact on battery efficiency. They used Mica2 hardware, commercial lithium-coin batteries and a mixture of different techniques, including measurements on actual batteries, on low level battery simulator Dual foil and battery emulation for battery discharge analysis. Furthermore, they studied critical trade-offs in sensor network design by experimentally analyzing the interactions between the WSN parameters and battery characteristics. Especially, they focused on the problem of selecting a suitable power level for transmitting sensor data and demonstrated that by using a battery-aware approach compared to an approach that assume ideal source of energy of the same capacity, leads to 52% increase in the amount of data which is sensed and transmitted before. Their results clearly demonstrated that battery-aware effect can significantly improve the battery lifetime of sensor nodes. In [54], the authors experimentally measured the battery performance using MicaZ motes and commercial alkaline batteries. They evaluated the key parameters which have the impact on the battery performance namely communication distance, working channel and operating power. They presented a real battery discharge measurements in [54] which can serve as a quantitative basis for future research.

2.6 Experimental Analysis of Sensor Node Lifetime

The wireless sensor node lifetime starts when it is operating and ends when it is not able to communicate or do its other basic tasks, that is, sensing, processing, and so forth. Ultimately, network lifetime depends on the individual lifetime of the sensor but can be specified in many other different ways too. Everyone in the sensor network research community agrees that battery lifetime is a key performance metric. Initial studies of researchers for the expected network lifetime were on simulation and real hardware, which was based on basic assumptions about the power consumption of different states (sleep, transmit, receive) of the onboard radio and about the duty cycle. But these studies failed to account for the nonlinear behavior of the batteries and the power consumption of sensing, processing, flash writing, and so forth. For example [55], the energy consumed by the sensor node in writing to the flash was often neglected compared to the energy consumption in the radio operation. In [55], the authors presented the results of an experimental study on the lifetime of sensor nodes (TelosB) with different application parameters.

They concentrated on flash usage and duty-cycled radio operation throughout various battery brands. In particular, they focused on high duty cycles and provided data for the battery consumption for real sensor nodes. For example, they observed that energy consumption in flash writing is very high and comparable to the power consumption in radio transmission. Second, they found that different battery brands can affect node lifetime up to 25%. Third, nodes are able to transmit data to others, but are not able to receive anything toward the end of their battery lifetime. They calculated half-dead duration and found that it was 27% of the total lifetime. Fourth, they observed that lower duty cycles of the radio can significantly increase the sensor node lifetime. Yet the increase was not linear. Fifth, they observed that if radio transmission uses less power than radio reception, then it consumes more energy.

2.6.1 Energy Measurements for MicaZ Node

A microcontroller, a transceiver, three light-emitting diodes (LEDs), and a few other components make up the MicaZ node. Some sort of energy-saving mechanism is present in all the components. To find out the operation time, the knowledge of two things is required: energy consumption of the mobile device and the capacity of the battery. The energy consumption varies as the device is in idle state, processing, transmitting, or in some energy-saving mode. It is very hard to save energy by using energy-saving techniques if the device is never idle. In [56], the authors proposed an elegant degradation of functionality when the battery is low. This means that when the battery is low, the system provides limited functionality rather than complete failure. For example, consider a device that has some core functions and some additional functions that are working. When the battery drains and gets low, the additional functions are degraded, meaning their operational time is reduced and then set to zero, but the core functions are still in operation with less runtime or reduced execution frequency. Shifting to energy-saving modes has some disadvantages, for example, stopping of timers or the CPU and discarding external events. Thus, it is necessary to consider the requirement of the application running on the device as well (Table 2.3).

Table 2.3   Energy Measurements for TelosB Motes

Modes	Program (TinyOS)	Program (TinyOS)	Current	Reference
Sleep mode	CountSleep Radio Program	2 s	0.6 mA	[57]
Normal mode	TestRadio	–	20 mA	[57]
Setup period	TestRadio	–	8 mA	[57]
Transmission mode	TestRadio	1 s	19.2 mA	[57]

2.6.2 Energy Analysis of Different MAC Protocols for WSNs

The radio embedded on a sensor mote uses the majority of a node's power. That is why developing medium access control (MAC) protocols that consume less power is important for wireless sensor networks. Idle listening, overhearing, and transmission collisions are the most common sources of energy waste in WSN radio communication. In carrier sense multiple access with collision avoidance (CSMA/CA) protocols, idle listening (i.e., listening to a wireless channel when no transmissions occur) happens often, but selecting when to receive the packet can improve the network life of WSNs. Sensor node waste receives state energy when it overhears a transmission aimed for another node. To reduce the receive state energy waste, advanced overhearing strategies are required. Extra mote energy is spent when coinciding transmissions on the same channel collide.

In [58], the authors gave a neutral evaluation of energy consumption over a single-hop network topology for four power-aware MAC protocols (Asynchronous-MAC [AS-MAC], Broadcasting Asynchronous Scheduled-MAC [BAS-MAC], crankshaft, and Scheduled Channel Polling-MAC [SCP-MAC]) employing three different standard WSN traffic patterns (broadcast, convergecast, and local gossip). The authors implemented AS-MAC, BAS-MAC, crankshaft, and SCP-MAC on 10 TelosB motes by using the MAC Layer Architecture (MLA) framework. The AS-MAC protocol, due to built-in overhearing avoidance, performs best under local gossip and convergecast traffic, but consumes the most energy under the broadcast traffic scenario. On the other hand, SCP-MAC performs best under broadcast traffic. Crankshaft and BAS-MAC performed likewise around all the experiments, but BAS-MAC consumed slightly less energy in most of the experiments. Not a single protocol is best under all conditions; the measurements provided by the authors lead to general strategies for minimizing energy consumption in future WSN MAC protocols.

2.6.3 Energy Model for Transmission in Telos Sensor Nodes

The major issue in deployment of WSNs is energy efficiency because sensor nodes run on small batteries. Most of the energy is consumed in transmission of information that can't be avoided. Researchers proposed most of their work to reduce the energy consumption for transmission. Other researchers proposed harvesting techniques to prolong network lifetime. Much simulation work has been proposed to test the energy consumption of WSNs, but it is still limited in real-world applications due to the size of the networks. Furthermore, these simulation results are not perfectly accurate. In [57], the authors proposed creating an energy model for data transmission based on the size of payloads in bytes for Telos sensor nodes. Their work focused on the exact energy usage of sensor nodes in different scenarios, which was measured by a digital oscilloscope.

Different settings for transmission power, sleeping mode reception, overhearing reception, and normal data reception were included in the scenarios. Based on the measurements done in different scenarios, the authors created an energy model for the sensor node, Telos, as a simple equation to ease the calculations while using network simulation tools.

2.7 Application-Specific Problems of WSN

2.7.1 Wireless Rechargeable Sensor Networks for Smart Grid

The smart grid is expected to utilize versatile communication technologies such as fiber-optic, satellite, power-line, cellular, mesh, and short-range wireless communication according to the location and requirements of the application [39]. Within all those communication technologies, short-range wireless communications are best for smart grid monitoring and diagnosing compared to the wired sensors.

WSNs are widely known as a monitoring tool and are used in a wide variety of fields, including health and defense, that require accurate solutions. Since WSNs are low cost, the smart grid can take advantage of this to cover more geographical regions with high redundancy. In [39], the authors proposed a sustainable wireless rechargeable sensor network (SuReSense) for smart grid monitoring. Furthermore, they proposed an optimization model to select the minimum number of landmarks according to the energy replenishment requirements and locations of the sensors. SuReSense provided long-term, timely, and reliable monitoring of the smart grid and used mobile wireless charger robots (MICROs) to recharge the batteries of the sensor nodes. First, a minimum number of landmarks were selected, and then landmarks were organized into clusters and each cluster was served by one MICRO according to the location and energy requirement. To reduce the path length of the MICRO, increase the duration spent on the docking station, and reduce the waiting time of each sensor, MICROs, following the shortest Hamiltonian cycle, visited the assigned landmarks in their cluster.

2.7.2 Sustainability Analysis and Resource Management for Wireless Mesh Networks

Traditional energy sources are more reliable and supplied from the electric grid station, while renewable energy sources are dynamic and unreliable in nature. Renewable energy sources depend on the atmospheric conditions and geography. For example, a wind turbine can deliver power if the weather is windy [59]. This dynamic nature of renewable sources created the need for an energy sustainable network design. In the early research in wireless communication, improving energy efficiency was the main issue, but now the design criteria and performance metric have shifted to energy sustainability. The authors of [59] were motivated by this fact and developed a mathematical model to study the energy sustainable performance of a wireless mesh network powered by renewable energy sources. They studied the transient evolution of an energy buffer and modeled the energy buffer as G/G/1/1 and G/G/1/N, infinite- and finite-energy buffer cases, respectively. For the purpose of examining the transient evolution, they applied a diffusion approximation. Based on this model, they proposed adaptive resource management and admission control schemes to improve the sustainable network performance by exploring the transient behavior of the energy buffer. They demonstrated that the distributed admission control strategy guaranteed high resource utilization and improved energy sustainability.

2.7.3 Wired and Wireless Sensor Networks for Monitoring Pipeline Infrastructures

The life and economy of most countries are strongly dependent on their gas, oil, and water pipelines. These pipelines are the veins of the economy for most countries. Protecting and maintaining the pipeline infrastructure is absolutely necessary for these countries. For example, there are around 500,000 miles of oil and gas pipelines in the United States [60]. Therefore, the economy of the United States is strongly dependent on them. To control and monitor pipelines, there are various technologies. Most of them depend on the sensors and networks for transferring the collected data from inside and outside pipelines to the central control stations. Throughout the pipelines, network components are spread out to transfer the collected data from distributed sensors that are scattered throughout the pipelines.

There have been several issues recently raised in the proposed network architecture using wireless sensor networks for pipeline monitoring, including the issues of power management and efficient routing. In [60], the authors proposed a fault-tolerant network architecture for monitoring pipelines based on integrated wired and wireless networks. The wired part of the network was considered a primary network by the authors, while the wireless part was used as a backup between sensor nodes if there was any failure in the wired network. This newly proposed architecture not only enhanced the reliability and performance, but also solved issues such as power management and efficient routing. Moreover, it also solved the problem of disabling the wired network by disconnecting the network cables due for any reason.

2.7.4 Solar-Powered WSN Framework for Aquatic Environmental Monitoring

To monitor underwater luminosity and temperature for the purpose of deriving the health status of a coralline barrier, WSN can be used. However, there are several complex challenges in the aquatic environment, for example, algae deposits on the sensor, possible water infiltrations in the unit, communication inefficiency due to RF reflections, and a contained elevation of the antenna with respect to the water surface [24]. In [24], the authors proposed an aquatic monitoring framework based on WSNs that was scalable, adaptive to the topological changes, and powered by an energy harvesting technique, that is, solar powered. This ad hoc system was based on clusters relying on star topology; it covers sensing activity and is able to transmit local and remote information from nodes to the gateway and the gateway to the control center, respectively. The authors deployed this system in Queensland, Australia, for monitoring luminosity and temperature, and it was able to monitor the marine environment at multiple scales, predict the occurrence of ecological phenomena, and provide quantitative indications related to cyclone formations in tropical areas based on the collected data.

2.7.5 Solar Biscuit: Battery-Less WSN System for Environmental Monitoring Applications

Self-discharging is an inherent problem of batteries, and the replacement of batteries is quite unrealistic for wireless sensor network applications [25]. In the case of military applications, WSNs can be assumed to be disposable, but in many other cases, disposable sensor nodes are not acceptable. The problem of battery replacement can be removed by using various energy conversion techniques, but the conversion devices usually provide 0.001–0.1 W, while the WSN node consumes

0.01–0.1 W [25]. The power consumption of a node depends on the clock speed of the micro-processor and the communication module. The power delivered by conversion devices can be increased if the physical size of these devices is increased, but this is not preferable due to practical reasons. Therefore, a sensor node should have to store the harvested energy in some storage device, such as supercapacitor or electric double-layer capacitor (EDLC), but the sensor node has to wait for sensing and communication until the stored energy is enough. In [25], the authors described the design and implementation of a battery-less WSN system in which EDLC was equipped with a small solar cell as its energy source. The authors designed a communication mechanism to solve the long-interval communication problem. The developed communication mechanism was suitable for battery-less WSN nodes based on an environmental monitoring applications scenario. Lastly, they implemented the designed communication mechanism on custom hardware and evaluated it via experimentation, but the performance of the system was less than expected due to the inefficiency of the newly designed communication protocol.

2.7.6 WPT for Structural Health Monitoring of Fiber-Reinforced Composite Materials

Structural health monitoring (SHM) has gained much attention in the research community for the purpose of developing new methods. To decrease the cost and increase the safety of structure, the composite materials glass fiber–reinforced polymer (GFRP) and carbon fiber–reinforced polymer (CFRP) are commonly used. Integration of SHM technologies in these complex structures is the major challenge according to [61]. The goal in [61] was to build a WSN for structural monitoring by enforcing the idea of inductively coupled coils. It was observed and experimented to determine the possibility of using inductive coils within composite materials. CFRP is known for lightweight structures, and it behaves as a conductor, which complicates the use of wireless communication techniques.

First, the authors of [61] studied the design of antenna coils for both low- and high-frequency bands for the purpose of integrating the antenna in composite materials to achieve wireless power and data transmission. They implemented an SHM system using lamb waves and piezo-wafer-active sensors (PWASs). They showed, with the help of experimental demonstration, that it was feasible to transmit enough power from the PWAS to a receiver covered with one layer of CFRP.

2.7.7 Wireless Underground Sensor Networks: Research Challenges

The authors of [62] introduced the concept of wireless underground sensor network (WUSN). WUSN has a variety of applications in the agricultural and environmental monitoring domains, such as soil properties (water and mineral content) and toxic substances. The technology previously used for underground monitoring consisted of buried sensors, and the collected data was stored in a data logger that was on the surface of earth. The data logger was connected to buried sensors with wired connections. This technology had several shortcomings, which are presented in [62], such as visibility, ease of deployment, timeliness of the data, reliability, and potential for coverage density. All these shortcomings are addressed by WUSN. WUSN has all of its required transmitting and sensing equipment underground and connected wirelessly, providing safety from damage and theft, unlike buried sensors and data loggers, which were connected via wired connections. WUSN has ease-of-deployment facility and can be deployed at any desired location within the range of other devices, unlike other sensing equipment, which require more data loggers and underground wiring. WUSN transmits sensed data in real time, while data loggers store the collected data for later retrieval. Similarly, WUSNs are self-healing, and in the case of failure, the failure report is routed to other devices and then to the network operator in real time, while data loggers have only a single point for the representation of failure, and tens of sensors are connected at one time.

Despite the above benefits described by [62], there are several challenges in the deployment of WUSNs. In [62], authors consolidated all the design challenges and presented a broad overview. The challenges included the methods for predicting path losses in an underground link and the communication protocol stack at each layer.

2.8 Conclusion

In this chapter, we presented a broad survey of energy replenishment and battery recharging techniques for wireless sensor networks used in the literature. First, the need of sensor nodes for an unlimited lifetime was discussed. Then, the classification of energy harvesting and wireless power transmission techniques were presented and discussed. Each technique was critically analyzed, and their challenges were discussed. Subsequently, experimental analyses of sensor node lifetimes were presented. In the end, application-specific problems of WSNs were described and solutions presented. In summary, this chapter provides an up-to-date survey of the battery recharging problem and the limited lifetime of wireless sensor nodes and consolidates all the information presented in the literature.

References

1. I. F. Akyildiz, W. Su, Y. Sankarasubramaniam, E. Cayirci, Wireless sensor networks: A survey, *Computer Networks* 38 (2002) 393–422.
2. M. G. L. Roes, J. Duarte, M. Hendrix, E. Lomonova, Acoustic energy transfer: A review, *IEEE Transactions on Industrial Electronics* 60 (1) (2013) 242–248.
3. W.-G. Seah, Z. A. Eu, H. Tan, Wireless sensor networks powered by ambient energy harvesting (WSN-HEAP)—Survey and challenges, in *1st International Conference on Wireless Communication, Vehicular Technology, Information Theory and Aerospace and Electronic Systems Technology 2009 (Wireless VITAE 2009)*, Aalborg, Denmark, 2009, pp. 1–5.
4. S. Roundy, P. K. Wright, J. Rabaey, A study of low level vibrations as a power source for wireless sensor nodes, *Computer Communications* 26 (11) (2003) 1131–1144.
5. H. S. Kim, J.-H. Kim, J. Kim, A review of piezoelectric energy harvesting based on vibration, *International Journal of Precision Engineering and Manufacturing* 12 (6) (2011) 1129–1141.
6. A. C. Valera, W.-S. Soh, H.-P. Tan, Survey on wakeup scheduling for environmentally-powered wireless sensor networks, *Computer Communications* 52 (2014) 21–36.
7. F. B. James M. Gilbert, Comparison of energy harvesting systems for wireless sensor networks, *International Journal of Automation and Computing* 5 (4) (2008) 334–347.
8. G. Anastasi, M. Conti, M. Di Francesco, A. Passarella, Energy conservation in wireless sensor networks: A survey, *Ad Hoc Networks* 7 (3) (2009) 537–568.
9. S. Sudevalayam, P. Kulkarni, Energy harvesting sensor nodes: Survey and implications, *IEEE Communications Surveys and Tutorials* 13 (3) (2011) 443–461.
10. C. P. Ankit Patel, Recharging mechanism of wireless sensor network: A survey, *IJSRD—International Journal for Scientific Research and Development* 1 (2) (2013) 56–58.
11. G. J. Pottie, W. J. Kaiser, Wireless integrated network sensors, *Communications of the ACM* 43 (5) (2000) 51–58.

12. A. Sinha, A. Chandrakasan, Dynamic power management in wireless sensor networks, *IEEE Design and Test of Computers* 18 (2) (2001) 62–74.

13. G. Bianchi, A. T. Capossele, C. Petrioli, D. Spenza, Agree: Exploiting energy harvesting to support data-centric access control in WSNs, *Ad Hoc Networks* 11 (8) (2013) 2625–2636.

14. P. Zhang, G. Xiao, H.-P. Tan, Clustering algorithms for maximizing the lifetime of wireless sensor networks with energy-harvesting sensors, *Computer Networks* 57 (2013) 2689–2704.

15. S. He, J. Chen, F. Jiang, D. Yau, G. Xing, Y. Sun, Energy provisioning in wireless rechargeable sensor networks, *IEEE Transactions on Mobile Computing* 12 (10) (2013) 1931–1942.

16. C.-K. Chau, M. H. Wahab, F. Qin, Y. Wang, Y. Yang, Battery recovery aware sensor networks, in *WiOpt*, Seoul, South Korea, 2009, pp. 1–9.

17. C.-K. Chau, F. Qin, S. Sayed, M. H. Wahab, Y. Yang, Harnessing battery recovery effect in wireless sensor networks: Experiments and analysis, *IEEE Journal on Selected Areas in Communications* 28 (7) (2010) 1222–1232.

18. N. Y.-Y. S. Xi Chen, S. Xu, 1.6 v nanogenerator for mechanical energy harvesting using PZT nanofibers, *Nano Letters* 10 (2010) 2133–2137.

19. S. Roundy, P. K. Wright, A piezoelectric vibration based generator for wireless electronics, *Smart Materials and Structures* 13 (2004) 1131–1142.

20. R. Abbaspour, A practical approach to powering wireless sensor nodes by harvesting energy from heat flow in room temperature, in *International Congress on Ultra Modern Telecommunications and Control Systems and Workshops (ICUMT)*, Moscow, Russia, 2010, pp. 178–181.

21. X. Lu, S.-H. Yang, Thermal energy harvesting for WSNs, in *IEEE International Conference on Systems Man and Cybernetics (SMC)*, Istanbul, Turkey, 2010, pp. 3045–3052.

22. D. Ramasur, G. Hancke, A wind energy harvester for low power wireless sensor networks, in *IEEE International Instrumentation and Measurement Technology Conference (I2MTC)*, Graz, Austria, 2012, pp. 2623–2627.

23. R. Myers, M. Vickers, H. Kim, S. Priya, Small scale windmill, *Applied Physics Letters* 90 (5) (2007) 054106.

24. C. Alippi, R. Camplani, C. Galperti, M. Roveri, A robust, adaptive, solar-powered WSN framework for aquatic environmental monitoring, *IEEE Sensors Journal* 11 (1) (2011) 45 –55.

25. M. Minami, T. Morito, H. Morikawa, T. Aoyama, Solar biscuit: A battery-less wireless sensor network system for environmental monitoring applications, in *Proceedings of the 2nd International Workshop on Networked Sensing Systems (INSS2005)*, San Diego, 2005.

26. M. L. D. Bouchouicha, F. Dupont, Ambient RF energy harvesting, in *International Conference on Renewable Energies and Power Quality (ICREPQ 10)*, Granada, Spain, March 23–25, 2010.

27. D. Dondi, S. Scorcioni, A. Bertacchini, L. Larcher, P. Pavan, An autonomous wireless sensor network device powered by a RF energy harvesting system, in *IECON 2012—38th Annual Conference on IEEE Industrial Electronics Society*, Montreal, QC, 2012, pp. 2557–2562.

28. M. R. Ayesha Feroz, An efficient and low-cost technique for charging nodes in wireless sensor network, *International Journal of Engineering Research and Technology* 2 (9) (2013) 2937–2941.

29. T. Ichihara, T. Mitani, N. Shinohara, Study on intermittent microwave power transmission to a ZigBee device, in *IEEE MTT-S International Microwave Workshop Series on Innovative Wireless Power Transmission: Technologies, Systems, and Applications (IMWS)*, Kyoto, Japan, 2012, pp. 209–212.

30. C. Park, P. Chou, Ambimax: Autonomous energy harvesting platform for multi-supply wireless sensor nodes, in *3rd Annual IEEE Communications Society on Sensor and Ad Hoc Communications and Networks, 2006 (SECON '06)*, Reston, Virginia, 2006, vol. 1, pp. 168–177.

31. M. Chetto, A. Queudet, A note on EDF scheduling for real-time energy harvesting systems, *IEEE Transactions on Computers* 63 (4) (2014) 1037–1040.

32. H. Ju, R. Zhang, A novel mode switching scheme utilizing random beamforming for opportunistic energy harvesting, *IEEE Wireless Communications* 13 (4) (2014) 2150–2162.

33. N. Bhatti, A. Syed, M. Alizai, Sensors with lasers: Building a WSN power grid, in *IPSN-14 Proceedings of the 13th International Symposium on Information Processing in Sensor Networks*, Berlin, Germany, 2014, pp. 261–272.

34. S. D. Ozeri, Ultrasonic transcutaneous energy transfer for powering implanted devices, *Ultrasonics* 50 (2010) 556–566.

35. S. S.-C.-C. W. Shaul Ozeri, D. Shmilovitz, Ultrasonic transcutaneous energy transfer using a continuous wave 650 kHz Gaussian shaded transmitter, *Ultrasonics* 50 (2010) 666–674.

36. A. Denisov, E. Yeatman, Ultrasonic vs. inductive power delivery for miniature biomedical implants, in *International Conference on Body Sensor Networks (BSN)*, Singapore, 2010, pp. 84–89.

37. Y. Peng, Z. Li, W. Zhang, D. Qiao, Prolonging sensor network lifetime through wireless charging, in *IEEE 31st Real-Time Systems Symposium (RTSS)*, San Diego, 2010, pp. 129–139.

38. A. Kurs, A. Karalis, R. Moffatt, J. D. Joannopoulos, P. Fisher, M. Soljacic, Wireless power transfer via strongly coupled magnetic resonances, *Science* 317 (5834) (2007) 83–86.

39. H. T. Melike Erol Kantarci, SuReSense: Sustainable wireless rechargeable sensor networks for the smart grid, *IEEE Wireless Communications* 12 (2012) 1536–1284.

40. L. Xie, Y. Shi, Y. Hou, A. Lou, Wireless power transfer and applications to sensor networks, *IEEE Wireless Communications* 20 (4) (2013) 140–145.

41. L. Xie, Y. Shi, Y. T. Hou, H. D. Sherali, Making sensor networks immortal: An energy-renewal approach with wireless power transfer, *IEEE/ACM Transactions on Networking* 20 (6) (2012) 1748–1761.

42. Y. Shi, L. Xie, Y. Hou, H. Sherali, On renewable sensor networks with wireless energy transfer, in *Proceedings of IEEE INFOCOM*, Shanghai, China, 2011, pp. 1350–1358.

43. C. Angelopoulos, S. Nikoletseas, T. Raptis, Efficient wireless recharging in sensor networks, in *IEEE International Conference on Distributed Computing in Sensor Systems (DCOSS)*, Cambridge, Massachusetts, 2013, pp. 298–300.

44. M. L. Wen Yao, M.-Y. Wu, Inductive charging with multiple charger nodes in wireless sensor networks, in *APWeb Workshops 2006*, ed. H. T. Shen et al., vol. 3842 of Lecture Notes in Computer Science, Springer-Verlag, Berlin, 2006, pp. 262–270.

45. M. Afzal, W. Mahmood, A. Akbar, A battery recharge model for WSNs using free-space optics (FSO), in *IEEE International Multitopic Conference (INMIC 2008)*, Karachi, Pakistan, 2008, pp. 272–277.

46. S. M. S. M. I. Afzal, W. Mahmood, S. Seoyong, Optical wireless communication and recharging mechanism of wireless sensor network by using CCRs, *International Journal of Advanced Science and Technology* 13 (2009) 49–60.

47. C. Wang, J. Li, F. Ye, Y. Yang, Netwrap: An NDN based real-time wireless recharging framework for wireless sensor networks, *IEEE Transactions on Mobile Computing* 13 (6) (2014) 1283–1297.

48. X. Li, R. Falcon, A. Nayak, I. Stojmenovic, Servicing wireless sensor networks by mobile robots, *IEEE Communications Magazine* 50 (7) (2012) 147–154.

49. L. Fu, P. Cheng, Y. Gu, J. Chen, T. He, Minimizing charging delay in wireless rechargeable sensor networks, in *Proceedings of IEEE INFOCOM*, Turin, Italy, 2013, pp. 2922–2930.

50. S. Guo, C. Wang, Y. Yang, Mobile data gathering with wireless energy replenishment in rechargeable sensor networks, in *Proceedings of IEEE INFOCOM*, Turin, Italy, 2013, pp. 1932–1940.

51. J. P. T. Z. Liang He, Y. Gu, On-demand charging in wireless sensor networks: Theories and applications, in *IEEE 10th International Conference on Mobile Ad-Hoc and Sensor Systems*, Hangzhou, China, 2013.

52. J. Torregoza, I.-Y. Kong, W.-J. Hwang, Wireless sensor network renewable energy source life estimation, in *First International Conference on Communications and Electronics 2006 (ICCE '06)*, Hanoi, Vietnam, 2006, pp. 373–378.

53. C. Park, K. Lahiri, Battery discharge characteristics of wireless sensor nodes: An experimental analysis, in *Proceedings of the IEEE Conference on Sensor and Ad-Hoc Communications and Networks (SECON)*, 2005.

54. W. Guo, W. Healy, M. Zhou, Battery discharge characteristics of wireless sensors in building applications, in *9th IEEE International Conference on Networking, Sensing and Control (ICNSC)*, Beijing, China, 2012, pp. 133–138.

55. H. A. Nguyen, A. Forster, D. Puccinelli, S. Giordano, Sensor node lifetime: An experimental study, in *IEEE International Conference on Pervasive Computing and Communications Workshops (PERCOM Workshops)*, Seattle, 2011, pp. 202–207.

56. M. Krämer, A. Geraldy, Energy measurements for micaz node, *5. GI/ITG KuVS Fachgespräch "Drahtlose Sensornetze"*, Stuttgart, Germany, (2006) 61.

57. Y. Panthachai, P. Keeratiwintakorn, An energy model for transmission in Telos-based wireless sensor networks, in *International Joint Conference on Computer Science and Software Engineering 2007 (JCSSE 2007)*, May 2–4, 2007.

58. B. Bates, A. Keating, R. Kinicki, Energy analysis of four wireless sensor network MAC protocols, in *6th International Symposium on Wireless and Pervasive Computing (ISWPC)*, Hong Kong, 2011, pp. 1–6.

59. L. Cai, Y. Liu, T. Luan, X. Shen, J. Mark, H. Poor, Sustainability analysis and resource management for wireless mesh networks with renewable energy supplies, *IEEE Journal on Selected Areas in Communications* 32 (2) (2014) 345–355.

60. N. Mohamed, I. Jawhar, A fault tolerant wired/wireless sensor network architecture for monitoring pipeline infrastructures, in *Second International Conference on Sensor Technologies and Applications 2008 (SENSORCOMM '08)*, Cap Esterel, France, 2008, pp. 179–184.

61. M. Salas, O. Focke, A. Herrmann, W. Lang, Wireless power transmission for structural health monitoring of fiber-reinforced-composite materials, *IEEE Sensors Journal* 14 (7) (2014) 2171–2176.

62. I. F. Akyildiz, E. P. Stuntebeck, Wireless underground sensor networks: Research challenges, *Ad Hoc Networks* 4 (6) (2006) 669–686.

WIRELESS BODY AREA NETWORKS

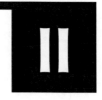

Chapter 3

Wearable Wireless Sensor Networks: Applications, Standards, and Research Trends

Muhammad Mahtab Alam and Elyes Ben Hamida

Contents

Abstract

Wearable wireless sensor networks (W-WSNs) aim to provide an attractive alternative for the conventional medical care system. It is an effective way to monitor patients within clinics, hospitals, and remotely from home, offices, and so forth. Several tiny sensors connected on the human body can effectively monitor the physiological status of a person. Wearable wireless sensor networks are a fast emerging technology with potential applications in various domains of daily life, such as health and physiological monitoring, sports, localization, and fashion. In this chapter, we present an extensive set of W-WSN applications along with their specific requirements and architectures. In this context, a number of different standards are used to realize the above-mentioned applications. We have highlighted the importance and significance of these different standards and their key features and also presented a comparison. Further, current research trends in terms of effective and reliable W-WSN specific routing, coexistence strategies, privacy and security, and hardware and software and simulating platforms are discussed. In this regard, several issues pertaining to effective W-WSN solutions are also highlighted.

3.1 Introduction

In this chapter, a newly emerging information and communication technology called wearable is introduced. The wearable computer is defined as a means of personal empowerment through human–computer interaction of user-programmable devices. The key characteristics of these devices are that they are always on and always available or ready for the user's interaction. Unlike portable or other smart devices, wearable devices don't need to turn on, and we can augment the reality of the physical world more powerfully from our surroundings.

The fundamental part of this technology is wireless sensor networks (WSNs). The wearable wireless sensor network (W-WSN) is a self-organizing network at the human body scale. It consists of heterogeneous smart devices that are low power, miniaturized, hardware constrained (with limited processing and storage capabilities), and attached to (or implanted inside) a human body. These devices can be sensors (to sense, transmit, and receive data), actuators (to react according to the perceived data), or coordinators (to act as a gateway for the external network). Typically sensors are connected on the body to collect signals about the human body's physiological signs (e.g., heartbeat and temperature), movement and activity (e.g., acceleration and orientation), and surrounding environments (e.g., temperature and toxic gases).

W-WSNs have a number of potential applications in various domains of daily life, such as health care (i.e., monitoring of physiological signals) in a hospital environment, ambient intelligence (mobile health and remote monitoring), sports and fitness, localization, augmented reality, and fashion. Wearable technology is also envisioned for applications such as worker safety and critical operations, as well as rescue and emergency management [1]. With regard to such applications, W-WSN technology can also play a vital role to not only save human lives but also protect critical and valuable assets.

Generally, W-WSN solutions are based on a number of different standards, such as ZigBee (i.e., IEEE 802.15.4, IEEE 802.15.4a, and IEEE 802.15.4j), Bluetooth (i.e., IEEE 802.15.1), and Wi-Fi (i.e., IEEE 802.11a, IEEE 802.11b, and IEEE 802.11g). More recently, the IEEE 802.15.6 standard [2] was introduced to specifically target the emerging W-WSN applications and meet their design requirements [1]. With reference to meeting the design constraints (of the various applications discussed above), such as reliability, quality of service, data transmission rate, and energy efficiency, key features of the physical and medium access control (MAC) layers of the IEEE 802.15.6 standard are very important to analyze. Therefore, all of the above-mentioned standards are discussed to understand their available potential and limitations. Moreover, various W-WSN architectures suited to different applications are presented.

The communication stack of W-WSNs is mainly governed by the application layer, which specifies a number of important parameters, such as the packet transmission rate, network topology, traffic patterns, and required quality of service (QoS). Moreover, the selection of appropriate routing strategies, feasible medium access and coexistence mechanisms, privacy and security, available physical layers, and channel and mobility modeling are important building blocks in the design of W-WSNs. This chapter provides the cross-layer perspective of the existing state of the art. The impact of designing a new communication and networking protocol, physical layer optimizations, or low-power strategies needs to be evaluated in the context of the global network performance. This performance evaluation through the development of a body area network (BAN) prototype or platform is a complex and time-consuming process. However, there are a number of prototypes and products that are available on the market and discussed in this chapter. Further, simulating tools that are generally considered for analyzing and evaluating new algorithms and protocols for the networking and communication stack are also discussed.

3.2 Applications

More than a decade ago, one of the natural and obvious applications of wearable wireless sensor networks was related to health care, as shown in Figure 3.1. This includes patient monitoring inside hospitals, that is, intensive care unit, hospital wards and rooms, and so forth. With this kind of W-WSN monitoring system, it was no longer necessary to keep checking patients regularly; instead, W-WSN-based systems provide various readings of the physiological signals and vital signs at any time, and alarms can be generated in case of emergency or abnormality. Later on, it became possible to remotely monitor from the home, offices, and so forth, as shown in Figure 3.1. In ambient intelligence, pervasive, ubiquitous, and context-aware human–computer interaction was made possible with an objective to bring emerging technologies much closer to human beings. In particular, the concept of ambient assistive living has enabled us to remotely keep track of the health patterns of sick and elderly patients, which has had several impacts. First, there are less crowded hospitals and fewer burdens on the staff inside the hospitals. Second, it provided a home environment to noncritical patients, which has a psychological impact on their well-being. Third, as an example, it was recently reported that the UK's National Health Service (NHS) could save up to £7 billion per year by using innovative technologies to deliver quality health care to the chronically ill with fewer hospital visits and admissions [1].

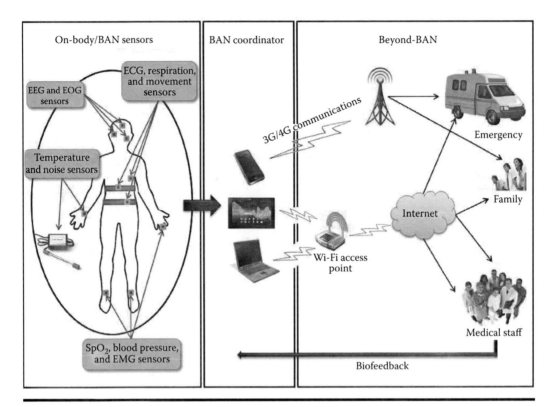

Figure 3.1   Wearable wireless sensor networks for health care and remote patient monitoring.

Within the last decade, W-WSN has also been utilized in sports and fitness, where athletes keep track of their movements, motion, muscles, heart rate, sweat control, and so forth, to perform much better with complete care and fitness. Further, in various games, children can physically interact with video screens using gestures, accelerometers, and gyrosensors. In the fashion industry and augmented reality, digital glass, wearable armbands, and various necklaces are becoming popular [2].

All of the above-mentioned applications are related to on-body or non-invasive networks, also called wearable wireless body area networks. One of the other aspects of health care revolution is through implant devices also called in-body or invasive body area networks. Sometimes in health monitoring it becomes necessary to not only observe physiological signals or vital signs, but also implant certain devices inside the body for repairing the malfunctioning of the body organs. This includes various implants such as heart, kidney, ear, birth control, and back pain. These devices facilitate the proper functioning of different organs. For example, hearing aid devices such as cochlear ear implants help deaf and elderly people hear comfortably [3].

Moreover, in the recent past there have been a number of other applications introduced that are still related to health care, but with a much broader context. For example, it is envisioned that wearable human assistive technologies will play an important role in managing, monitoring, and ensuring the safety of humans during mission-critical operations [4], as illustrated in Figure 3.2. The scenarios of these mission-critical operations include disasters (such as building collapse due to earthquake) or emergency (such as fire outbreak in a building). In such cases, the existing

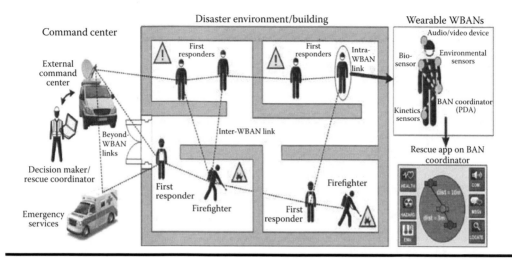

Figure 3.2 **Wearable wireless sensor networks for safety and mission-critical applications.**

communication infrastructure is either completely damaged or oversaturated. Consequently, W-WSN is targeted to create a new network to facilitate the first responder to effectively coordinate and communicate to reach the casualties as soon as possible. In this context, the health of the first responder is also very critical and is monitored using W-WSN.

Furthermore, W-WSNs can also be used for the monitoring of workers' safety and protection in a harsh environment. To that end, each worker can be fitted with a set of wearable sensing devices to continuously monitor his or her (1) physiological parameters (e.g., energy expenditure, heart pulse rate, and stress level), (2) location (e.g., GPS position and relative proximity to hazards), (3) body movement (e.g., body posture and orientation, activity, and fall detection), and (4) surrounding environment (e.g., toxic gases, CO_2 or CO, and heat). All of the collected information is then processed locally or remotely to better anticipate and protect the deployed mobile workforces from accidents and hazards, thus reducing injuries and fatalities and improving working conditions.

3.3 Design Requirements and Architecture

Concerning the design requirements, all of the applications explained above have different and specific requirements in terms of devices, sensors and their types, data rates and traffic patterns, miniaturization, and so on. As an example, inside hospitals having classical health care patient monitoring systems, such as for vital signs, ECG, or EEG, the coordinating node does not necessarily need to be multistandard, multifunctional, and very powerful; instead, having access to the nearest access point (e.g., Wi-Fi) is sufficient. For the case of mobile or remote monitoring systems, the coordinator should be able to support different standards and functionalities with minimum battery constraints.

With respect to the application requirements in sports, games, and fitness, generally two to three specific sensors, such as accelerometers, gyroscopes, heartbeat, and sweat, are used, which can operate standing alone without any external communications. Often there is no requirement of a more powerful coordinating node. On the other hand, rescue and critical applications require more resources, much better quality of service, and reliability. In such systems, body-to-body and off-body communications are necessary, and they add explicitly different requirements and constraints. For example, four to five sensors are typically used to monitor the health, orientation, and movement of firefighters, along with the support of multiple functions and a multistandard-based coordinator.

To represent all the applications in terms of their requirements, we present a global view based on a number of different parameters. Table 3.1 shows various application layer parameters (such as used devices, sensor types and traffic patterns, system coordination, and location awareness) and their corresponding requirements. Further, in Table 3.2, we classify the target applications as medical and nonmedical. It covers a broad range of applications, as discussed earlier, and presents a corresponding data range of various physiological signals, frequency, accuracy, and data rates.

In order to develop or understand the architecture of the above applications along with their requirements (as mentioned in Tables 3.1 and 3.2), a generic architecture is presented in Figure 3.3, which includes three main parts: the on-body communications part (or intra-BAN), the body-to-body communications part (or inter-BAN), and the off-body communications part (or beyond-BAN).

In the on-body communications part, various sensors are placed on (or inside) the human body, and all interact together to forward the sensed data to the coordinating device. A coordinator normally controls and coordinates all the communication, for example, scheduling the transmission of every sensor in a synchronized manner such that every sensor can easily transmit its data. A coordinator is also a much more powerful device in comparison to the other sensors; it has fewer power constraints, and often it is strengthened with multiple standard and multiple communication modes. Depending on the considered application, a coordinator device may act as a gateway with respect to the external network (cf. remote patient monitoring application in Figure 3.1).

Table 3.1 Application Layer Parameters and Corresponding Requirements

Parameters	Requirements
Used devices	Sensor based, actuator based, smart phones, base station.
Sensor types	Vital signs: Heartbeat/breathing rate, stress, sweating, temperature, blood pressure, oxygen saturation (SiO_2). Surrounding environment: Pressure, heat, humidity, light intensity, carbonic gases. Body movement: Orientation, acceleration, fall detection.
Node type (data)	Source, sink, router, multistandard gateway.
Traffic pattern	Periodic, event driven, burst traffic.
Traffic type	Audio, video (e.g., coordinator can be used as vision sensor), data, time series (signals). Information can be raw, encoded.
System coordination	Centralized: Coordinator is controlling the communication on the body (intra-BAN shall be centralized). Distributed: For interbody communications.
Location awareness	Absolute or relative.

Table 3.2   Typical BAN Sensors and Data Rates

Applications	Signals	Data Range	Frequency (Hz)	Accuracy (bits)	Data Rate
Medical/ health	Glucose concentration	0–20 mM	40	12	480 bps
	Blood flow	1–300 ml/s	40	12	480 bps
	ECG	0.5–4 mV	500	12	6 kbps
	Respiratory rate	2–50 breaths/ min	20	12	240 bps
	Pulse rate	0–150 BPM	4	12	48 bps
	Blood pressure	10–400 mm Hg	100	12	1.2 kbps
	Blood pH	6.8–7.8 pH units	4	12	48 bps
	Body temperature	32–40°C	0.2	12	2.4 bps
Nonmedical	High-quality audio	–	–	–	1.4 Mbps
	Voice	–	–	–	100 kbps
	Video	–	–	–	1–2 Mbps
	GPS positions	–	1	32	96 bps
	Motion sensor	–	100	16	4.8 kbps

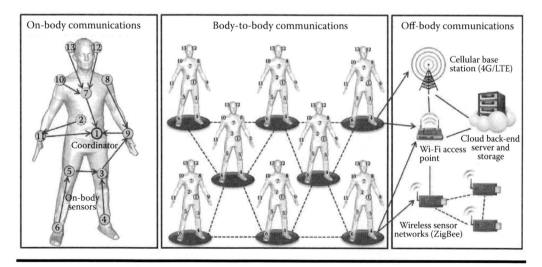

Figure 3.3   Generic architecture for wearable wireless sensor networks.

In this case, off-body communications are needed to ensure the delivery of the collected data toward remote cloud back-end servers for further data analysis and decision making, via either cellular networks (e.g., 3G and 4G/Long-Term Evolution [LTE]) or preexisting infrastructures (e.g., Wi-Fi access points or ZigBee-based sensor networks). However, in certain applications (cf. safety and mission-critical applications in Figure 3.2), the existing network infrastructures might be unavailable, damaged, or out of range. In this case, the coordinators or on-body sensors can exploit cooperative and multihop body-to-body (or inter-BAN) communications to extend the end-to-end network connectivity, thus forming a self-organizing and dynamic wireless body-to-body network (BBN).

3.4 Key Enabling Standards and Technologies

Over the last decade, various low-power standards have been used in BAN research as well as for commercial applications, where most of them partly satisfy the requirements for life signs monitoring, as already discussed in Section 3.2. Some low-power standards designed to support low-power sensing have been adapted for health care applications. These standards include personal area network (PAN) technologies, such as Bluetooth (IEEE 802.15.1) [5] and Bluetooth low energy (BLE) [6]; wireless sensor network (WSN) technologies, such as ZigBee (IEEE 802.15.4) [7], ultrawideband (IEEE 802.15.4a) [8], and an alternate physical layer extension to support medical body area networks (IEEE 802.15.4j) [9]; and wireless local area network (WLAN) technologies, such as Wi-Fi (IEEE 802.11a/b/g/n) [10]. More recently, a specific BAN standard, that is, IEEE 802.15.6 [11], was proposed to meet the increasing demand for such applications. Below we will provide the necessary details of these standards.

3.4.1 IEEE 802.11 Standard (Wi-Fi)

The 802.11 family of standards [10] is widely known as Wi-Fi. After the introduction of Wi-Fi with 802.11a and 802.11b, the standard became very popular. In particular, 802.11b got more attention because of its compatible speed and low price in comparison with other variants operating at 5 GHz. IEEE 802.11b specifies one MAC and several physical (PHY) layers, including frequency hopping spread spectrum (FHSS), direct-sequence spread spectrum (DSSS), and infrared. Then, in order to provide a higher speed to 802.11a, a new standard was introduced called 802.11g. At the 2.4 GHz band, 802.11b uses a DSSS modulation scheme, whereas 802.11g uses both DSSS and the orthogonal frequency division multiplexing (OFDM) method. More improved variants have since been proposed, with a major difference being the higher-data-rate support. With reference to W-WSN, 802.11g is sufficient, as it can satisfy the requirement of an access point.

In several W-WSN applications, Wi-Fi-based access points (APs) are responsible for the stable and reliable interconnection and extension of the network. Further, an access point also helps in scheduling the transmissions and keeping the nodes synchronized, consequently allowing the devices to save energy. Generally, wireless connections can be made in ad hoc mode or infrastructure mode. Ad hoc mode, also known as peer to peer, is simply a group of devices that interact wirelessly with each other without an access point (AP), and therefore they are limited in range and functionality. On the other hand, infrastructure mode uses one AP to connect multiple clients. Both of these modes can be used in different contexts of W-WSN applications. Below, a brief overview of the IEEE 802.11g standard is presented.

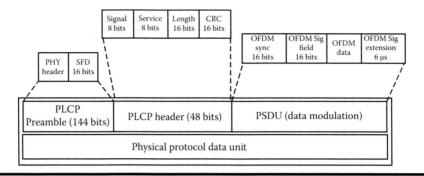

Figure 3.4   Physical layer OFDM-based frame format of the IEEE 802.11g standard.

3.4.1.1 IEEE 802.11g Physical Layer

The PHY layer is responsible for meeting the desired data rates, time, and processing constraints. The components of the physical layer frame format are shown in Figure 3.4. It consists of the physical layer convergence protocol (PLCP) preamble (144 bits), header (48 bits), and protocol service data unit (PSDU), which constitutes the physical protocol data unit (PPDU). IEEE 802.11g provides a maximum raw data throughput of 54 Mbps. In order to provide resilience against multipath effects while also being able to carry the high data rates, the main modulation method chosen for 802.11g was OFDM. It is a form of transmission that uses a large number of close-spaced carriers that are modulated with a low data rate. Normally these signals would be expected to interfere with each other, but by making the signals orthogonal to each other, there is no mutual interference.

3.4.1.2 IEEE 802.11g Medium Access Layer

The MAC layer controls the communication between the nodes in a network to effectively utilize the bandwidth and at the same time optimize the energy consumption. To provide fairness to every user, the carrier sense multiple access with collision avoidance (CSMA/CA) scheme is considered by default. Distributed coordinated function (DCF) and point coordinated function (PCF) communication modes were proposed in the protocol. In DCF, all the nodes listen before talking. If they find the channel free, they transmit their data; otherwise, they back off for a certain duration of time. A four-way handshake approach is used. It includes request to send (RTS), clear to send (CTS), data transmission, and acknowledgment. The receive node sends back an acknowledgment on the successful reception of the data. Since all the nodes in the network behave like a distributed system where each node is independent from the other, packet collisions can take place and therefore degrade the performance in terms of both latency and energy consumption. In the PCF mode, the AP periodically sends beacon frames, which communicate network identification and management parameters. The important difference between DCF and PCF is that PCF splits the time in such a way that each superframe includes first a contention-free period (CFP) and then possibly a contention period (CP). DCF mechanisms are used during the contention period, whereas in the contention-free period, a polling technique is used for transmission under the control of AP.

3.4.2 IEEE 802.15.1 Standard (Bluetooth)

The Bluetooth technology [5] was designed as a short-range wireless communication standard and later widely used for connecting a variety of personally carried devices to support data and voice applications. This wireless personal area network (WPAN) technology operates at the 2.4 GHz industrial, scientific, and medical (ISM) band, in which two and up to eight Bluetooth devices form a short-range network called piconet. These devices are synchronized to a common clock and hopping sequence in the same physical channel. The key features of this wireless technology include robustness, low power consumption, and low cost. It has several variants: (1) Bluetooth basic rate (BR) with an optional enhanced data rate (EDR); (2) Bluetooth 3.0, or generic alternate MAC and PHY (AMP) extensions (which have a provision of an additional radio); and (3) Bluetooth low energy (LE), or Bluetooth 4.0 [6]. Although classic Bluetooth operating at the basic data rate can reach up to 721.2 kbps, and at the enhanced data rate up to 2.1 Mbps, Bluetooth LE packets can only transfer data at up to 1 Mbps. However, they offer a simpler link layer design, ultra-low-power idle mode operation, simple device discovery, and reliable point-to-multipoint data transfer with advanced power savings and secure encrypted connections. Bluetooth LE technology supports very short data packets (i.e., 8 octet minimum and up to 27 octets maximum), whereas classic Bluetooth supports a maximum packet of 372 octets, which includes 8–9 octets of access code, 7 octets of header, and a payload of up to 342 octets. The Bluetooth LE network is mostly based on a star topology. It uses fewer channels for pairing devices, and the synchronization can be achieved in a few milliseconds. Further, with all of the above-mentioned features and support, Bluetooth LE is starting to get attention in many W-WSN applications and platforms [12].

3.4.2.1 IEEE 802.15.1 Physical Layer

Bluetooth LE operates in the frequency range of 2400–2483.5 MHz. There are 40 radio frequency (RF) channels, with each one having a bandwidth of 2 MHz, and the center frequencies are $f_c = 2402 + i \cdot 2$ MHz, $i = 0, \ldots, 39$. Gaussian frequency-shift keying (GFSK) modulation is used with a modulation index between 0.45 and 0.55. The maximum achievable symbol rate is 1 MS/s, and correspondingly, 1 Mbit/s of bit rate. For a transmitter, the output power level at the maximum power setting will be between the minimum output power (–20 dBm) and the maximum output power (+10 dBm). Table 3.3 shows the physical protocol data unit (PPDU) frame format; in total, it consists of 48 octets, including up to 38 octets of payload and only 2 octets of header.

The RF channels are allocated into two different types: advertising physical channels and data physical channels. The advertising channels are used for discovering devices, initiating a connection, and broadcasting data. The data physical channels are used for communication between connected devices during normal operation. In both cases, channels are subdivided into time units known as events: advertising events on the advertising channels and connection events on the data channels. The allocation of channel type to each RF channel is shown in Table 3.4.

Table 3.3 Frame Format of IEEE 802.15.1 PPDU

Preamble	Access Address	PDU Header	PDU Payload	CRC
1 octets	4 octets	2 octets	Variable (0–37 octets)	3 octets

Table 3.4   Bluetooth Low-Energy RF Channel Allocations

RF Channel	RF Center Frequency	Channel Type	Data Channel Index	Advertising Channel Index
0	2402 MHz	Advertising channel		37
1	2404 MHz	Data channel	0	
2	2406 MHz	Data channel	1	
...	...	Data channels	...	
11	2424 MHz	Data channel	10	
12	2426 MHz	Advertising channel		38
13	2428 MHz	Data channel	11	
14	2430 MHz	Data channel	12	
...	...	Data channels	...	
38	2478 MHz	Data channel	36	
39	2480 MHz	Advertising channel		39

3.4.2.2 IEEE 802.15.1 Link Layer

The functioning of the link layer can be described through a state machine having multiple states. These include standby, advertising, scanning, initiating, and connection states. The transitions from one state to another are directed by the host. The functionalities of different states are explained below.

- The standby state is the default state, in which it is not possible to send or receive packets. The link layer may leave the standby state to enter the advertising state, scanning state, or initiating state, and the standby state can be entered from any other state.
- A device in the advertising state is known as an advertiser. The link layer in the advertising state transmits advertising packets in advertising events and possibly listens to and answers responses triggered by these packets. A device in the scanning state is known as a scanner.
- The link layer in the scanning state listens for advertising channel packets from advertisers. There are two types of scanning, determined by the host: passive and active. In passive scanning, the link layer can only receive the packets. In active scanning, the link layer listens for advertising packets and, depending on the advertising type, may request an advertiser to send additional information.
- A device in the initiating state is known as an initiator. The link layer in the initiating state listens for advertising channel packets from specific devices and responds to these packets to initiate a connection with the other devices.
- The connection state can be entered either from the initiating state or from the advertising state. The link layer in the connection state sends or receives data packets in connection events. A device in the connection state can be either a master or a slave. When entered

from the initiating state (after sending a connection request), the connection state will be in the master role, whereas when entered from the advertising state (after receiving a connection request), the connection state will be in the slave role. The master defines the timing of transmissions in connection events.

3.4.3 IEEE 802.15.4 Standard (ZigBee)

IEEE 802.15.4 [7] is another standard that is very popular for wireless sensor networks (WSNs), as well as used for W-WSN. ZigBee is a low-power, short-distance wireless communication standard designed by ZigBee Alliance based on the IEEE 802.15.4 low-rate (LR) WPAN standard. It uses the license-free ISM bands, either 2.4 GHz or 868/915 MHz. The network standard allows unicast, broadcast, and group-cast messaging in ad hoc self-created networks. Based on that messaging scheme, ZigBee defines mesh, cluster tree, and star network topologies.

Nodes within a ZigBee network can be in three different roles: coordinator, router, and end devices. A coordinator stores information about the network and provides connection to the other networks. A router is capable of running application functions and relaying data to and from other devices. The end devices only communicate the information generated among the neighbors with less critical roles in the network topology.

3.4.3.1 IEEE 802.15.4 Physical Layer

The standard provided three PHY layers, based on three different frequency bands, modulation, and spreading formats, as summarized in the Table 3.5. The maximum achievable data rate is 250 kbps. The physical protocol data unit (PPDU) is composed of a synchronization header (SHR); the PHY header (PHR), which contains frame length information; and the PHY payload (PSDU), which is at most 128 octets.

Table 3.5 IEEE 802.15.4 Physical Layer Specification

Frequency Range (MHz)	Spreading Parameters		Data Parameters		
	Chip Rate (Kchip/s)	Modulation	Bit Rate (kbps)	Symbol Rate (kilo symbol/s)	Symbols
868–868.6	300	BPSK	20	20	Binary
868–868.6	400	ASK	250	12.5	20-bit PSSS
868–868.6	400	O-QPSK	100	25	16-array orthogonal
902–928	600	BPSK	40	40	Binary
902–928	1600	ASK	250	50	5-bit PSSS
902–928	1000	O-QPSK	250	62.5	16-array orthogonal
2450–2483.5	2000	O-QPSK	250	62.5	16-array orthogonal

Note: ASK, amplitude shift keying; BPSK, binary phase shift keying; O-QPSK, orthogonal quadrature phase shift keying; PSSS, parallel sequence spread spectrum.

3.4.3.2 IEEE 802.15.4 Medium Access Layer

The responsibilities of the IEEE 802.15.4 MAC layer include generating network beacons (coordinator), synchronizing these beacons, supporting MAC association and disassociation, supporting MAC encryption, employing unslotted and slotted CSMA/CA mechanisms for channel access, and handling guaranteed time slot (GTS) allocation and management. IEEE 802.15.4 defines four frame structures: beacon frame, data frame, acknowledgment frame, and MAC command frame. Data transfers are completely controlled by the devices rather than by the coordinator. Based on the application-defined rate, the device either transfers data to the coordinator or polls the coordinator to receive data, which helps in optimizing the energy consumption.

Two modes are provided for IEEE 802.15.4 multiple access schemes: the beacon-enabled and non-beacon-enabled modes. In a beacon-enabled mode, a superframe structure is used. A superframe is divided into two portions: active and inactive. During the inactive portion, devices may enter a low-power mode according to the requirement of its application. The active portion consists of the contention access period (CAP) and contention-free period (CFP). During the CAP phase, multiple devices have to compete with other devices using a slotted CSMA/CA mechanism, while the CFP contains guaranteed time slots where no contention exists. However, if a coordinator does not prefer to use the beacon-enabled mode, it may turn off the beacon transmissions, and the unslotted CSMA/CA algorithm can be used. In this case, both downlink and uplink compete for the same resources.

The MAC protocol data unit is composed of the MAC header (MHR), which contains, in particular, the frame type indication (beacon, data, acknowledgment, and MAC command); the MAC payload, whose maximum length is 128 octets; and the MAC footer (MFR).

3.4.4 IEEE 802.15.4a Standard (IR-UWB)

The IEEE 802.15 Low-Rate Alternative PHY Task Group (TG4a) for wireless personal area networks was initially created to propose an amendment to the low-data-rate IEEE 802.15.4 standard, aiming to define an alternative physical layer [8]. The main objectives were to provide high-throughput, ultralow power consumption, scalable bit rates, longer achievable ranges, low cost, compliance with worldwide regulation, and the possibility for different receiver architectures (hence tolerating trade-offs between performance and complexity). In this enhanced standard, only the physical layer was modified, and some of the important features are presented below.

IEEE 802.15.4a provides two physical layers: infrared ultrawideband (IR-UWB) and chirp spread spectrum (CSS) in the unlicensed band. The format of the PHY protocol data unit (PPDU) is slightly the same as a classical IEEE 802.15.4 structure, including a synchronization header (SHR), which is composed of a preamble, start-of-field delimiter (SFD), and PHY header (PHR). In comparison with IEEE 802.15.4, three features have been added included: variable preamble length, variable data rates, and ranging support. Consequently, new fields have been included in the PHR, which includes preamble length, data rate, and ranging flag.

3.4.5 IEEE 802.15.4j Standard

IEEE802.15.4j [9] is another extension of the IEEE 802.15.4 standard for a band reserved for medical body area networks (MBANs). The standard proposed an alternate PHY specified for 2360–2400 MHz, as well as defined MAC modifications to support the PHY implementation. The band between 2360 and 2390 MHz is reserved only for in-hospital patient monitoring applications, whereas in the range between 2390 and 2400 MHz, remote health care monitoring and ambulance monitoring is possible. The physical layer is based on direct-sequence spread spectrum (DSSS), employing offset quadrature phase-shift keying (O-QPSK) modulation, operating in 780, 868, 915, 2380, and 2450 MHz. The additional MAC functionalities include a channel switch command in which the coordinator may require devices to switch their operating channel at a specific time. The device is required to vacate the channel at the scheduled time and move to another channel that does not interfere with the primary user. Finally, multiperiodic guaranteed time slots (GTSs) are allocated to each device at every *N* superframe.

3.4.6 IEEE 802.15.6 Standard

In order to address a wide range of W-WSN applications, it is important to have a generic standard that leads to uniformity toward the development of specific solutions and products. Therefore, in 2012, the Institute of Electrical and Electronics Engineers (IEEE) formally released a wireless body area network specific standard called IEEE 802.15.6 [11]. The IEEE 802.15.6 standard intends to complete the IEEE 802.15 family for body area networks. In the following sections, we will explore the relevant details of the standard and its important features.

3.4.6.1 IEEE 802.15.6 Physical Layer

The IEEE 802.15.6 standard has proposed three different alternatives for the PHY layer based on the specific applications and their requirements. It includes human body communications (HBC), narrowband (NB) PHY, and ultrawideband (UWB) PHY. The physical layer protocol data unit (PPDU) represents the information that is sent through the propagation medium to the receiver device. It is composed of the physical layer convergence protocol (PLCP) preamble, PLCP header, and physical layer service data unit (PSDU), as illustrated in Figure 3.5, and is briefly explained below:

■ PLCP preamble: The purpose of the preamble is to aid the receiver in packet detection, timing synchronization, and carrier-offset recovery. Two unique preambles are defined in order to mitigate false alarms due to other networks operating on adjacent channels. The preamble

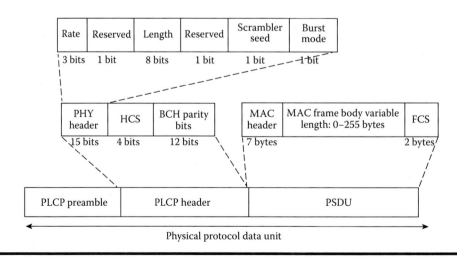

Figure 3.5 IEEE 802.15.6 physical frame format.

is transmitted at the symbol rate for the desired band of operation and will be encoded using the same modulation parameters as defined for different physical types.

■ PLCP header: It is added to convey information about the PHY and MAC parameters that are needed at the receiver side in order to decode the PSDU. Details on the different parts composing the PLCP header and how to properly set the bits of each field can be found in Section 8.3 of [11].

■ PSDU: It is formed by concatenating the MAC header with the MAC frame body and frame check sequence (FCS). The PSDU is then scrambled and optionally encoded by a Bose, Ray-Chaudhuri, Hocquenghem (BCH) code. The PSDU is transmitted using any of the available data rates in the operating frequency band.

3.4.6.2 IEEE 802.15.6 Medium Access Layer

The MAC protocol data unit (MPDU) is an ordered sequence of fields delivered to or from the PHY service access point (PHY SAP). The MAC frame consists of a fixed-length MAC header (seven octets), a variable-length MAC frame body, and a fixed-length FCS field (two octets), as shown in Figure 3.6. The MAC frame body has an octet length L_FB such that $0 \leq L\_FB \leq pMax$-$FrameBodyLength$, and is present only if it has a nonzero length, where $pMaxFrameBodyLength$ is the maximum frame body length at the physical layer. The Low-order security sequence number and message integrity code (MIC) fields are not present in unsecured frames. Management, control, and data type are the three MAC frames that are described in detail in the standard. Each of them implies a different composition of the MAC frame body, in particular for the frame payload field.

The standard provides greater flexibility to the users to adapt MAC according to their requirements, for example, at the MAC level, carrier sense multiple access with collision avoidance (CSMA/CA), scheduled and unscheduled time division multiple access (TDMA) access, and scheduled and unscheduled polling and posting. In CSMA/CA, priorities are assigned based on the traffic; for example, emergency traffic has higher priority than normal W-WSN traffic.

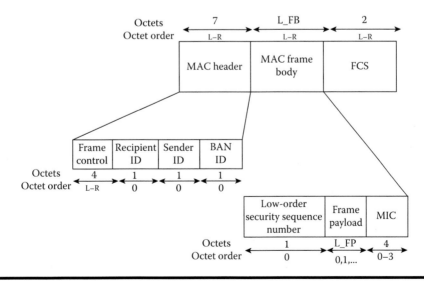

Figure 3.6   IEEE 802.15.6 MAC frame format.

Further, based on these priorities, contention window sizes and back-off mechanisms are proposed [11]. In scheduled access, TDMA slots can be allocated by the coordinator to specific nodes based on their requirements. To compare these two schemes, scheduled access has very high performance in terms of packet delivery ratio, but at the expense of synchronization overheads and the cost of keeping the nodes always in sync, which is very critical for the complete network performance. Moreover, in time-critical applications and due to the dynamic nature of W-WSN traffic, a scheduled-based approach can be degraded significantly. As far as the CSMA/CA-based MAC approach is concerned, it consumes more energy, but it can be improved, provided the packet collisions are minimized and the packet delivery is maximized.

With regard to medium access mechanisms, the BAN coordinator can decide to operate two different access modes. The first mode corresponds to the beacon mode with superframe boundaries. A beacon is transmitted at the beginning of every beacon period. In this way, a common time base is established, to enable time-referenced allocations. There are two exclusive access phases (EAPs) and two random access phases (RAP),* which can be configured together based on the application requirements as EAP1 and EAP2, respectively. For the medium access, only a contended allocation is possible, which means either CSMA/CA or Slotted Aloha can be used during EAP, RAP, and CAP periods, whereas during the managed access phase (MAP), there are scheduled, unscheduled, and improvised access options.

The second mode is the nonbeacon mode with superframe boundaries. In this mode, beacons are not transmitted, but superframe and allocation slot boundaries are established, which provides time referencing to the medium access. In the nonbeacon mode with superframe boundary only, managed access is possible. In the nonbeacon mode without superframe boundaries, beacons are not transmitted, and superframe and allocation slot boundaries are not established, because there is no time reference involved in accessing the medium. It is based on unscheduled access with type II polled uplink and postallocation for downlink by using the CSMA/CA mechanism.

3.4.7 Comparison of Key Enabling Standards

Despite the specific constraints and requirements of W-WSN, as discussed in Section 3.2, currently there are many standards that can be used for various applications. A comparison based on several metrics is presented in Table 3.6. The following main differences can be highlighted between W-WSN and existing PAN-, WSN-, and WLAN-related standards:

- The power consumption of existing standards is too high in comparison to the specific requirements of W-WSN applications. As shown in Table 3.6, power levels are in the range of 800 mW for Wi-Fi, 100 mW for Bluetooth, and 50 mW for ZigBee, whereas W-WSN requires lower power levels of around 0.1–1 mW to increase the battery and network lifetime, especially for on-body communication.
- Existing communication standards were not optimized for body-centric communications, in particular regarding W-WSN inherent intra- and intermobility patterns and the RF propagation around, or inside, the body.
- The communication stacks of several W-WSN used standards do not meet the specific requirements of applications in terms of reliability, low power consumption, high data rates, and robustness against external interferences.

---

* RAP is used for regular nonemergency traffic.

Table 3.6 Comparisons of the Key Enabling Standards for Wearable Wireless Sensor Networks

	IEEE 802.11 a/b/g/n (Wi-Fi)	*IEEE 802.15.1 (Bluetooth)*	*IEEE 802.15.1 (Bluetooth LE)*	*IEEE 802.15.4 (ZigBee)*	*IEEE 802.15.4a (UWB)*	*IEEE 802.15.4j (MBAN)*	*IEEE 802.15.6 (WBAN Standard)*
Modes of operation	Ad hoc, infrastructure	Ad hoc	Ad hoc	Ad hoc	Ad hoc	Ad hoc	Ad hoc
Physical (PHY) layers	NB	NB	NB	NB	UWB	NB	NB, UWB, HBC
Radio frequencies (MHz)	2400, 5000	2400	2400	868/915, 2400	75–724, 3000–5000, 6000–10,000	2360–2490/2390–2400	402–405, 420–450, 863–870, 902–928, 950–956, 2360–2400, 2400–2438.5
Power consumption	High (≈800 mW)	Medium (≈100 mW)	Low (≈10 mW)	Low (≈60 mW)	Low (<50 mW)	Low (≈50 mW)	Ultralow (≈1 mW at 1 m distance)
Maximal signal rate	Up to 150 mbps	Up to 3 mbps	Up to 1 mbps	Up to 250 mbps	Up to 27.24 mbps	Up to 250 kbps	10 kbps to 10 mbps
Communication range	Up to 250 m (802.11n)	100 m (class 1 device)	>100 m	Up to 75 m	Up to 30 m	Up to 75 m	Up to 10 m (nominal ~2 m)

(Continued)

Table 3.6 (Continued)   Comparisons of the Key Enabling Standards for Wearable Wireless Sensor Networks

	IEEE 802.11 a/b/g/n (Wi-Fi)	IEEE 802.15.1 (Bluetooth)	IEEE 802.15.1 (Bluetooth LE)	IEEE 802.15.4 (ZigBee)	IEEE 802.15.4a (UWB)	IEEE 802.15.4j (MBAN)	IEEE 802.15.6 (WBAN Standard)
Networking topology	Infrastructure based	Ad hoc very small networks	Ad hoc very small networks	Ad hoc, peer to peer, star, mesh	Ad hoc, peer to peer, star, mesh	Ad hoc, peer to peer, star	Intra-WBAN: 1/2-hop star Inter-WBANs: Nonstandardized
Topology size	2007 devices for structured Wi-Fi BSS	Up to 8 devices per piconet	Up to 8 devices per piconet	Up to 65,536 devices per network	Up to 65,536 devices per network	Up to 65,536 devices per network	Up to 256 devices per body and up to 10 WBANs in 6 m³
Target applications	Data networks	Voice links	Health care, fitness, beacons, security, etc.	Sensor networks, home automation, etc.	Short-range and high data rates, localization, etc.	Short-range medical body area networks	Body-centric application
Target BAN architectures	Off body	On body	On body	Body to body, off body	Body to body	On body	On body

Note: HBC, human body communication; NB, narrowband; UWB, ultrawideband.

To conclude, most of the above-mentioned communication standards were designed with other applications in mind and not specifically W-WSN. Therefore, they do not always meet the specific requirements and constraints of W-WSN applications [4]. For example, there are major limitations in terms of peak power consumption, achieved data rates, communication range, generated RF interferences, and efficient on-body and body-to-body communications.

On the other hand, existing communication and radio technologies and standards are currently still used for the design of short-term and ready-to-use W-WSN solutions [13]. This is mainly due to the current unavailability of commercial and off-the-shelf IEEE 802.15.6-compliant devices.

However, it is necessary to highlight that different standards can still be used and combined to design an end-to-end complete W-WSN architecture. For example, for on-body communications Bluetooth low energy, IEEE 802.15.4j, and IEEE 802.15.6 can be used, though IEEE 802.15.6 seems to be more suitable, adaptable, and optimized. For the context of body to body, we might use IEEE 802.15.4, IEEE 802.15.1, or IEEE 802.11 in ad hoc mode, as they are much better in terms of network connectivity due to their higher transmit power. Finally, for the off-body communication, infrastructure and cellular networks, such as 3G/4G/LTE, Wi-Fi, and satellite, have to be used.

3.5 Research Trends, Opportunities, and Challenges

The standards presented in Section 3.4 provide details about the PHY and MAC layers. Although there are further enhancements at both layers, those extensions are beyond the scope of this chapter. In this section, we present the current research trends and future opportunities with reference to routing protocols, coexistence, privacy and security, and hardware and software platforms for body area networks (BANs).

3.5.1 Routing Protocols

A body-to-body network (BBN) is required to extend the interaction between multiple bodies, which is required to realize several applications, as discussed in Sections 3.2 and 3.3. BBNs can be further seen as centralized, structured, and distributed, as depicted in Figure 3.7. In centralized BBNs, one of the many BANs can be selected as a leader to easily coordinate and control the interaction between multiple BANs, whereas in the structured approach, many clusters can be formed, in which an individual cluster head can manage all the activities of the cluster and between the clusters as a gateway to enable end-to-end connectivity between all BANs. Finally, a distributed approach can also be used where every single BAN is responsible for its own communication and in a competition to interact with the surrounding BANs.

The classification and state-of-the-art routing protocols can be divided into several important subcategories, which include multipath routing, QoS-aware routing, cluster-based routing, and mission-critical reliable routing. Detailed comparison and description about the individual categories can be found in [4]. The networking layer is responsible for providing a global end-to-end network connectivity between multiple BANs that are operating in the same vicinity. Further, the design of innovative topology management and addressing capabilities and self-organizing intra/inter-BAN routing/relaying strategies are the important issues at the network level. Typically, W-WSN is envisaged as only an on-body network in which star (i.e., one-hop) or two-hop topology is sufficient. However, considering the broader range of W-WSN applications, as discussed in

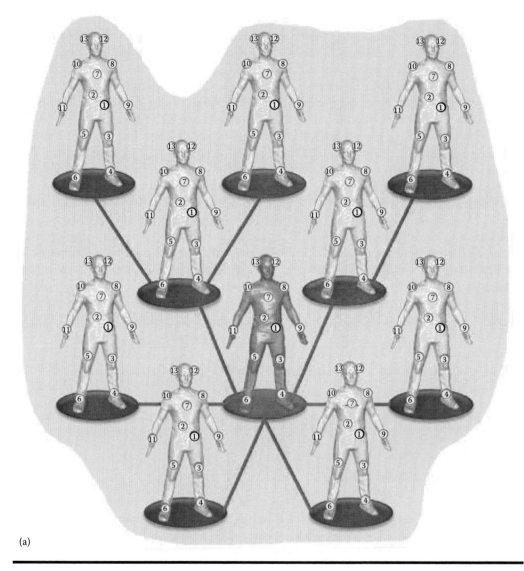

(a)

Figure 3.7 Body-to-body network architectures. (a) Centralized. *(Continued)*

Section 3.2, multihop routing becomes necessary to realize effective system performance in some applications.

Self-organization for wireless multihop systems can be based on reactive on-demand solutions, which are adapted to low-energy, low-traffic W-WSN. The authors of AnyBody [14] have shown that despite the relatively high cost to build and maintain a topology, a cluster-based approach is particularly suited for body area networks. There are five steps of the AnyBody protocol: neighbor discovery, density calculation, cluster head construction, backbone setup, and routing path setup. AnyBody uses similar principle of low-energy adaptive clustering hierarchy (LEACH) [15], which selects the cluster head at regular time intervals to balance the energy consumption, and the cluster head aggregates the data and sends it to the remote station. In LEACH, it is assumed

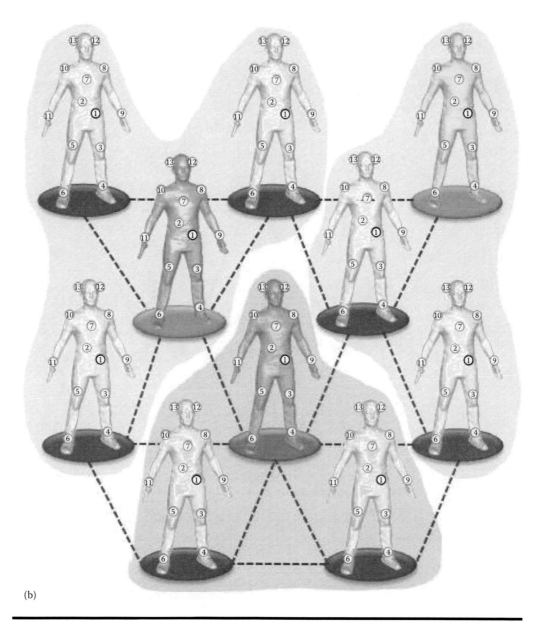

(b)

Figure 3.7 (Continued)   Body-to-body network architectures. (b) Structured.   *(Continued)*

that all nodes are within the range of the remote base station, whereas AnyBody addresses this problem by using a density-based cluster head selection method and these cluster heads to build a backbone network. The main advantage of AnyBody [14] is that the number of clusters remains almost constant even when the number of nodes in the network is increased, whereas in LEACH, as the number of nodes increases, the number of clusters increases as well. This results in a bigger cluster size of AnyBody compared to LEACH. Furthermore, AnyBody also reduces the cost of setting up the cluster.

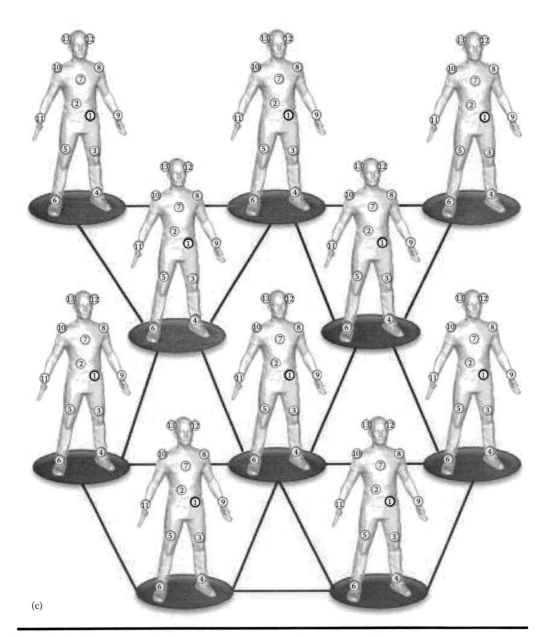

(c)

Figure 3.7 (Continued)   Body-to-body network architectures. (c) Distributed.

In infrastructure based Multipath routing protocols which is also necessary for certain W-WSN application, in particular, for off-body communication. In such protocols, the main objective is to construct and maintain multiple paths from source to destination. An infrastructure provides reliable and fast data transmission because every intermediate data routing node has its next hop set up in advance. It also provides the protocol for reducing failure recovery time by assigning the alternative route, which is also discovered in advance [16]. Several techniques are exploited for reliable, secure, and energy-balanced designs; this includes energy-aware and hierarchy-based multipath routing algorithms, which are explained in much more detail in [16]. However, the focus here is to highlight multipath routing in general, with reference to an overall overview of routing protocols.

There have been so far only a few research studies related to the issue of self-organizing, cooperative, and multihop communications between multiple W-WSNs (i.e., hybrid on-body and body-to-body communications) in infrastructure-less environments. All of these major research issues have yet to be resolved in order to enable the emergence of future ubiquitous communication and monitoring systems enabling health care, life-critical and rescue operations, public safety, and preparedness.

To conclude, one of the important constraints in designing routing protocols is the impact of mobility. Other than classical hospital monitoring, most of the W-WSN applications have a dynamic mobility environment, and for effective on-body communication, opportunistic and mobility-aware routing is very critical. For example, it can be seen in works such as [17] that the performance of traditional Dijkstra algorithm–based routing can be enhanced significantly with the addition of some important characteristics, such as accurate link estimation of the mobile W-WSN.

3.5.2 *Interference Mitigation and Coexistence Schemes*

Most of the W-WSN-related standards and technologies operate on the same frequency ISM bands, which results in significant interruption to each other. In particular, those devices that transmit their information at low transmit power suffer more. As for W-WSN devices, power consumption is one of the important constraints due to limited battery capacities, and therefore often these devices have to operate at ultralow power. Consequently, on-body sensors can be interfered with by neighboring Bluetooth or Wi-Fi devices that are significantly more powerful.

There are different sources of interference, such as intra-BAN or inter-BAN nodes operating in the same frequency band, that is, co-channel interference, or in different frequencies bands, that is, adjacent channel interference. Figure 3.8a shows an interference model where any node k can potentially generate interference at a given receiver j. For example, consider a given transmission between two nodes i and j; since other transmissions can also happen concurrently (within the same frequency band), the interference and noise level at receiver j can vary during the time, as shown in Figure 3.8b.

Generally, the interference mitigation is classified into collaborative and noncollaborative coexistence techniques. In collaborative methods, multiple nodes interact with each other to manage coexistence, whereas in noncollaborative methods, multiple nodes manage coexistence without any interaction. The initial research studies on BAN interference mainly concentrate on the impact from other technologies (e.g., adjacent channel interference), such as IEEE 802.11 and IEEE 802.15.1. It is clear from the various research studies, such as [18–21], that there is a dominant interference from other networks in BAN. These approaches of interference analysis are only enough for intra-BAN communication, where each node is synchronized with its coordinator and configured at the same transmit power. However, with the advent of body-to-body communications, inter-BAN interference and its mitigation is a new and unresolved problem.

There are several approaches that are used to coexist with other technologies. This includes noncollaborative schemes such as time sharing and channel hopping, whereas for the case of collaborative schemes, the MAC enhanced temporal algorithm (META) [22] and TDMA-based approach [23] are well known. However, with reference to W-WSN, there were no standardized approaches until the IEEE 802.15.6 standard was formulated and released. The standard has

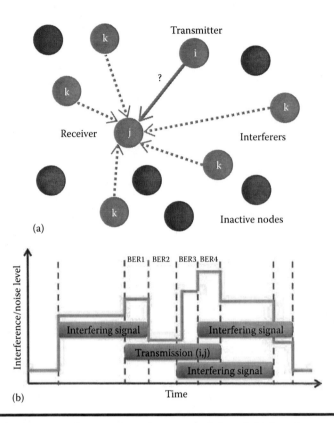

Figure 3.8 Interference overview. (a) Interference model and (b) interference/noise level with reference to time. i, transmitter; j, receiver; k, interfering nodes.

proposed three methods for coexistence: beacon shifting, channel hopping, and adaptive superframe interleaving. The performance of these techniques is still yet to be evaluated for many diverse applications, as discussed in Section 3.2.

3.5.3 Privacy and Security

The security and privacy of patient-related data are two essential components for the system security of W-WSNs. Privacy means that the data is accessed by only certain people who are authorizated to access it. Through a user-defined access control, a fine-grained data access policy can be enforced to prevent unauthorized access to patient-related data generated by the W-WSN.

The security is divided into five parts: integrity, availability, confidentiality, authentication, and authorization [24]. Data security is ensured by keeping it stored and transferred in a safe manner. Data confidentiality, dependability, and integrity are the three most important requirements for distributed data storage in W-WSNs. In order to enhance the dependability of the data, error correcting code techniques can be employed, which add extra redundancy in the information and enable the data to be more dependable. To provide confidentiality, the data is normally encrypted by public keys, which are authorized by storage nodes. However, data integrity is not ensured whenever the number of errors is more than the detecting capability [25].

Given the specific requirements and constraints of W-WSN solutions, in terms of limited resource capabilities and harsh environmental conditions, the design of new security and privacy mechanisms remains an important challenge yet to be achieved. Unlike traditional communication networks, there has been relatively little research so far on solving the security and privacy issues in W-WSN [25,26].

IEEE 802.15.6 [11] already proposed three basic security levels for W-WSN systems: unsecured communications (level 0), authenticated but not encrypted communications (level 1), and authenticated and encrypted communications (level 2). In the last security level, all the messages are transmitted within authenticated and encrypted frames, thus enabling the authenticity, integrity, confidentiality, privacy, and replay defense of the exchanged messages. However, these traditional encryption and authentication mechanisms remain not perfectly suitable for W-WSN systems due to their stringent requirements in terms of limited power consumption, computation capability, heterogeneous intra/inter/beyond-BAN communications, infrastructure-less deployments, and so on.

In this context, new lightweight security and privacy paradigms are currently being developed, such as the efficient generation and management of cryptographic keys using biometric techniques [27] (i.e., based on collected physiological signals, such as ECG) and physical channel–based approaches [28] (i.e., based on radio channel impulse responses or receive signal strength indicator [RSSI] values). Another promising research trend is the development of new schemes to solve WBAN's dependability issues [26], defined as the conjunction of reliability, security, and availability, from both an intra- and an inter-BAN communication perspective.

3.5.4 Hardware and Software Platforms and Simulators

There are a number of hardware platforms available for typical health care environments of body area networks. Code Blue is one of the earliest platforms developed mainly for vital sign monitoring in emergency situations [29]. In this system, different sensors connected to the body send their information through a coordinator or sink node to a database that is used to store, secure, and manage the data, as well as trigger an alarm in case of any danger. After it, several similar platforms were introduced with specific enhancements and features. These include iCare [30], Medical MoteCare [31], and DexterNet [32], specifically dedicated to the elderly and infirm patients monitoring wirelessly from their homes at any time. Further, MEMSWear [33] monitors vital signs and detects falls among the elderly, whereas IBM Personal Care Connect (PCC) [24] and ALARM-NET [34] are home-based platforms. AMON provides continuous medical care to high-risk cardiac and respiratory patients [35]. The system includes continuous collection and evaluation of multiple vital signs, intelligent multiparameter medical emergency detection, and a cellular connection to a medical center. A more extensive list of platforms can be found in [13].

In the recent past, with the emergence of wearable technologies, smarter, cost-effective, easy-to-wear application-specific watches, armbands, and necklaces have been introduced. Companies such as Vuzix, Samsung, LG, Sony, and Razer have introduced several products for medicine and fitness, entertainment, augmented reality, and more. It is predicted that wearable technologies could be a key booster for the consumer electronics industry. According to the wearable technology database [36], there are currently 41 medical, 77 fitness, 117 lifestyle, and 26 entertainment devices already available on the market.

To sum up different hardware and software platforms, the availability of these platforms is one of the major issues, since most of them were developed under projects for academic purposes. Another limitation for the comparison is that some of them follow quality of service as a major requirement, while others use energy efficiency and reliability as a metric. Further, security must be implemented in terms of shielding the communication channels between the networking devices, as well as the data stored in an integrity database. Also, one of the aspects that is missing in most of these platforms is the management of dynamic mobility. As pointed out by [13], most of the platforms are dealing with either limited or no mobility. In various applications, discussed in Section 3.2, accurate radio-link and mobility modeling is necessary, and efforts such as [17] and [37] will help to enhance the performance of the W-WSN systems.

The impact of designing a new communication and networking protocol, physical layer optimizations, or low-power strategies needs to be evaluated in the context of the global network performance. This performance evaluation through the development of a body area network (BAN) prototype or platform is a complex and time-consuming process. Thus, simulating tools are generally considered to assist in and be used for evaluating new BAN algorithms and protocols.

Some simulators can be used for measuring the performance of a routing protocol with respect to the shortest-path-finding mechanism, while a few of them focus on power efficiency and energy reduction in the MAC protocol. In general, most of them allow the development of new protocols and algorithms for all layers of the protocol communication stack. This can be easily achieved by following certain rules, and then new protocols can be integrated with built-in modules and easily tested and verified.

NS-2, NS-3, OMNet, OPNet, and WSNet are a few examples of different network simulators that are often used. Tables 3.7 and 3.8 present several important features of different simulators. There are a number of simulators that are available as open source for research and academic purposes; however, the simulators with real W-WSN characteristics, features, and modules are much less. A SystemC-based fast simulator was presented in [38,39], which helps to provide accurate timing and the energy estimated for health monitoring applications on wireless BANs. These

Table 3.7 Simulator Features and Comparison

Simulators		Programming Languages	Platforms	License	Document Support	Extension
NS-2		C++ and Tcl	Linux/GNU, Solaris, MAC, Windows with Cygwin support	Yes (GNU-GPL)	Good	Yes
NS-3		C++ and Python	Linux/GNU, Windows support through visual studio	Yes (GNU-GPL)	Good	Yes
OMNet ++	Castalia	C++, Java, C#, and SystemC	Linux/GNU, Eclipse-based IDE	Yes (GNU-GPL)	Good	Yes
	MiXiM	C++, Java, C#, and SystemC			Little	N/A
WSNet		C and XML	Linux/GNU	Yes (CeCILL)	Little	Very easy

Note: GNU-GPL, general public license; IDE, integrated development environment; XML, extensible markup language.

Table 3.8 Simulator Features and Comparison

Simulators		Standards Supported	Wireless Channel	Physical Layer	MAC Support	Network Support
NS-2		Wi-Fi, WSN	Friis propagation loss and two-ray ground reflection, lognormal shadowing	Omnidirectional antennas, 802.11 PHY parameters	IEEE 802.11, single-hop TDMA	Unicast, multicast, hierarchical, etc. Protocols: geographical, greedy forward, AODV, DSDV, DSR, etc.
NS-3		Wi-Fi, WiMAX, LTE, LR-WPAN	Models: Friis, building propagation loss, log distance, Rayleigh, Nakagami, ITU, Jakes, etc.	Parabolic, cosine, isotropic antenna models, 802.11 PHY parameters	Wi-Fi MAC, 802.11, DCF	6LowWPAN, Nix-Vector, DSDV, DSR, OLSR, etc.
OMNet ++	Castalia	Wi-Fi, ZigBee, IEEE 802.15.4/15.6	Lognormal, free space, body shadowing, path loss measurements, temporal variations, interference	Narrowband, FSK, PSK, DBPSK, DQPSK, etc. (OMNet++)	CSMA: Tunable 802.15.4 TDMA: TMAC, SMAC Hybrid: Baseline 802.15.6	Simple tree routing, multipath ring, bypass routing
	MiXiM	ZigBee, IEEE 802.15.4	Jakes model, path loss, fading and shadowing, random walk	Narrowband, FSK, PSK, DBPSK, DQPSK, etc. (OMNet++)	CSMA, SMAC, TMAC, adaptive MAC (OMNet++)	OMNet++
WSNet		Wi-Fi, ZigBee/ IEEE 802.15.4, IEEE 802.15.6	Free space, disk models, lognormal, ITU, Nakagami, IEEE 802.15.6, etc.	IEEE 802.15.4, IEEE 802.15.6, narrowband	802.11 DCF, BMAC, XMAC, IEEE 802.15.4 CSMA, IEEE 802.15.6 CSMA/beacon enabled	Greedy, geographic, file static, AODV, etc.

Note: AODV, Ad hoc on demand distance vector; BMAC, Berkeley medium access control; DBSPK, differential binary phase shift keying; DQPSK, differential quadrature phase shift keying; DSDV, destination-sequenced distance-vector; DSR, dynamic source routing; ITU, International Telecommunication Union; OLSR, optimized link state routing; SMAC, sensor-medium access control; TMAC, time-out medium access control; XMAC, a low power MAC protocol for wireless sensor networks (WSNs).

simulators were able to obtain timing and energy measurements for each function in the program, as well as for each module in the hardware, which is then used to strategize the optimizations toward critical functions and components. One of the major contributions in terms of W-WSN simulators comes from Castalia [40], which is a framework of OMNet++ simulator. It is based on realistic wireless channel and radio models, with a realistic node behavior especially relating to radio access. Castalia can also be used to evaluate different platform characteristics for specific applications, since it is highly parametric and can simulate a wide range of platforms.

3.6 Conclusion

To conclude, in this chapter, current state-of-the-art wearable wireless sensor networks were presented with an objective to give a technical overview of the complete W-WSN system. A number of new and emerging, as well as classic, W-WSN applications were explored, along with their key requirements, which it is necessary to understand. Before 2012, there was no specific wireless body area network standard, and therefore various existing WPAN and low-power standards were used to design W-WSN solutions. Some of those standards are still very much valid, in particular to realizing end-to-end communication. Therefore, most of those standards were discussed with their main features, and finally, a comparison was presented. This comparison reflects that even though IEEE 802.15.6 has been released and is available, we still require other technologies in several applications to meet their requirements. Further, several research tracks are briefly touched upon, including effective and reliable routing, interference mitigation and coexistence schemes, privacy and security, and different platforms. Finally, there are number of open research opportunities, such as effective on-body and body-to-body communication and channel modeling, accurate radio-link and mobility modeling, and reliable communication and routing.

Acknowledgments

This publication was made possible by National Priority Research Program (NPRP) grant 6-1508-2-616 from the Qatar National Research Fund (a member of Qatar Foundation). The statements made herein are solely the responsibility of the authors.

References

1. S. Curtis, Silver surfers demand digital health services, *The Telegraph*, http://www.telegraph.co.uk/technology/internet/10631423/Silver-surfers-demand-digital-health-services.html?fb (accessed 2014).
2. http://www.wearable-technologies.com/2015/01/meet-the-wt-wearable-technologies-heroes-of-the-year/ (accessed January 7, 2015).
3. Cochlear, 2014, http://www.cochlear.com/wps/wcm/connect/au/home/understand/hearing-and-hl/hl-treatments/cochlear-implant (accessed January 14, 2015).
4. M. M. Alam, E. Ben Hamida, Surveying wearable human assistive technology for life and safety critical applications: Standards, challenges and opportunities, *Sensors*, 14(5), 9153–9209, 2014.
5. IEEE (Institute of Electrical and Electronics Engineers), IEEE standard for local and metropolitan area networks—Part 15.1: Wireless medium access control (MAC) and physical layer (PHY) specifications for wireless personal area networks (WPANs), IEEE 802.15.1, IEEE, New York, 2005.
6. Bluetooth SIG, http://www.bluetooth.com/Pages/Bluetooth-Smart.aspx (accessed 2015).

7. IEEE (Institute of Electrical and Electronics Engineers), IEEE standard for local and metropolitan area networks—Part 15.4: Low-rate wireless personal area networks (LR-WPANs), IEEE 802.15.4, IEEE, New York, 2011.

8. IEEE (Institute of Electrical and Electronics Engineers), Amendment to 802.15.4-2006: Wireless medium access control (MAC) and physical layer (PHY) specifications for low-rate wireless personal area networks (LR-WPANs), IEEE 802.15.4a, IEEE, New York, 2007.

9. IEEE (Institute of Electrical and Electronics Engineers), IEEE 802.15.4j, IEEE, New York, 2013, http://standards.ieee.org/findstds/standard/802.15.4j-2013.html.

10. IEEE (Institute of Electrical and Electronics Engineers), Wireless LAN medium access control (MAC) and physical layer (PHY) specifications, IEEE 802.11, IEEE, New York, 2012.

11. IEEE (Institute of Electrical and Electronics Engineers), IEEE standard for local and metropolitan area networks—Part 15.6: Wireless body area networks, IEEE 802.15.6, IEEE, New York, 2012.

12. CSR, Bluetooth 4.0: Low energy, http://chapters.comsoc.org/vancouver/BTLER3.pdf.

13. B. Alghamdi, H. Fouchal, Wireless body area network platforms evaluation, in *9th International Wireless Communications and Mobile Computing Conference (IWCMC)*, Sardinia, Italy, July 2013, 1348–1352.

14. T. Watteyne, I. Augé-Blum, M. Dohler, D. Barthel, AnyBody: A self-organization protocol for body area networks, in *ICST 2nd International Conference on Body Area Networks (BodyNet)*, Florence, Italy, 2007, 1–7.

15. W. Heinzelman, A. Chandrakasan, H. Balakrishna, An application-specific protocol architecture for wireless microsensor networks, *IEEE Wireless Communications*, 1(4), 660–670, 2002.

16. K. Sha, J. Gehlot, R. Greve, Multipath routing techniques in wireless sensor networks: A survey, *Wireless Personal Communications*, 70(2), 807–829, 2013.

17. E. Ben Hamida, M. M. Alam, M. Maman, B. Denis, Short-term link quality estimation for opportunistic and mobility aware routing in wearable body sensors networks, in *IEEE 10th International Conference on Wireless and Mobile Computing, Networking and Communications (IEEE WiMob 2014)*, Larnaca, Cyprus, 2014.

18. F. Martelli, R. Verdone, Coexistence issues for wireless body area networks at 2.45 GHz, in *18th European Wireless Conference (EW)*, Poznan, Poland, 2012, 1–6.

19. D. M. Davenport, F. Ross, B. Deb, Coexistence of WBAN and WLAN in medical environments, in *70th VTC Conference*, Anchorage, AK, 2009, 1–5.

20. T. Hayajneh, G. Almashaqbeh, S. Ullah, A. Vasilakos, A survey of wireless technologies coexistence in WBAN: Analysis and open research, *Wireless Networks*, 20(8), 2165–2199, 2014.

21. D. Jie, D. Smith, Coexistence and interference mitigation for wireless body area networks: Improvements using on-body opportunistic relaying, 2013, http://arxiv.org/abs/1305.6992.

22. J. Lansford, MEHTA: A method for coexistence between co-located 802.11b and Bluetooth systems, 2000, http://www.ieee802.org/15/pub/TG2.html.

23. S. Shellhammer, Collocated collaborative coexistence mechanism: TDMA of 802.11 and Bluetooth, 2001, http://www.ieee802.org/15/pub/TG2.html.

24. M. Blount, V. Batra, A. Capella, M. Ebling, W. Jerome, S. Martin, M. Nidd, M. Niemi, S. Wright, Remote health-care monitoring using Personal Care Connect, *IBM Systems Journal*, 46(1), 95–113, 2007.

25. M. Li, W. Lou, Kui Ren, Data security and privacy in wireless body area networks, *IEEE Wireless Communications*, 17(1), 51–58, 2010.

26. Y. Hovakeemian, K. Naik, A. Nayak, A survey on dependability in body area networks, in *Proceedings of the 5th International Symposium on Medical Information and Communication Technology (ISMICT)*, Montreux, Switzerland, 2011.

27. H. Wang, H. Fang, L. Xing, M. Chen, An integrated biometric-based security framework using wavelet-domain HMM in wireless body area networks (WBAN), in *Proceedings of IEEE International Conference on Communications (ICC)*, Kyoto, Japan, 2011.

28. S. Hamida, J. Pierrot, B. Denis, C. Castelluccia, B. Uguen, On the security of UWB secret key generation methods against deterministic channel prediction attacks, in *Proceedings of IEEE Vehicular Technology Conference (VTC Fall)*, Quebec, Canada, 2012.

29. D. Malan, F. J. Thaddeus, M. Welsh, S. Moulton, Code-Blue: An ad hoc sensor network infrastructure for emergency medical care, in *Proceedings of the MobiSys 2004 Workshop on Applications of Mobile Embedded Systems*, Boston, 2004, 1–4.
30. L. Ziyu, X. Feng, W. Guowei, Y. Lin, C. Zhikui, iCare: A mobile health monitoring system for the elderly, in *IEEE/ACM International Conference on Cyber, Physical and Social Computing (CPSCom)— Green Computing and Communications (GreenCom)*, Hangzhou, China, 2010, 699–705.
31. K. F. Navarro, E. Lawrence, B. Lim, Medical MoteCare: A distributed personal healthcare monitoring system, in *International Conference on eHealth, Telemedicine, and Social Medicine*, Cancun, Mexico, 2009, 25–30.
32. P. Kuryloski, A. Giani, R. Giannantonio, K. Gilani, R. Gravina, V.-P. Seppa, E. Seto, V. Shia, C. Wang, P. Yan, A. Yang, J. Hyttinen, S. Shankar Sastry, S. Wicker, R. Bajcsy, DexterNet: An open platform for heterogeneous body sensor networks and its applications, in *Sixth International Workshop on Wearable and Implantable Body Sensor Networks*, Berkeley, CA, 2009, 92–97.
33. F. E. Tay, D. G. Guo, L. Xu, M. N. Nyan, K. L. Yap, MEMSWear—Biomonitoring system in a body area, in *International Semiconductor Conference*, Sinaia, Romania, 2006, 207–214.
34. A. Wood, G. Virone, T. Doan, Q. Cao, L. Selavo, Y. Wu, L. Fang, Z. He, S. Lin, J. Stankovic, ALARM-NET: Wireless sensor networks for assisted-living and residential monitoring, *IEEE Journal of Networks*, 22(4), 26–33, 2008.
35. U. Anliker, J. Ward, P. Lukowicz, G. Troster, F. Dolveck, M. Baer, F. Keita, E. Schenker, F. Catarsi, L. Coluccini, A. Belardinelli, D. Shklarski, M. Alon, E. Hirt, R. Schmid, M. Vuskovic, AMON: A Wearable multiparameter medical monitoring and alert system, *IEEE Transactions on Information Technology in Biomedicine*, 8(4), 415–427, 2004.
36. Vandrico Solutions, Inc., 2008, http://vandrico.com/database (accessed 2014).
37. M. M. Alam, E. Ben-Hamida, Towards accurate mobility and radio link modeling for IEEE 802.15.6 wearable body sensor networks, in *IEEE 10th International Conference on Wireless and Mobile Computing, Networking and Communications (WiMob)*, Larnaca, Cyprus, 2014.
38. I. Cutcutache, T. Dang, W. K. Leong, S. Liu, K. D. Nguyen, L. T. X. Phan, E. J. Sim, Z. Sun, T. B. Tok, L. Xu, F. E. H. Tay, W. F. Wong, BSN simulator: Optimizing application using system level simulation, in *Body Sensor Networks (BSN)*, Berkeley, CA, 2009, 9–14.
39. K. Nguyen, I. Cutcutache, S. Sinnadurai, Fast and accurate simulation of biomonitoring applications on a wireless body area network, in *Body Sensor Networks (BSN)*, Hong Kong, 2008, 145–148.
40. NICTA, https://castalia.forge.nicta.com.au/index.php/en/documentation.html.

Chapter 4

Revisiting Routing in Wireless Body Area Networks

Zuneera Aziz, Umair Mujtaba Qureshi, Faisal Karim Shaikh,
Nafeesa Bohra, Abdelmajid Khelil, and Emad Felemban

Contents

Abstract

Wireless body area networks (WBANs) are yet another milestone achieved in automating the human environment for pervasive applications. The small, miniaturized design of these sensors is giving way to many applications, including e-health, lifestyle, gaming, and disaster recovery. WBANs are expected to increase life expectancy by performing preventive tasks on time. The patient could be treated in time due to continuously being in touch with the medical health service. This can also lead to decreasing the cost of medical expenses by avoiding extreme conditions. However, the point to ponder about these networks is the transfer of information, which is dealt with by the routing strategies. In order to combat and maintain a trade-off between the reliability and energy efficiency, a number of routing algorithms have been designed that work in conjunction with a prescribed topology. This chapter aims to provide a survey on the existing routing protocols for WBANs in terms of intra-WBAN and inter-WBAN communication. We classify intra-WBANs as thermal aware, cluster, cross-layer, delay tolerant, location based, and quality of service (QoS) based. For inter-WBAN communication, a list of available techniques on the physical, data link, and network layers is discussed.

4.1 Introduction

Wireless sensor networks (WSNs) are a technology that enables remote sensing and off-sight monitoring. WSNs are greatly used in many fields for monitoring, such as seismology, agriculture, medicine, construction, and irrigation, to prevent disasters. The sensed data can provide timely information to prevent great losses [1].

One of the most promising fields of WSNs is the wireless body area networks (WBANs), which are specifically designed to sense human vital signs such as temperature, acceleration, motion, location, electrocardiography (ECG), and electroencephalography (EEG). The sensed data can give rise to numerous applications in the domain of e-health, disaster recovery, lifestyle, and gaming [2,3].

There are two paradigms in WBAN communications: intra-WBAN and inter-WBAN communications. Intra-WBAN communications correspond to communication between different sensors in the same WBAN, whereas inter-WBAN communications correspond to communication between different WBANs [4]. It could be from base station to base station, sensor to senor, or base station to sensor, or vice versa. Figure 4.1 shows how routing is done in both paradigms of WBANs.

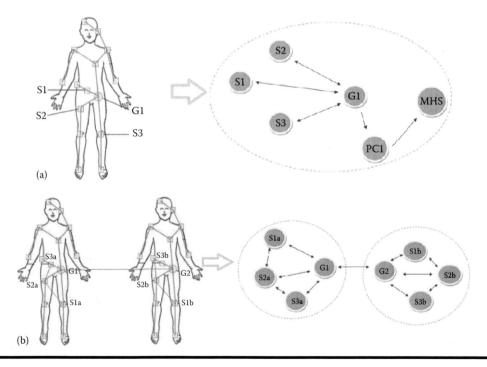

Figure 4.1   Data transmission in WBANs.

As shown in Figure 4.1a, in intra-WBANs, these networks are deployed on the human body in order to collect details about their vital signs. Here, S1, S2, and S3 are the sensing nodes. These vital signs are then routed to the central node, G1, which is also called the base station. The base station routes this data to the medical health service (MHS) through a third-party carrier, for instance, a laptop connected to the Internet (PC1). After the data reaches the MHS, the doctor can analyze the data and instruct the patient accordingly [5]. In Figure 4.1b, the inter-WBAN paradigm is shown. It can be seen that the two base stations (G1 and G2) are communicating to exchange information. Such communications are favorable in applications such as crowd counting and disaster recovery.

The critical nature of e-health applications demands integrity and correct synchronization of the data, without any latency in routing the information. Hence, there is a need for routing protocols to be designed keeping in view two important factors. First, they must utilize the least amount of possible resources to achieve maximum data transfer efficiency; second, they must be energy-efficient to satisfy WBAN requirements [6]. Energy efficiency algorithms are implemented in WBANs, but this accounts for higher computation most of the time. A topological aspect also plays a key role in defining proper data transmission techniques, which can further cut down the energy consumption. Apart from that, there are different reliability parameters that can be measured to test the deployed network. This chapter not only highlights the existing routing protocols, but also classifies them into subcategories. The rest of the chapter is organized as follows: Section 4.2 specifies the existing routing protocols and the hierarchy followed for WBANs. Section 4.3 discusses thermal-aware routing protocols. Section 4.4 enunciates routing algorithms based on clusters. Section 4.5 specifies the cross-layer routing algorithms. In Section 4.6, delay-tolerant network routing algorithms are discussed. Section 4.7 discusses location-based routing, whereas quality of service–based routing is discussed in Section 4.8. The protocols in the inter-WBAN category are discussed in Section 4.9. A thorough discussion with respect to the classification of Intra WBAN protocols is given in Section 4.10. Finally, Section 4.11 concludes the chapter and gives direction for future work.

4.2 Classification of Existing Routing Protocols in WBANs

A number of protocols exist to handle data routing in WBANs. The routing protocols can be classified as intra-WBAN and inter-WBAN paradigms. Figure 4.2 classifies WBAN protocols

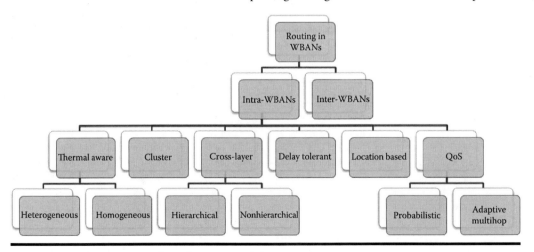

Figure 4.2 Classification of routing protocols in WBANs.

in terms of intra-WBANs and inter-WBANs. Intra-WBAN routing protocols have been summarized, and a list of Inter-WBAN cross-layer techniques is given in Table 4.1.

The classification illustrated in Figure 4.2 and Table 4.1 is based on the routing metrics and the techniques that these protocols follow. Moreover, each protocol enlisted in Table 4.1 is analyzed with respect to the following parameters:

- **Forwarding methodology:** This parameter defines how the packet is forwarded or routed. For example, the packet can be forwarded on the basis of shortest hops, remanent energy, maximum energy, and least bandwidth.
- **Packet discarding mechanism:** This parameter defines whether the protocol contains a packet discarding mechanism. If so, how is the packet discarded?
- **End-to-end delay:** The time data transmission takes from the node to the base station.
- **Topology:** This parameter reflects which topology the protocol follows during the data transmission phase, that is, star, multihop, or hybrid.
- **Periodic message:** This parameter defines periodic messages that are exchanged within a protocol to join a network.

4.3 Thermal-Aware Routing

Biomedical sensors deployed or implanted inside the human body communicate with each other using radio frequency (RF) transmissions (*in vivo* sensors). The data processing and communication causes heat dissipation and a rise in temperature. As a result, human tissues may be damaged because they can only sustain a certain amount of temperature or heat. The specific absorption rate (SAR) defines the amount of heat absorbed by the tissue and is expressed by Equation 4.1 [7].

$$\text{SAR} = \frac{\sigma |E|^2}{\rho} \text{ W/kg} \tag{4.1}$$

where σ is the amount of electrical conductivity of the tissue, E is the electric field of radiation, and ρ is the density of tissue. A SAR of 4 W/kg consistent for 15 min or more can cause damage to tissues in the human head [7]. The protocols that fall in this category take SAR as one of the cost metrics and design the routing strategy accordingly. Thermal-aware routing protocols are further classified into protocols for heterogeneous sensors and homogeneous sensors and are summarized in Table 4.1.

It is observed that most of the protocols are under the heterogeneous category. The reason is that the sensors deployed in WBANs are usually deployed to measure different parameters.

The routing protocols under the heterogeneous and homogeneous categories are discussed in the following subsections.

4.3.1 Thermal-Aware Routing Algorithm

The thermal-aware routing algorithm (TARA) [7] has been developed to cope with overheating due to communication and sensing in *in vivo* sensors. This routing algorithm reduces the possibilities to route information through overheated sensors or "hotspots." TARA [7] assumes that all the

Table 4.1 Routing Protocols in WBANs

Thermal Aware			Cross-Layer		Delay Tolerant	Location Based	QoS Based			Inter-WBANs
Heterogeneous	Homogeneous	Cluster	Hierarchical	Nonhierarchical			QoS	Adaptive	Probabilistic	
TARA [7]	RAIN [8]	AnyBody [9]	CICADA [10]	TICOSS [11]	EDSR [12]	RLOCF [13]	ZEQoS [14]	EAR [15]	PRPLC [16]	Ali and Khan [17]
LTR [18]		HIT [19]	CICADA-S [20]	BIOCOMM [21]	OPPT [22]	OBSFR [13]	DMQoS [23]	AMR [24]	DVRPLC [13]	WARP [25]
ALTR [18]		WASP [26]			RAND [27]	EER [28]	TMQoS [29]	OLSR [30]	PSR [31]	Fang et al. [32]
LTRT [33]					DOR [34]		LOCALMOR [35]		BAPR [36]	
LR [37]							RLQRP [38]		PRoPHET [39]	
M-ATTEMPT [40]									PRMPL [41]	
RE-ATTEMPT [42]										
HPR [43]										
TSHR [44]										

nodes forward packets to the base station and the base station is responsible for aggregating the data and forwarding it to the nearest MHS. Initially, the temperature rise and SAR are calculated using the finite-difference time-domain (FDTD) method and Pennes bioheat equation [45]. As soon as the temperature of each node is calculated, TARA is applied for transmission of packets. In this protocol, the hotspots are identified by neighbors, and this information is communicated to other nodes by withdrawing packets.

The protocol features smaller maximum temperature rise, smaller average temperature rise, higher transmission delay, and less traffic congestion, and it balances temperature rise and ultimately network load (load balancing).

4.3.2 Least Temperature Routing

Least temperature routing (LTR) [18] always chooses the neighbor node with the lowest temperature for routing the data. Every packet keeps a record of the hops it has passed. The maximum hop count is predetermined and denoted by the keyword *MAXHOPS*. When hop count exceeds MAXHOPS, the current packet is discarded and is rerouted using an alternate route. Also, in this routing algorithm, each node maintains a routing table, which keeps track of all the recently passing packets. If the lowest-temperature node to which the packet is forwarded is already in the table, the packet is forwarded to the second lowest-temperature node that is not in the table to avoid choosing the same route. The difference between TARA and LTR is that if the information is destined to the neighboring node, TARA buffers the packet, whereas in LTR, the packet is directly forwarded to the destination.

4.3.3 Adaptive Least Temperature Routing

Adaptive least temperature routing (ALTR) [18] is a variant of LTR. The prefix *adaptive* is added because the protocol is adaptive to topological changes. LTR keeps a record of all the least temperature nodes, whereas ALTR employs the shortest-path routing algorithm to keep a subset of all those routes. It also utilizes a strategy called proactive delay. In this strategy, if two nodes are encountered that have high temperatures, then the algorithm makes the network wait for some time, until the individual nodes cool down. This somehow increases the end-to-end delay. But the positive part of employing proactive delay is that the overall temperature of the network decreases.

In ALTR, when the hop count exceeds MAX-HOP-ADAPTIVE, rather than discarding packets, it send packets to the destination in a minimum hop count using shortest-hop routing (SHR). If neighboring nodes have a comparatively high temperature, the current node will wait a unit of time for the sake of cooling down the temperature.

4.3.4 Least Total Route Temperature

TARA [7], LTR [18], and ALTR [18] consume more power and cause delay. To counter these problems, another thermal-aware routing protocol, known as least total-route temperature (LTRT) [33], has been proposed. LTRT selects a least temperature route from every possible route from the sender node to a destination, not always choosing the least temperature sensor nodes. LTRT works on Dijikstra's shortest-path-first algorithm and choose routes that have the least temperature from the source to the destination, and hence save bandwidth by reducing hop count. The protocol converts the temperature of the nodes into weights and then associates the weights to the route between two sensor nodes. After that, the shortest-path algorithm is applied in order to get the weights of all the network nodes and routes in the network. Since the network nodes can heat up within short durations, the route is updated periodically to account for temperature changes.

4.3.5 Lightweight Temperature Scheduling Routing Algorithm

With reference to the problems faced by most of the thermal-aware routing protocols, such as end-to-end delay, building hotspot entities, and consumption of memory due to hotspot creation, the lightweight temperature scheduling routing algorithm (LR) [46] has been proposed. For example, in TARA [7], there is a temperature threshold set. If the node has a temperature beyond that threshold, the sensing stops, and as soon as the node cools down, the routing starts again. This produces a delay. There are similar problems in LTR [18], ALTR [18], and LTRT [33]. LR [46] counters this problem by dividing the whole network into small clusters. Furthermore, there is a rendezvous node (RN), which is responsible for collecting information regarding subscription, broker, and publisher nodes in any cluster. Initially, all the sensors in the network are informed of the roles of the other sensors (subscriber, publisher, broker, etc.). Then, in each cluster, the subscriber node has to be informed of the broker node, so it takes all the information associated with a broker node. The broker node then communicates with the rendezvous node and provides information regarding the subscriber nodes. The subscribers and publisher nodes are requested to start or stop their service after this.

4.3.6 M-ATTEMPT

M-ATTEMPT [40] is applicable to heterogeneous WBANs, and it divides the whole working strategy into two domains; one deals with critical data and the other one with the usual routing data. The authors suggest placing nodes in descending order of their data rates. Initially, all nodes in the network broadcast the HELLO packets. This way, all the nodes have information of the other nodes in the network in terms of hop count and neighborhood information. In the routing phase, the shortest path is selected. In case the data is critical, the sensor node communicates with the base station directly through a star connection. Since it is a thermal-aware protocol, if the sensor nodes at any time have a higher temperature than the threshold, the sensor nodes break that particular link. If a sensor node heats up after receiving a packet, the sensor returns the same packet to the previous node. This protocol supports mobility and posture changes.

4.3.7 RE-ATTEMPT

In this protocol [42], the position of the base station is assumed to be at the center of the body. Similar to M-ATTEMPT [40], the data transmission is performed through star connection if the data has a higher priority, whereas shortest-path multihop connection is adopted for other data. Initially, all the nodes exchange HELLO packets, this HELLO packet is distributed to other nodes of the network, additionally having information regarding the hop counts from the node to the base station. Then the priority of data is analyzed and the data with the highest priority is sent directly to the base station, whereas the data with less priority is sent through the shortest multihop routes. Once the node information is obtained, the base station allocates time slots to the nodes in the scheduling phase. This makes all the data transmission possible on the same frequency with different time slots. The high-priority data is sent through the contention-free period (CFP) in the medium access control (MAC) superframe structure. The data transmission begins afterwards, and data is routed to its specified destination.

4.3.8 Hotspot Preventing Routing

The hotspot preventing routing (HPR) [43] prevents hotspots for delay-sensitive applications. This routing protocol sends packets via the shortest path if there is no hotspot on the way. The number of hops is also taken into consideration. If the number of hops exceeds the number of MAX_HOPs, the packet is discarded and a new route is established. An interesting feature is that HPR maintains the list of routes that are the most visited. This is done to avoid any loops in the network.

4.3.9 Routing Algorithm for Networks of Homogeneous and ID-Less Biomedical Sensor Nodes

The routing algorithm for networks of homogeneous and ID-less biomedical sensor nodes (RAIN) [8] is applicable to sensors that do not have any IDs and are homogeneous in nature. Initially, all nodes randomly generate their own IDs. ID = 0 is given to the base station. Later, a HELLO packet is sent from all nodes to each other in order to negotiate the IDs. Later, the packets are routed. During the routing phase, the hop count of each packet is compared with the time to live (TTL). If the hop count is less than TTL, the packet will be routed; otherwise, it will be dropped. One of the most significant features of RAIN is that it is fault tolerant. The fault-tolerant nature is achieved by the property that whenever a node has consumed 98% of the power, the node will shut off by itself.

4.3.10 Thermal-Aware Shortest-Hop Routing

The thermal-aware shortest-hop routing (TSHR) [44] routing protocol has been designed for high-priority data. It retransmits the packet if it is dropped due to any reason. It takes into consideration the SHR for establishing routes to forward packets. There are two types of temperatures considered for hotspot avoidance: T_{Dn}, which considers the temperature of neighboring nodes, and T_y, which defines the threshold value that should not be exceeded by a node.

Thermal-aware routing protocols are summarized in Table 4.2.

4.4 Cluster-Based Routing

In cluster-based routing, sensor nodes group together to form clusters, each having a leader (head) in them. Energy saving is accomplished in this protocol by requiring only the leader nodes to make the long-distance transmission to the base station.

4.4.1 AnyBody

The AnyBody [9] protocol features a cluster-based transmission in a WBAN. The specialty of the protocol is that the cluster heads are swapped from time to time. This swapping is done because a cluster head involves a lot of computation, which costs energy consumption. Initially, during the discovery of neighbor nodes, each node broadcasts a HELLO1 for sending and a HELLO2 for relaying in separate

Table 4.2 Comparison of Thermal-Aware Routing Protocols

Protocol	Forwarding Methodology	Packet Discarding Mechanism	End-to-End Delay	Topology	Periodic Messages
TARA	Lowest-temperature path	TTL	Very high	Multihop	Not present
LTR	Coolest neighbor	MAX_HOPS	High	Multihop	Not present
ALTR	Coolest-neighbor SHR when hop counts > MAX_HOPs_ ADAPTIVE	Not present	High	Multihop	Not present
LTRT	Minimum hops using Dijkstra's algorithm	Not present	Low	Multihop	Not present
LR	Forwards packets if subscribed to RN only	Not present	Moderate	Multihop	Not present
M-ATTEMPT	Direct for critical data, SHR for normal data	Not present	Less for critical data, moderate for normal data	Star for critical data, multihop for normal data	HELLO
RE-ATTEMPT	Direct for critical data, SHR for normal data	Not present	Less for critical data, moderate for normal data	Star for critical data, multihop for normal data	HELLO
HPR	SHR	MAX_HOPS	Low	Multihop	Not present
RAIN	Forwards packets to coolest neighbor	TTL	Low	Multihop	HELLO
TSHR	SHR	Not present	High	Multihop	Not present

time frames. As soon as the information regarding the number of nodes and neighborhood nodes is determined, the density is calculating using Equation 4.2 by sending the HELLO3 message.

$$\text{Density} = \frac{\text{Number of links}}{\text{Number of nodes}} \tag{4.2}$$

Later, each node sends information regarding itself and the one-hop neighbors to the neighbor with the highest density, ultimately making it to the highest-density node. This forms clusters in the network and cluster heads are selected. In the last stage, the routes are formed for packet routing. If a node has to route information from one point to another, it has to contact the cluster head. The cluster head routes the information to the destination cluster head, ultimately transferring the information to the destination cluster.

The most significant feature of AnyBody is that it keeps changing the cluster heads, which results in equal computation on each node periodically. However, a major concern is the end-to-end delay, which is very high.

4.4.2 Hybrid Indirect Transmission

The hybrid indirect transmission (HIT) protocol [19] is hybrid because it is a combination of LEACH [47] and PEGASIS [48] protocols. It is an indirect protocol because the sender node makes an indirect or multihop connection to the base station. The protocol has two features combined. The first one is that it works in clusters, so that each base station can be reached in fewer hops. Second, the transmission can be done in parallel even though other clusters exist at the same time. Initially, cluster heads are selected according to various schemes [19]. Next, the cluster heads inform all the nodes regarding their status with a fixed power level. Each node in the cluster then calculates the distance $D(H, j)$, where H is the cluster head and j is the ID of the node. Next, all the nodes in a cluster subscribe for membership within the cluster and bear the parameters $d(i, j)$, which is the distance from node i to any node j, and $d(j, H)$, which is the distance from node j to the cluster head H of the same node. A communication distance is then calculated employing the carrier sense multiple access with collision detection (CSMA/CD) technique. Next, a blocking set is calculated. Node i blocks any node j if $d(i, ui) > d(i, uj)$. The nodes then communicate with each other using the cluster heads by employing a time division multiple access (TDMA) schedule setup scheme. Finally, the data transmission occurs.

The most significant feature of HIT is that it tends to reduce the number of data connections directly to the base station. However, one of the major shortfalls is that if the number of nodes is small, HIT deliberately puts the network on a high amount of computation.

A summary of the important features of cluster-based routing algorithms is enunciated in Table 4.3.

4.5 Cross-Layer Routing Protocols

Cross-layer routing algorithms are designed so that different layers can cooperate with network layers to decide the routing path. The information received from the MAC layer specifically incorporates scheduling and timing information that can lead to designing energy-efficient routing protocols [49]. This protocol is divided into hierarchical and nonhierarchical categories. Hierarchical protocols follow a hierarchy (leveled) structure, whereas nonhierarchical protocols do not follow any hierarchy.

4.5.1 Cascading Information Retrieval by Controlling Access with Distributed Slot Assignment Protocol

The cascading information retrieval by controlling access with distributed slot assignment (CICADA) protocol [10] has been specifically designed for WBANs keeping in consideration the low power characteristic. This is a multihop protocol based on TDMA and is an improvement to the WASP [26] protocol. At first, CICADA generates a spanning tree. Every (child) node

Table 4.3 Comparison of Cluster-Based Routing Protocols

Protocol	Forwarding Methodology	Packet Discarding Mechanism	End-to-End Delay	Topology	Periodic Message
AnyBody	Cluster to cluster	Not present	High	Hybrid	HELLO1, HELLO2, HELLO3
HIT	Cluster to cluster	Not present	Low	Hybrid	Not applicable

connected to the parent node is assigned a time to transmit or receive. This is done by a slot assignment mechanism in which the time axis is given a number of cycles and each cycle represents a particular number of slots. A node can send all the data in one cycle. This does not give birth to complicated routing strategies. Time slotting eliminates the problem of mixing of signals and unnecessary listening. Each node instructs its underlying nodes to send data according to the allocated time scheme. Initially, all the upper-level nodes in the hierarchy send the time slot information or the scheme to the underlying nodes. New nodes can join by sending the JOIN-REQUEST. Each cycle of the time slot is divided into control and data parts. The control part sends the scheme information to the underlying nodes, whereas the data part incorporates data transmission control part works from upper-level nodes into lower-level nodes, whereas data transmission happens from lower-level nodes to upper-level nodes.

The most significant feature of CICADA is that it has a small overhead, and this decreases the end-to-end delay. The delay is further decreased because all the data is sent in one cycle. However, one of the drawbacks of this protocol is that it can give access to unrecognized nodes to join the network.

4.5.2 Cascading Information Retrieval by Controlling Access with Distributed Slot Assignment Protocol—Secure

Cascading information retrieval by controlling access with distributed Slot assignment protocol—secure (CICADA-S) [20], the secure version of the CICADA protocol, has a built-in security mechanism. A spanning tree is set up to route the data, which consists of different levels. The higher-level nodes are called parents, and the remaining nodes are called children. CICADA-S goes through the following procedure: First, in the start-up phase, a sensor contacts a designated server to ensure authorization. This can be a very energy-hungry process. To avoid that, an independent device can be attached to communicate with the designated server. Second, the cryptographic key is communicated to the gateway. A new key is generated randomly to avoid eavesdropping. As soon as the keys are generated, a JOIN-REQUEST is sent by each node in the network. This JOIN-REQUEST is actually a HELLO message containing the information regarding the identity of the node and the CICADA-S counter value given under CTRA. The key is updated periodically to facilitate the security under this scheme. When a node leaves the network, the parent node (a level above the children node) informs the base station that the node has left.

4.5.3 Time Zone Coordinated Sleep Scheduling

In time zone–coordinated sleep scheduling (TICOSS) [11], the network is slotted down into time zones by means of a V-table, which is taken from the MERLIN [50] protocol.

Initially, the protocol relates itself to the 802.15.4 standard, a low-rate wireless personal area network. During this phase, each node in the network is given a MAC address of 64 bits, which is later resolved into a short 16-bit address. Later, a beacon request (BR) is sent to all the network nodes. After the reception of the beacon request, a personal area network (PAN) ID replies, containing a superframe specification and guaranteed time slot (GTS) field. After this, an association request is sent to all the network nodes. Each node then replies with a data request, containing the long MAC ID. After reception at the administrator end, a short 16-bit ID is sent back. These shorts IDs are unique locally, so if a network is large, the IDs cannot conflict with each other.

After the initialization phase, the network is segmented into time zones. In this phase, an algorithm is implemented to allow the node to transfer data in the minimum number of hops. Initially, the PAN coordinator sends a message to the underlying nodes, having information about the time zones. Every node receives this information and sets the prescribed time zone. If a new node joins or any existing node leaves the network, time zones are updated periodically. Finally, in the V-scheduling phase, scheduling is done so that each node in a time zone can transmit in the same slot. Apart from that, sending and receiving can be done simultaneously. After the scheduling phase, routing is performed with nodes having the lowest number of hops to the PAN coordinator.

TICOSS [11] is a key enabler protocol that provides the WBAN networks with elimination of hidden nodes. This protocol also helps to save energy by implementing coordinated sleeping between the nodes. However, a major drawback is that TICOSS has a very high node computation in terms of initial ID exchanges.

4.5.4 Wireless Autonomous Spanning Tree Protocol

The wireless autonomous spanning tree protocol (WASP) [26] deals with time domain slotting. The synchronization of timing is done from the base station node through frequently transmitting beacon request (BR) messages. It is also a spanning tree routing protocol that takes coordinating information from the MAC layer and decides the routing strategy. The routing decision is communicated by a WASP scheme. The WASP scheme is communicated from the parent to the children and from the children to the parent. At the start of each cycle, a WASP scheme is sent from the base station to the children nodes. This WASP scheme communicates to the children nodes when it can transmit. In response, the children can send a WASP scheme to the parent. In this way, all the WASP schemes are communicated to each node from the remaining nodes.

One of the most significant features of WASP is that it is a double-sided communication protocol. The real problem with utilizing this scheme is that the end-to-end delay increases as the number of levels between the parent and children increase. However, for a low number of levels, this protocol results in a decreased end-to-end delay.

4.5.5 Cross-Layer MAC and Routing Protocol Codesign for Biomedical Sensor Networks

Cross-layer MAC and routing protocol codesign for biomedical sensor networks (BIOCOMM) [21] optimizes routing by incorporating a cross-layer messaging interface (CMI) to coordinate between the MAC and the network layer. The CMI sends information regarding the status of the MAC layer or network layer. If there is a free space F in the buffer, the network layer sends this message to the MAC layer through the CMI. If it is not free, the blocked B message is sent. As soon as a free space is found, the MAC layer assigns that place to the congested nodes. The routing is performed with the lowest number of hops to the base station. This protocol also implements the packet discarding mechanism to avoid loopholes through Equation 4.3.

$$\text{Hop\_Count} > \text{TTL} \tag{4.3}$$

where *TTL* is the time to live that is set by the network administrator.

The cross-layer protocol algorithms are summarized in Table 4.4.

4.6 Delay-Tolerant Routing

The routing protocols that are designed specifically for the kind of networks that are more prone to having link breakages or unpredictable behavior fall in this category [51].

4.6.1 Enhanced Dynamic Source Routing

Enhanced dynamic source routing (EDSR) [12] is an enhanced version of the dynamic source routing (DSR) [52] protocol. EDSR is an on-demand, on-call routing protocol that creates routes on request by flooding route request packets in the entire network by attempting a quick route when no route is known to be efficient and also reduces the control overhead.

In addition to the DSR, the following new features are included in EDSR:

- ▪ Starting and ending times of the route request packet
- ▪ Path selection time of the route
- ▪ Trust time for the route

The EDSR sets up a route on demand. It consists of five steps: (1) route request, in the case where no route is present; (2) route reply, for building a route upon receiving a route request; (3) caching overheard routing information, which is information on any route present if and only

Table 4.4 Comparison of Cross-Layer Routing Protocols

Protocol	Forwarding Methodology	Packet Discarding Mechanism	End-to-End Delay	Topology	Periodic Messages
CICADA	Spanning tree	Not present	Low	Multihop	JOIN-REQUEST
CICADA-S	Spanning tree	Not present	Low	Multihop	JOIN-REQUEST
TICOSS	Minimum hops to PAN coordinator	Not present	Moderate	Multihop	BR
WASP	Spanning tree	Not present	Variable	Multihop	BR
BIOCOMM	Minimum hops	Hop count > TTL	Moderate	Multihop	CMI

if there is any route; (4) route request hop, using cached routes and route request hop limits, which gives the total number of hops; and (5) route maintenance, using route maintenance feature of EDSR during communication.

In EDSR, the node that has data packets to send checks to see whether the route cache has the route to the destination. If it a route exists, the node sends data packets through that route; otherwise, it initiates the route discovery process by broadcasting the RREQ packet to other nodes. The node that receives the RREQ packets checks its destination address with its node address. Trust time depends on the route request period, route cache time, and broadcast jitter. If the node address is equal to the destination address, the processing time for the RREQ packets is calculated for the route. The node sends RREP only if the processing time of the route is found to be less than that of the trust route. The route that has a processing time greater than the trust time is discarded. The route for which the processing time is less than the trust time is the efficient route, as it reduces the time needed for routing the packets. The routes for which the processing time is greater than the trust time take more time to route the packet, leading to increased energy consumption. Hence, EDSR, when compared to DSR, has low delay and low energy consumption with a high packet delivery ratio and throughput. This proves that EDSR is energy-efficient compared to DSR.

4.6.2 Opportunistic Routing

Opportunistic routing (OPPT) [22] assumes that the base station is attached to the wrist and is moving forward and backward with respect to a walking position. Another node that can relay the information is attached to the waist. The sensor node is placed on the chest. There are two cases that can be followed. In the first one, when the wrist is in the front, the sensor node can send the data through a line-of-sight (LOS) communication. On the other hand, when the wrist is behind, the data can be relayed through the relay installed on the waist through a non-line-of-sight (NLOS) communication. Equal probabilities are considered for NLOS and LOS communication.

If the node sends data through LOS, it sends a request to send (RTS), and an acknowledgment (ACK) signal is received if the LOS communication is possible. On the other hand, if LOS is not possible, NLOS communication is possible. This happens when the RTS is generated and there is no ACK received. In that case, once the timer expires, a wake-up signal is generated.

One of the major drawbacks of this protocol is that it has been designed specifically keeping in mind either one-hop or two-hop communication. The end-to-end delay is less for LOS, whereas it is variable for NLOS, as the delay depends on the timer that is set by the network administrator.

4.6.3 Random Routing Protocol

The random routing protocol (RAND) [27] is also designed for a network where the links are short, disturbed, or interrupted. The working mechanism of the node is the same as that of the opportunistic routing algorithms; that is, a node in the WBAN network transmits the packet data only when it encounters the node nearby. Later, depending on the application scenario, either a single copy of the data packet is transmitted or multiple copies are distributed. The random routing protocol differs from the opportunistic one in a manner that when a node intends to transmit the data packet, it randomly chooses the node, rather than choosing the first encountered one or flooding the copy to all its neighbors. The decision to randomly choose one or a limited number of nodes to which the copies are transmitted is based on spatial Euclidian coordination. Therefore, the random routing protocol adheres to the features of being robust, compared to single-copy opportunistic routing, and avoids flooding, which creates unnecessary congestion, along with an increase in bandwidth and energy efficiency compared to multiple-copy opportunistic routing.

4.6.4 Delay-Tolerant Opportunistic Routing

Delay-tolerant opportunistic routing [34] counters the flooring approach. First, a node sends a packet to the destination only when it is in the vicinity; that is, a direct link is available. If the sender node and the destination node are not in the vicinity of each other, the packet is temporarily stored in a buffer space.

One of the major drawbacks of this protocol is that when the sender and destination are not in communication range, the packet delivery ratio (PDR) can go up very high. However, this protocol is very easy to implement. The end-to-end delay is variable, as when the two nodes are in the vicinity of each other, the delay is less, and the opposite is also true.

4.7 Location-Based Routing

These protocols use information of the location of different nodes that are in the WBAN. This information can be used to decide the best route possible based on energy comparison [53].

4.7.1 Relative Location-Based Forwarding

The relative location-based forwarding (RLOCF) [13] protocol depends on routing packets with respect to the distance between the sending and receiving nodes. For a network consisting of N nodes, a list of underlying nodes is created, such as $[n_1, n_2, n_3, ...]$. The first node is prescribed as the gateway node. The node makes a connection with the first node in the list of underlying nodes, which is closest to the base station node. If there is no node in the vicinity of the base station or sink node, the gateway buffers the data for as long as it remains isolated. This makes it possible for the closer nodes, with respect to the base station, to do the forwarding. The same logic is applied to the nodes afterwards. In order to increase the functionality to multipoint sources to multipoint destinations, a node can maintain as many sets as the number of nodes available [13].

In this way, RLOCF gives the capability to the network to have multiple links, which gives rise to reliability. This can further be extended to many-to-many connections, but in this manner, each node has to keep a larger data set. However, one of the major drawbacks is that this protocol is designed for multipoint-to-one communication.

4.7.2 On-Body Store and Flood Routing

The on-body store and flood routing (OBSFR) protocol modifies the existing RLOCF [13] and PRoPHET [39] protocols. It tends to modify the initial route selection methodology in a way that the WBAN is first flooded. When the WBAN is flooded, multiple copies of a single set of data reach the destination at different times. The data, properly designated as test data, is further evaluated. The first copy to reach the destination follows the shortest path. This shortest path is then used for RLOCF and PRoPHET operations [13].

In OBSFR, all the data that is being routed also include information about their routes. A modified flooding strategy is used in this protocol to save energy. This protocol tends to increase the probability of packet transfers for small networks, such as a WBAN. The packet delay of OBSFR tends to be low compared to PRoPHET [39] and RLOCF [13].

4.7.3 Energy-Efficient Location Routing

The energy-efficient location routing (EER) protocol [28] uses the geographical location of the sensor nodes along with the energy levels. In this protocol, there is a one-sided transmission that is done from the node to the base station. This protocol involves three phases. In the first phase, all nodes make an advertisement. The objective of this phase is to limit the area in which routing is done. In the second phase, the conditional reply phase, the nodes adjacent to the advertisement node decide whether to reply to the advertisement. There is a condition on which the decision is made, and this condition is based on the direct distance that a node can make to the base station. Finally, in the route selection phase, the best route is selected on the basis of the condition [28]. One of the major drawbacks of EER is that it is a one-sided communication protocol.

Location-based protocols are compared in Table 4.5.

4.8 Quality of Service–Based Routing

Quality of service (QoS)–based protocols have stringent requirements in terms of delay and packet loss tolerance. QoS protocols are further divided into QoS-aware routing protocols, probabilistic routing protocols, and adaptive multihop routing protocols.

4.8.1 QoS-Aware Routing Protocols

4.8.1.1 Zahoor Energy and QoS-Aware Routing Protocol

The Zahoor energy and QoS-aware routing (ZEQoS) [14] protocol assumes that the WBAN is associated with an indoor (hospital) environment, which is called the ZK-BAN framework [54]. ZEQoS divides the packet stream into three categories: ordinary packets (OPs), delay-sensitive packets (DSPs), and reliability-sensitive packets (RSPs) For OPs, the protocol computes the communication cost (Ci) that is calculated, taking into consideration the geographical and energy information of the nodes in the neighborhood. For DSPs, the lowest end-to-end delay path is

Table 4.5 Comparison of Location-Based Routing Protocols

Protocol	Forwarding Methodology	Packet Discarding Mechanism	End-to-End Delay	Topology	Periodic Message
RLOCF	Closest node to base station	Not present	Low	Multihop	Not present
OBSFR	Shortest path	Not present	Low	Multihop	Not present
EER	Shortest distance to base station	Not present	Low	Multihop	Not present

selected, which is based on the next-hop (NHD) parameter. For RSPs, ZEQoS initially calculates the possible paths, further short-lists these paths to three exactly paths, and later finds the redundancy among these paths. Finally, it chooses the most reliable end-to-end path.

4.8.1.2 Data-Centric Multiobjective QoS-Aware Routing Protocol

The data-centric multiobjective QoS-aware routing (DMQoS) [23] protocol assumes that the WBAN is divided into a set of clusters. Each cluster head in a cluster broadcasts a HELLO packet as soon as some significant change occurs in the network. As soon as the HELLO packet is received, the cluster heads put information regarding the neighbors in their routing tables. If a HELLO packet is not received during the frequent updates, there is a time-out and the entry is deleted from the routing table.

One of the most significant features of DMQoS is that it distributes energy evenly among the available sensor nodes. However, a drawback is that if the update frequency is increased, the computation energy is also increased.

4.8.1.3 Thermal-Aware Multiconstrained Intrabody QoS Routing

The thermal-aware multiconstrained intrabody QoS (TMQoS) [29] protocol gathers information regarding the routing table by transmitting beacon messages. These beacon messages are sent to neighboring nodes. This protocol includes different modules, such as the MAC receiver module, routing table constructor module, delay estimator module, and reliability estimator module. The first two modules are responsible for updating the routing table entries in regard to reliability, delay, hop count, and temperature of a particular sensor node. The latter are responsible for estimating the hop-to-hop delay and the reliability of the link with regard to its neighborhood nodes. Another module, called the temperature estimator module, is responsible for measuring the temperature of the node, which means if a node has a higher temperature, it is characterized as a hotspot.

4.8.1.4 New QoS and Geographic Routing

New QoS and geographic routing (LOCALMOR) [35] characterizes the data packets into regular, delay-sensitive, reliability-sensitive, and critical data. The protocol further suggests two types of base stations for the WBAN: primary base station and secondary base station. Each base station receives the data individually. Additionally, four different modules, namely, the reliability-sensitive module, power efficiency module, neighborhood manage module, and delay-sensitive module, are present. The reliability-sensitive module makes sure that the two base stations get the individual copies of data reliability. The power efficiency module keeps a balance between the transmission power and the residual energy. The neighborhood manager module sends or receives the HELLO packets to update the neighborhood information in the routing table. Last, the delay-sensitive module takes care of delay-sensitive data.

4.8.1.5 Reinforcement Learning-Based Routing Protocol with QoS Support for Biomedical Sensor Networks

In the reinforcement learning-based routing protocol with QoS support for biomedical sensor networks (RLQRP) [38], the network offers QoS support by incorporating the geographical information and Q-learning algorithm. This results in selecting optimal routes that are found through previous experiences and rewards. The reward is positive or negative, depending on the data transfer to the neighborhood node. This award is used in the future for forwarding the packets.

The protocols under QoS-aware routing are compared in Table 4.6.

4.8.2 Adaptive Routing

Adaptive routing incorporates routing algorithms that can account for changes in different conditions in the WBAN.

1. *Environment-adaptive routing*: The environment-adaptive routing (EAR) [15] algorithm consists of the following three modules: routing table constructor, fault detector, and path selector. EAR builds its routing algorithm when a node in WBAN transmits a broadcast message to all other nodes. Every node, on reception of the message, executes the routing table constructor module to build its routing table, which consists of the information of the routing table of that node. Similarly, once the routing tables are established with the help of routing table constructors, the data is transmitted. The node that transmits the data uses the path selector module. If the node is faulty or the link between the nodes is partially damaged or there is a breakdown or detection of congestion, the fault detector module, on the basis of the fault detected, transmits the data through a neighboring node; hence, the data transmission is made. In this entire algorithm, the important information laid out in the routing table, which is constructed for a node. It consists of five fields containing the ID, cost, energy, level, and flag. The ID field characterizes the ID of the next hop to the node in transmission. The cost field specifies the

Table 4.6 Comparison of QoS-Aware Routing Protocols

Protocol	Forwarding Mechanism	Packet Discarding Mechanism	End-to-End Delay	Topology	Periodic Message
ZeQoS	Ci for OPs, NHD for DSPs, most reliable path for RSPs	Not present	Moderate for OPs and DSPs, low for RSPs	Multihop	HELLO
DMQoS	Cluster to cluster	Time-out	Depends on update frequency	Multihop	HELLO
TMQoS	Based on different modules	Not present	Moderate	Multihop	Beacon
LOCALMOR	Based on different modules	Not present	Low	Multihop	HELLO
RLQRP	Geographic position and Q-learning algorithms	Not present	Low	Multihop	Not present

communication cost, that is, the routing metric, for the path from one node to another. The level field indicates the level of the device and is different based on the characteristic of the node. The energy field represents the residual energy of the node, and the flag field is a Boolean value representing whether a path contains faulty nodes. Each node records information of neighbor nodes in its routing table based on the broadcast message from the coordinator. The above-mentioned fields make this algorithm event driven and environment adaptive in the sense that if a flag indicates the presence of any sort of fault, the decision for selection of an alternative route considers all the above fields and the best decision is made.

2. *Adaptive multihop tree-based routing*: Adaptive multihop tree-based routing (AMR) [24] is a tree-based routing protocol. This protocol combines the receive signal strength indicator (RSSI), hop count, and remanent energy in a node by fuzzy logic. In this protocol, there are message exchanges such as HELLO, JOIN, ACCEPT, LEAVE, and DATA. Initially, a HELLO is sent to all the nodes in the network. The nodes that recognize the HELLO message reply back with a JOIN message to request access. If the nodes are inside the network, the sink node replies back with an ACCEPT message. When nodes are registered in the network, the registering node broadcasts a HELLO message. As soon as all the HELLOs have been received by all the nodes, the upper-level hierarchy is formed with the help of routing metrics. After the upper and lower hierarchies are formed, a DATA message is sent to initiate the transmission.

3. *Optimized link state routing protocol*: The optimized link state routing (OLSR) protocol is aided by multipoint relays. These multipoint relays avoid redundant packet retransmission. All the nodes in the network communicate with each other and share the topology information. Nodes individually broadcast information relating the topology to the neighboring nodes. This is done in a periodical manner. This periodical message transfer is done to calculate the link cost. In this protocol, the intermediate nodes are called multipoint relays (MPRs). These MPRs help to construct routes when flooding is employed. OLSR is different from link state routing because it modifies the functionality a bit. It takes into account the delays achieved at individual nodes to account for a better or optimized routing strategy.

The main advantage of this routing protocol is the minimum size of updates. Since only the neighborhood information is transferred, the update size has only a minimum set of information. However, if the nodes are scarcely separated, OLSR might not work properly. A densely populated network is required for the proper working of OLSR.

The protocols in the adaptive routing category are compared in Table 4.7.

Table 4.7 Comparison of Adaptive Routing Protocols

Protocol	Forwarding Methodology	Packet Discarding Mechanism	End-to-End Delay	Topology	Periodic Message
EAR	Lowest communication cost	Not present	High	Multihop	Routing table
AMR	Combination of RSSI, hop count, and remanent energy	Not present	Moderate	Multihop	HELLO
OLSR	Fewer delay nodes	Not present	Low	Multihop	Topology information

4.8.3 Probabilistic Routing

Probabilistic routing accounts for routing data packets with respect to some reference to the best route possible to forward data. The best route is decided on the basis of some metric, for example, latency, link outage, and energy consumption [55].

The packet delivery ratio (PDR) for this technique is very high, as the links are updated time to time and the best data path is decided with reference to a metric.

However, with the said technique, it consumes a lot of energy in order to maintain the routing tables and to decide the best path.

4.8.3.1 Probabilistic Routing with Postural Link Costs

Probabilistic routing with postural link costs (PRPLC) [16] associates a probability P_{ij}^t of connection between two nodes i and j. Equation 4.4 specifies the probability value if links between i and j are connected, and Equation 4.5 represents the probability if the links are disconnected.

$$P_{ij}^t = \left\{ P_{ij}^{t-1} + 1 - P_{ij}^t w \right\} \tag{4.4}$$

$$P_{ij}^t = \left\{ P_{ij}^{t-1} w \right\} \tag{4.5}$$

where w is a quantity that depends on the link quality between the two nodes. Since the link status can change, a period HELLO message is sent to update the nodes with the link quality. If there are multiple routes to the same destination, the link with the highest probability, depending on the link quality, is followed.

One of the most significant features of PRPLC is that it maintains a high network lifetime, but the end-to-end delay is high due to calculating probabilities by their link qualities.

4.8.3.2 Distance Vector Routing with Postural Link Costs

Distance vector routing with postural link costs (DVPRLC) [56] proposes a strategy similar to that of PRPLC [16], but instead of the probabilities, a cost factor C_{ij}^t is defined to associate the routing. C_{ij}^t is given for connected links in Equation 4.6.

$$C_{ij}^t = \left\{ C_{ij}^{t-1} \left(1 - w_{i,j}^t \right) \right\} \tag{4.6}$$

Here $w_{i,j}^t$ varies according to the link quality. DVRPLC has features similar to those of PRPLC.

4.8.3.3 Prediction-Based Secure and Reliable Routing

In prediction-based secure and reliable routing (PSR) [31], a matrix is maintained that contains a list of each node n and shows the link quality of a particular node with any other node. This link quality is recorded for a past time p. The link quality is determined through experience by measuring the received signal. Initially, the data is sent to all the nodes to measure the received signal. Only the destination replies with an ACK message, through which the matrix is updated.

One of the major drawbacks of this protocol is that it has a large overhead, and therefore it is only feasible to use for low-scale WBANs with a small number of nodes.

4.8.3.4 Behavior-Aware Probabilistic Routing

In behavior-aware probabilistic routing (BAPR) [36], the postural movements and historical link statistics are taken into consideration. A routing table is broadcasted by each node that consists of destination D, next hop N, and connectivity cost C in the form (D, N, C). The node with the highest C is selected for reliable transfer of information. The most significant feature of this protocol is that it has the lowest packet loss, but the energy consumption is too high because of selecting the link with the highest connectivity cost.

4.8.3.5 Probabilistic Routing Protocol Using History of Encounters and Transitivity

The probabilistic routing protocol using history of encounters and transitivity (PRoPHET) [39] is a probability-based epidemic routing protocol. Epidemic routing is done by continuous flooding of the data packets in an uninterrupted manner. Hence, no attempt is made to stop the replication, which gives rise to overheads or unnecessary congestion. However, in PRoPHET, this is mitigated by calculating the delivery predictability using history of encounters between the nodes. The nodes in a network duplicate the data packets continuously and flood it to only those nodes that do not have the packets' copy. In order to select the next best hop, PRoPHET utilizes the history of encounters and computes the predictability of the data packet delivery for the nodes along with the information of transitivity. PRoPHET thus has a high data packet delivery ratio, as the algorithm intelligently exploits and maintains a set of successful delivery predictabilities of the node to all destinations and only distributes the copies of the data packet to those that do not possess a copy. Continuous flooding creates varying delays and large jitter in the network performance for transmission of data packet delivery from source to destination.

4.8.3.6 Probabilistic Routing with Multiscale Postural Locality

In probabilistic routing with multiscale postural locality (PRMPL) routing [41], the route to the nodes are introduced by its probabilistic design matrix, which is by postural link cost (PLC). PLC traces the location of the nodes and calculates link localities on multiple timescales. With multiple time meanings, on different timings synchronously, the nodes keep calculating the postural link localities and assign a PLC value to them, establishing the link. With movement of the human body or node mobility, if the link is disconnected, the PLC automatically assigns and adjusts the values to the nodes once the link is reestablished.

Table 4.8 compares the probabilistic routing techniques.

4.9 Inter-WBAN Protocols

4.9.1 Secure Cluster Formation in Inter-WBAN Communication

In [17], it is assumed that in a geographical area, each WBAN makes a cluster with the closest WBAN (geographically). One of the WBAN base stations is selected as the cluster head (CH),

Table 4.8 Comparison of Probabilistic Routing Protocols

Protocol	Forwarding Methodology	Packet Discarding Mechanism	End-to-End Delay	Topology	Periodic Message
PRPLC	Depends on probability	Not present	High	Multihop	HELLO
DVRPLC	Depends on cost factor	Not present	High	Multihop	Not present
PSR	Best link quality route	Not present	High	Multihop	ACK
BAPR	Highest connectivity cost	Not present	Low	Multihop	D, N, C
PRoPHET	Past history	Not present	High	Multihop	Flooding
PRMPL	PLC	Not present	Moderate	Multihop	Not present

and the remaining base stations of other WBANs are called cluster members (CMs). The CH is selected on the basis of remanent energy and distance from a fixed remote base station (RBS). If a CM or CH has to send the data to a RBS, the cost is the total number of paths taken from the node to the RBS. After the CMs and CH are selected, each base station broadcasts a beacon solicitation message to see the members in that particular cluster.

4.9.2 Distributed Internetwork Interference-Aware Power Control Algorithm

This algorithm [32] is based on game theory, and the power control problem of WBANs is considered in it. Initially, the network's capacity is estimated using Shannon's capacity law. A set of N WBANs is considered. Each WBAN is in the transmission range of the other WBANs. Therefore, an interference pattern is expected. However, if a TDMA contention scheme is considered, the internetwork interference can be avoided. This work proposes a power control mechanism in which each WBAN is considered a player and the action is considered in place of the transmission power. The payoff function is given by Equation 4.7 [32].

$$\pi_i(p_i,\ p_{-i}) = \log\left(\frac{h_{ii}p_i}{\frac{1}{b}\sum h_{ji}p_j + n_o} - c_i p_i\right) \tag{4.7}$$

Here p_i represents the transmit power by player i. This algorithm allows the user to set the WBAN specifications in terms of QoS, power, and so forth, considering the interference to and from the player with regard to its actions.

4.9.3 Weighted Random Value Protocol

The weighted random value protocol (WARP) [25] is a multiuser WBAN protocol used for controlling the bandwidth in inter-WBAN communication. This protocol assigns resources with regard to their bandwidth requirement and risk index. Since WBAN applications can be intended for crucial data, two types of bandwidth schemes are allowed: high risk and low risk. The bandwidth for a high-risk scheme contends to acquire the desired bandwidth: BW_{desire}. A low-risk scheme contend for the required bandwidth: BW_{require}. The risk index can be 0 or 1, depending on the priority of the application. A weighted value is generated by any WBAN considering the BW_{desire}, BW_{require}, and risk index. Contention is done according to these factors.

4.10 Discussion

Thermal-aware routing protocols have their own packet forwarding methodologies and network topologies, but most of these protocols (TARA, LTR, ALTR, LTRT, LR, HPR, and TSHR) lack a periodic message mechanism. Therefore, in the case of any topological changes, the routing tables are not updated. This ultimately results in high end-to-end delays in the network. Also, the protocols (ALTR, LTRT, LR, and TSHR) do not have packet discarding mechanisms, also contributing to end-to-end delay. Thus, most of the thermal-aware routing protocols (TARA, LTR, ALTR, LTRT, LR, HPR, RAIN, and TSHR) need to be optimized in terms of prioritizing types of data and adapting to different network topologies by formulating a packet forwarding mechanism and eliminating end-to-end network delays.

In *cluster-based routing*, there are only two (AnyBody and HIT) protocols. Both of these protocols lack a packet discarding mechanism, which results in unnecessary packet processing time and overall end-to-end network delay. Thus, there is high need for packet discarding routing mechanisms. In the future, more cluster-based routing protocols will be required. IPv6 convergence will result in all IP networks. Thus, routing between cluster-to-cluster networks will require a more comprehensive, intelligent, and energy-efficient cluster-based protocol.

For cross-layer routing protocols, most of the protocols (CICADA, CICADA-S, TICOSS, WASP, and BIOCOMM) have their own packet forwarding methodologies and periodic message mechanism, which enables them to route the data packets and also encounter any change in the topology of the network, especially in a multihop scenario. Also, all cross-layer routing protocols exhibit end-to-end delays in the network, which is critical for the data packet interpretation, and also lack a packet discarding mechanism. Therefore, introducing a packet discarding mechanism is important and can be a significant parameter in reducing the network delays. Moreover, WBAN researchers need to develop data packet criticality based on cross-layer routing protocols that should be able to adapt in terms of the network topologies, having less overhead.

There are only three location-based routing protocols (RLOCF, OBSFR, and EER), having a distinct data packet forwarding mechanism, each one of them exhibiting low end-to-end network delays. All three protocols work in a multihop network topology. There is no packet discarding mechanism or periodic message mechanism in these protocols. One of the major flaws in location-based routing protocols is that they are not able to adapt to any change in the location of the nodes. The change in the location of the nodes can be due to possible postural movements of the body, such as in the case of walking, moving, or taking turns. This can be countered by developing a periodic message mechanism that can update the routing tables periodically, making the protocol aware of the current locations of the nodes. Also, developing a packet discarding mechanism can further counter the low end-to-end delays of the network. Thus, there is a significant need for a location-based routing protocol that is adaptive to different body movements and intelligent enough to send the data quickly and reliably.

For *QoS-aware routing protocols*, one of the most significant aspects is that all the protocols possess moderate or less end-to-end delay. This means that the protocols take into consideration the delay as one of the QoS parameters. However, the overhead due to calculating different parameters and differentiating traffic in terms of their criticality is high. Therefore, these protocols need to be optimized in terms of their energy efficiency.

There are only three (EAR, AMR, and OLSR) *adaptive routing protocols* with defined packet forwarding mechanisms, all applicable to multihop network transmission topology and exhibiting low end-to-end delays. Each of these routing protocols lacks a packet discarding mechanism. Thus, if there is any change in the network topology, the packet forwarding mechanism takes more than the required time to deliver the packet. This also increases the end-to-end delay in the network. Therefore, adaptive routing protocols can be optimized by incorporating a packet discarding mechanism, thus making them more adaptive routing protocols.

Each of the *probabilistic routing protocols* listed has a different packet forwarding mechanism to forward the data packet and periodic message to be updated with any change in the multihop network topology. However, nearly all probabilistic routing protocols exhibit very high network delays. One of the reasons is that none of the probabilistic routing protocols have a packet discarding mechanism. This end-to-end delay can also affect the overall lifetime and performance of the network. One method to reduce the delay can be by formulating and incorporating the packet discarding mechanism. This can eliminate any unnecessary, nonprioritized data packet looping inside a network.

For *inter-WBANs*, some techniques are illustrated. Since with inter-WBANs the WBAN-WBAN communication can possess threats to privacy, the techniques majorly focus on the security aspects. There is a need for cross-layering in inter-WBAN protocols, especially in the interworking of the data link and network layers.

4.11 Conclusion

This chapter highlighted the major routing protocols in WBANs, categorized as thermal, cluster, cross-layer, delay tolerant, location based, QoS, and some techniques in Inter-WBAN communication. This chapter gave light to the different aspects of the discussed protocols and the applications they can be employed in. Alongside, this chapter also highlighted the available topological options in WBANs. Each protocol is designed to fit in a category of topology. However, there exists a list of protocols that can switch their interworking and are based on the dynamic movement of the human body, which are listed as hybrid. All the routing protocols were compared on the basis of their forwarding methodology, packet discarding mechanism, end-to-end delay, topology, and periodic messages. Each parameter has significant influence on the overall network performance. The protocols need to be adaptive, intelligent, and energy-efficient for a proficient WBAN. Some protocols take into account topological changes, postural changes, and real-time and non-real-time data transfers. There is also a need to build up a connection between other layers and take their data into account for even better routing decisions.

Acknowledgments

This work is partially supported by the Mehran University of Engineering and Technology, Jamshoro, Pakistan, and by grant number 10-INF1236-10 from the Long-Term National Plan for Science, Technology and Innovation (LTNPSTI), the King Abdul-Aziz City for Science and Technology (KACST), Kingdom of Saudi Arabia. We also thank the Science and Technology Unit and TCMCORE at Umm Al-Qura University for their continued logistics support.

References

1. I. F. Akyildiz, W. Su, Y. Sankarasubramaniam, and E. Cayirci. Wireless sensor networks: A survey. *Computer Networks*, 38(4):393–422, 2002.
2. R. Cavallari, F. Martelli, R. Rosini, C. Buratti, and R. Verdone. A survey on wireless body area networks: Technologies and design challenges. *IEEE Communications Surveys Tutorials*, 16(3):1635–1657, 2014.
3. S. Movassaghi, M. Abolhasan, J. Lipman, D. Smith, and A. Jamalipour. Wireless body area networks: A survey. *IEEE Communications Surveys Tutorials*, 16(3):1658–1686, 2014.
4. A. Ali and F. Khan. Energy-efficient cluster-based security mechanism for intra-WBAN and inter-WBAN communications for healthcare applications. *EURASIP Journal on Wireless Communications and Networking*, 2013(1):216, 2013.
5. B. Latre, B. Braem, I. Moerman, C. Blondia, and P. Demeester. A survey on wireless body area networks. *Wireless Networks*, 17(1):1–18, 2011.
6. G. R. Tsouri, A. Prieto, and N. Argade. On increasing network lifetime in body area networks using global routing with energy consumption balancing. *Sensors*, 12(10):13088–13108, 2012.
7. Q. Tang, N. Tummala, S. K. S. Gupta, and L. Schwiebert. TARA: Thermal aware routing algorithm for implanted sensor networks. In V. K. Prasanna, S. S. Iyengar, P. G. Spirakis, and M. Welsh, eds., *Distributed Computing in Sensor Systems*, vol. 3560 of Lecture Notes in Computer Science. Springer, Berlin, 2005, pp. 206–217.
8. A. Bag and M. A. Bassiouni. Routing algorithm for network of homogeneous and ID-less biomedical sensor nodes (RAIN). In *Sensors Applications Symposium 2008 (SAS 2008)*, Atlanta, GA, February 2008, pp. 68–73.
9. T. Watteyne, I. Augé-Blum, M. Dohler, and D. Barthel. AnyBody: A self-organization protocol for body area networks. In *Proceedings of the ICST 2nd International Conference on Body Area Networks (BodyNets '07)*, Brussels, Belgium, 2007, pp. 6:1–6:7.
10. B. Braem, B. Latre, C. Blondia, I. Moerman, and P. Demeester. Improving reliability in multi-hop body sensor networks. In *Second International Conference on Sensor Technologies and Applications 2008 (SENSORCOMM '08)*, 2008, pp. 342–347.
11. A. G. Ruzzelli, R. Jurdak, G. M. P. O'Hare, and P. Van Der Stok. Energy-efficient multi-hop medical sensor networking. In *Proceedings of the 1st ACM SIGMOBILE International Workshop on Systems and Networking Support for Healthcare and Assisted Living Environments (HealthNet '07)*, New York, 2007, pp. 37–42.
12. R. Venkateswari and S. Rani. Design of an energy efficient and delay tolerant routing protocol for wireless body area network. *International Journal on Computer Science and Engineering*, 04(05):694–701, 2012.
13. M. Quwaider and S. Biswas. On-body packet routing algorithms for body sensor networks. In *First International Conference on Networks and Communications 2009 (NETCOM '09)*, 2009, pp. 171–177.
14. Z. A. Khan, S. Sivakumar, W. Phillips, and B. Robertson. ZEQoS: A new energy and QoS-aware routing protocol for communication of sensor devices in healthcare system. *International Journal of Distributed Sensor Networks*, 2014:627689, 2014.
15. D.-Y. Kim, W. Y. Kim, J. Cho, and B. Lee. EAR: An environment-adaptive routing algorithm for WBANs. In *Fourth International Symposium of Medical Information and Communication Technology 2010*, Taipei, Taiwan, 2010.
16. M. Quwaider, M. Taghizadeh, and S. Biswas. Modeling on-body DTN packet routing delay in the presence of postural disconnections. *EURASIP Journal on Wireless Communications and Networking*, 2011(3), 2011.
17. A. Ali and F. A. Khan. Energy-efficient cluster-based security mechanism for intra-WBAN and inter-WBAN communications for healthcare applications. *EURASIP Journal on Wireless Communications and Networking*, 2013(1), 2013.
18. A. Bag and M. A. Bassiouni. Energy efficient thermal aware routing algorithms for embedded biomedical sensor networks. In *2006 IEEE International Conference on Mobile Adhoc and Sensor Systems (MASS)*, Philadelphia, PA, 2006, pp. 604–609.

19. J. Culpepper, L. Dung, and M. Moh. Hybrid indirect transmissions (HIT) for data gathering in wireless micro sensor networks with biomedical applications. In *Proceedings of 2003 IEEE 18th Annual Workshop on Computer Communications (CCW 2003)*, 2003, pp. 124–133.

20. D. Singelée, B. Latré, B. Braem, M. Peeters, M. Soete, P. Cleyn, B. Preneel, I. Moerman, and C. Blondia. A secure cross-layer protocol for multi-hop wireless body area networks. In *Proceedings of the 7th International Conference on Ad-Hoc, Mobile and Wireless Networks (ADHOC-NOW '08)*, Sophia Antipolis, France, 2008, pp. 94–107.

21. A. Bag and M. A. Bassiouni. Biocomm—A cross-layer medium access control (MAC) and routing protocol co-design for biomedical sensor networks. *International Journal of Parallel, Emergent and Distributed Systems*, 24(1):85–103, 2009.

22. A. Maskooki, C.-B. Soh, E. Gunawan, and K. S. Low. Opportunistic routing for body area network. In *2011 Consumer Communications and Networking Conference (CCNC)*, Las Vegas, NV, 2011, pp. 237–241.

23. Md. A. Razzaque, C. S. Hong, and S. Lee. Data-centric multiobjective QoS aware routing protocol for body sensor networks. *IEEE Sensors*, 11(10):917937, 2011.

24. A. M. Ortiz, N. Ababneh, N. Timmons, and J. Morrison. Adaptive routing for multihop IEEE 802.15.6 wireless body area networks. In *2012 20th International Conference on Software, Telecommunications and Computer Networks (SoftCOM)*, Split, Croatia, 2012, pp. 1–5.

25. C. Y. Huang, M. L. Liu, and S. H. Cheng. WRAP: A weighted random value protocol for multi-user wireless body area network. In *2010 IEEE 11th International Symposium on Spread Spectrum Techniques and Applications (ISITA)*, Taichung, Taiwan, October 2010, pp. 116–119.

26. B. Braem, B. Latre, I. Moerman, C. Blondia, and P. Demeester. The wireless autonomous spanning tree protocol for multihop wireless body area networks. In *3rd Annual International Conference on Mobile and Ubiquitous Systems—Workshops 2006*, San Jose, CA, 2006, pp. 1–8.

27. M. Quwaider and S. Biswas. Delay tolerant routing protocol modeling for low power wearable wireless sensor networks. *Network Protocols and Algorithms*, 4(3):15–34. 2012.

28. K. Kim, I.-S. Lee, M. Yoon, J. Kim, H. Lee, and K. Han. An efficient routing protocol based on position information in mobile wireless body area sensor networks. In *First International Conference on Networks and Communications 2009 (NETCOM '09)*, 2009, pp. 396–399.

29. M. M. Monowar, M. M. Hassan, F. Bajaber, Md. A. Hamid, and A. Alamri. Thermal-aware multiconstrained intrabody QoS routing for wireless body area networks. *International Journal of Distributed Sensor Networks*, 2014:676312, 2014.

30. T. Clausen and P. Jacquet. Optimized link state routing protocol (OLSR). No. RFC 3626. 2003.

31. X. Liang, X. Li, Q. Shen, R. Lu, X. Lin, X. Shen, and W. Zhuang. Exploiting prediction to enable secure and reliable routing in wireless body area networks. In *2012 Proceedings of IEEE INFOCOM*, Orlando, FL, March 2012, pp. 388–396.

32. G. Fang, E. Dutkiewicz, K. Yu, R. Vesilo, and Y. Yu. Distributed inter-network interference coordination for wireless body area networks. In *2010 IEEE Global Telecommunications Conference (GLOBECOM 2010)*, Miami, FL, December 2010, pp. 1–5.

33. D. Takahashi, Y. Xiao, F. Hu, J. Chen, and Y. Sun. Temperature-aware routing for telemedicine applications in embedded biomedical sensor networks. *EURASIP Journal on Wireless Communications and Networking*, 2008:26:1–26:26, 2008.

34. L. Sassatelli, A. Ali, M. Panda, T. Chahed, and E. Altman. Reliable transport in delay-tolerant networks with opportunistic routing. *IEEE Transactions on Wireless Communications*, 13(10):5546–5557, 2014.

35. D. Djenouri and I. Balasingham. LOCALMOR: Localized multi-objective routing for wireless sensor networks. In *IEEE 20th International Symposium on Personal, Indoor and Mobile Radio Communications 2009*, September 2009, Washington, DC, pp. 1188–1192.

36. S. Yang, J.-L. Lu, F. Yang, L. Kong, W. Shu, and M.-Y. Wu. Behavior-aware probabilistic routing for wireless body area sensor networks. In *2013 IEEE Global Communications Conference (GLOBECOM 2013)*, Atlanta, GA, 2013, pp. 444–449.

37. R. Kamal, M. O. Rahman, and C. S. Hong. A lightweight temperature scheduling routing algorithm for an implanted sensor network. In *2011 International Conference on ICT Convergence (ICTC)*, Seoul, Korea, 2011, pp. 396–400.

38. X. Liang, I. Balasingham, and S.-S. Byun. A reinforcement learning based routing protocol with QoS support for biomedical sensor networks. In *First International Symposium on Applied Sciences on Biomedical and Communication Technologies 2008 (ISABEL '08)*, Kingston, Jamaica, October 2008, pp. 1–5.

39. A. Lindgren, A. Doria, and O. Scheln. Probabilistic routing in intermittently connected networks. In P. Dini, P. Lorenz, and J. N. Souza, eds., *Service Assurance with Partial and Intermittent Resources*, vol. 3126 of Lecture Notes in Computer Science. Springer, Berlin, 2004, pp. 239–254.

40. N. Javaid, Z. Abbas, M. S. Fareed, Z. A. Khan, and N. Alrajeh. M-ATTEMPT: A new energy-efficient routing protocol for wireless body area sensor networks. *Procedia Computer Science*, 19(0):224–231, 2013.

41. M. Quwaider and S. Biswas. On-body packet routing algorithms for body sensor networks. In *First International Conference on Networks and Communications 2009 (NETCOM '09)*, Chennai, India, 2009, pp. 171–177.

42. A. Ahmad, N. Javaid, U. Qasim, M. Ishfaq, Z. A. Khan, and T. A. Alghamdi. RE-ATTEMPT: A new energy-efficient routing protocol for wireless body area sensor networks. *International Journal of Distributed Sensor Networks*, Article ID 464010, 9 pp., 2014. doi:10.1155/2014/464010.

43. A. Bag and M. A. Bassiouni. Hotspot preventing routing algorithm for delay-sensitive biomedical sensor networks. In *IEEE International Conference on Portable Information Devices 2007 (PORTABLE07)*, Orlando, FL, May 2007, pp. 1–5.

44. M. Tabandeh, M. Jahed, F. Ahourai, and S. Moradi. A thermal-aware shortest hop routing algorithm for in vivo biomedical sensor networks. In *Sixth International Conference on Information Technology: New Generations 2009 (ITNG '09)*, Las Vegas, NV, April 2009, pp. 1612–1613.

45. H. H. Pennes. Analysis of tissue and arterial blood temperatures in the resting human forearm. *Journal of Applied Physiology*, 1(2):93–122, 1948.

46. R. Kamal, M. O. Rahman, and C. S. Hong. A lightweight temperature scheduling routing algorithm for an implanted sensor network. In *2011 International Conference on ICT Convergence (ICTC)*, Seoul, Korea, 2011, pp. 396–400.

47. M. J. Handy, M. Haase, and D. Timmermann. Low energy adaptive clustering hierarchy with deterministic cluster-head selection. In *4th International Workshop on Mobile and Wireless Communications Network 2002*, Stockholm, Sweden, 2002, pp. 368–372.

48. S. Lindsey and C. S. Raghavendra. PEGASIS: Power-efficient gathering in sensor information systems. In *Aerospace Conference Proceedings 2002*, Big Sky, Montana, 2002, vol. 3, pp. 3-1125–3-1130.

49. C. F. Chou and K. T. Chuang. COLANET: A cross-layer design of energy-efficient wireless sensor networks. In *Proceedings of Systems Communications 2005*, Montreal, Canada, 2005, pp. 364–369.

50. A. G. Ruzzelli, G. M. P. O'Hare, M. J. O'Grady, and R. Tynan. MERLIN: A synergetic integration of MAC and routing protocol for distributed sensor networks. In *2006 3rd Annual IEEE Communications Society on Sensor and Ad Hoc Communications and Networks 2006 (SECON '06)*, Reston, VA, 2006, vol. 1, pp. 266–275.

51. Z. Feng and K.-W. Chin. A survey of delay tolerant networks routing protocols. *Computing Research Repository*, 2012, abs/1210.0965.

52. D. B. Johnson and D. A. Maltz. Dynamic source routing in ad hoc wireless networks. In T. Imielinski and H. F. Korth, eds., *Mobile Computing*, vol. 353 of Kluwer International Series in Engineering and Computer Science. Springer, Berlin, 1996, pp. 153–181.

53. K. Kim, I.-S. Lee, M. Yoon, J. Kim, H. Lee, and K. Han. An efficient routing protocol based on position information in mobile wireless body area sensor networks. In *First International Conference on Networks and Communications 2009 (NETCOM '09)*, Chenai, India, 2009, pp. 396–399.

54. Z. A. Khan, S. Sivakumar, W. Phillips, and N. Aslam. A new patient monitoring framework and energy-aware peering routing protocol (EPR) for body area network communication. *Journal of Ambient Intelligence and Humanized Computing*, 5(3):409–423, 2014.

55. A. Lindgren, A. Doria, and O. Schelén. Probabilistic routing in intermittently connected networks. *SIGMOBILE Mobile Computing and Communications Review*, 7(3):19–20, 2003.

56. S. Movassaghi, M. Abolhasan, and J. Lipman. A review of routing protocols in wireless body area networks. *Journal of Networks*, 8(3), 2013.

Chapter 5

Thermal-Aware Communication Protocols for Body Sensor Networks

Muhammad Mostafa Monowar

Contents

Abstract

The massive proliferation of microelectronics and integrated circuits, system-on-chip design, and the wide adoption of wireless communication technologies have created significant opportunities for promoting numerous e-healthcare applications. The body sensor network (BSN) is such a recent technology in this regard. A BSN is formed by integrating miniature, intelligent, and low-power wireless communication-enabled devices that are placed on, near, or within a human body to monitor the biological activities and surrounding environment. An implant biosensor node, also known as an *in vivo* node, is a special type of sensor node that detects and collects the desired biometric data of a certain physiological change inside the body and transmits the data to a coordinator node exploiting wireless communication. One of the major problems caused by continuous sensing of the *in vivo* sensor node is the heat produced due to wireless communication. The increased heat could result in thermal damage to the human tissue inside the body if the communication is prolonged for a long time, which could be a threat to the human life. This thermal effect of an implant biosensor node thus imposes significant challenges in designing sensor hardware and communication protocols that need to be addressed effectively and carefully. In this chapter, the thermal effects of biomedical implants are extensively discussed. The contemporary solutions regarding thermal-aware communication protocols are also studied in detail, and a comparative study of the protocols is provided. This chapter intends to be a comprehensive reference of the thermal-aware communication protocols of BSNs for academicians and can be used as a suitable textbook for postgraduate students, academics, and networking professionals.

5.1 What Is a Body Sensor Network?

5.1.1 Definition

Thanks to the Moore's law, the continuous decrease in size of electronic devices, while increasing capacities, made it inevitable for the development of tiny and portable devices that could communicate with each other around the human body. This substantial development has caused a growing interest of researchers, system designers, and application developers in a new type of network architecture generally known as the body sensor network (BSN) or wireless body area network (WBAN).* A BSN is thus a special type of network that integrates miniaturized, intelligent, low-power sensor nodes in, on, or around a human body and continuously monitors the body functions and surrounding environment [1]. Some sensor devices are wearable, and some can be implanted in the body. All of the sensor nodes are intelligent in the sense that they have the capability of capturing, processing, and forwarding data to a nearby base station for further diagnosis. The nodes communicate with each other through a wireless link. The sensors provide

---

* Throughout this chapter, the terms *BSN*, *WBAN*, and *BAN* have been used interchangeably.

continuous and real-time feedback to the user or healthcare professional. Moreover, the captured data can be recorded for longer periods of time, improving the quality of the measured data [2].

BSNs possess some unique characteristics:

- Lossy channel: The wireless communication in BSNs takes place either in or on a human body. However, propagation loss around the human body is very high. Generally, in wireless networks, it is known that the transmitted power drops off with d^n, where d represents the distance between the sender and the receiver and n the path loss coefficient. In free space, n has a value of 2. Several researchers have been investigating the path loss around the human body [3–7]. All of them concluded that the radio signals experience large losses as n reaches a value between 4 and 7.

- Highly resource constrained: Sensor nodes in BSNs are battery powered; hence, these are highly energy constrained. In some applications, a sensor should operate while supporting a battery lifetime of months or even years without intervention. For example, a pacemaker or a glucose monitor would require a lifetime lasting more than 5 years. Considering the implanted devices, the lifetime is very crucial. It is not desirable and practical to frequently change the node or replace the battery for such implant devices [8]. Hence, the energy resources and, consequently, processing and memory resources are very limited.

- Nonredundancy: The sensor nodes in BSNs are deployed only when they are needed. All devices are equality important. BSNs do not employ redundancy to cope with diverse types of failures or increasing reliability.

- Low transmit power: The transmit power of the sensors must be kept as low as possible to minimize the interference and cope with health concerns [9].

- Mobility: The sensors are placed in the human body, which can move around. So, the devices share same mobility patterns. The mobility could cause frequent changes in network topology.

- Variable data rates: Due to the diverse application types, data rates of the sensors strongly vary, ranging from a few kilobits per second to megabits per second.

- Stringent security requirements: BSNs possess high security requirements, especially during the exchange of a patient's healthcare data between sensor nodes and the base station to the base station and the Internet. All of the data must be encrypted effectively to ensure the patient's privacy.

5.1.2 Communication Architecture

BSNs consist of nodes that are placed on, in, or near the human body and are capable of wireless communication to transmit continuously monitored patient physiological data or data related to other body activities to the responsible healthcare professional for further diagnosis. Min Chen et al. [10] presented a three-tier communication architecture of BSN, as illustrated in Figure 5.1.

The communication architecture employs three types of communication in three different tiers: *intra-BAN*, *inter-BAN*, and *beyond-BAN*. This tiered architecture covers different aspects, ranging from low-level to high-level design issues that promote the design of a modular and efficient BSN system for a wide range of applications.

The term *intra-BAN communication* refers to the radio communications among the body sensors of about 2 m in, on, or around the human body. As the figure shows, diverse types of sensors can be attached to a human body. For instance, electrocardiogram (ECG), electroencephalography (EEG), electromyography (EMG), blood pressure sensors, motion sensors, temperature

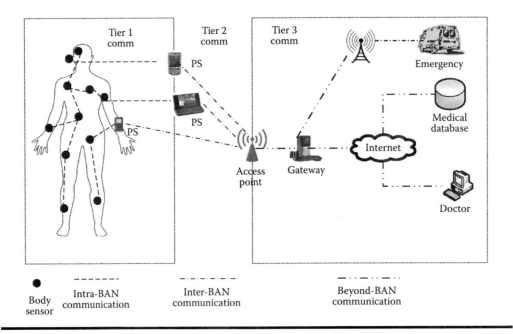

Figure 5.1   Three-tier-based BSN communication architecture.

sensors, and so forth, can be implanted or placed on the human body for transmitting biological data. In intra-BAN communication, the body sensors transmit the sensed data to a local personal server (PS). The communication may take place directly in the PS or may employ multihop communication, where the sensors act as a forwarder and the sensed data eventually reaches the PS. The multihop communication is often desirable for BSNs due to the lower transmission power of the sensors and high propagation loss inside the human body.

The inter-BAN communication occurs between the PS and one or more access points (APs). The APs can be deployed strategically to handle the radio communication in dynamic environments for addressing emergency situations. Through the inter-BAN communication, the BSNs can be connected with different types of networks, such as a wireless local area network (WLAN), cellular network, or Internet.

Tier 3 exploits the beyond-BAN communication to connect the BSNs with the outside environment. A gateway device is used that acts as a bridge between inter-BAN and beyond-BAN communication. As depicted in Figure 5.1, the beyond-BAN communication enables the healthcare data to be transmitted to the healthcare professionals (i.e., doctor or nurse), who can remotely access the data, or the data can be stored in a medical database for offline processing or readily transmitted to the emergency unit for taking prompt action. However, the design of beyond-BAN communication is application dependent and should address the user demands.

This chapter focuses on some challenging issues of intra-BAN communication. In particular, wireless communication among implant devices inside the body has some biological impacts on the human body, such as temperature increase. This chapter mainly discusses the thermal effects caused during intra-BAN communication of the implant devices, as well as the contemporary solutions offered while designing communication protocols that address the thermal issues.

5.1.3 Application Domain

In recent years, the BSN has been considered one of the striking technologies for promoting e-healthcare. BSNs have tremendous potential for various new, practical, and innovative applications to improve e-healthcare and quality of life. Some of the potential applications of BSN include [10]

- Remote health monitoring: BSNs are mainly designed for health monitoring applications. The implant or on-body sensors monitor the biological information in real time and transmit to nearby devices for diagnosis or storage purposes. This data can be further processed by off-site doctors. Some of the health monitoring applications include ECG, EEG, EMG, glucose monitoring, blood saturation monitoring, cancer cell monitoring, temperature monitoring, and pressure monitoring [11–14].
- Sports training: Sports players may wear motion sensors on both their hands and elbows for accurate feature extraction of their movements.
- Interactive gaming: Interactive game applications such as boxing and shooting may exploit BSNs for improving entertainment experiences. The body sensors can provide feedback of the actual body movement to the gaming console, thus allowing the player to directly interact with the game.
- Battlefield: BSNs can be used in a battlefield to connect soldiers and report their activities, such as running, digging, and firing, to the commander [1].
- Personal information sharing: BSNs might be used for storing personal information or business information in body sensors that can be used later for several daily life applications, such as shopping or information exchange with clients.
- Secure authentication: The potential problems of the existing authentication schemes have motivated research for new physical and behavioral characteristics of the human body, such as EEG and gait, and a multimodal biometric system, that is, facial pattern recognition, iris and fingerprint recognition, and so forth. BSNs can be effectively used in such applications.

5.2 Thermal Effects of Biomedical Implants

Significant advancements in microelectromechanical (MEMS) [15] and nanoelectromechanical (NEMS) system [16] technology have promoted the rapid proliferation of biomedical sensors for e-healthcare. The biomedical sensors can be wearable or implanted inside the human body. An implant biosensor node, also known as an *in vivo* node, is a special type of sensor node that detects and collects the desired biometric data of a certain physiological change inside the body and transmits the data to a practitioner terminal, such as a personal digital assistant (PDA) or tablet PC, exploiting wireless communication. Currently, the biomedical sensors are applied to glucose-level monitoring, artificial retinas, cancer detection, control of bladder function, monitoring pacemakers and cardiac defibrillators, and so forth [17,18].

5.2.1 The Problem

Biomedical implant devices usually do not cause any problem if they perform limited processing and limited communication with the external world. In those situations, the processing functions and communication operations are so limited that they do not produce significant heat inside the human body [19]. For event-based low-rate applications in which the devices remain silent most of the time, the processing and communication operations performed by the implants have limited power dissipation and do not change the temperature of the surrounding body tissue.

However, for periodic and high-rate applications where the implant devices continuously monitor and transmit the physiological data, significant heat might be produced in the surrounding tissue. Obviously, this thermal increase of the implant devices could be extremely dangerous for the neighboring tissues, and a high temperature may damage them during long-term monitoring [20,21]. Some organs are very sensitive to temperature rise due to a lack of blood flow to them and are prone to thermal damage even with modest heating [22].

5.2.2 Reasons for Thermal Increase

There are various parameters that affect the thermal increase related to the operation of an implant device; these include the current requirements of the stimulating electrodes, the number of electrodes, the power requirements of the implanted electronics, and the characteristics of the telemetry [19]. However, the major sources that are predominant for a thermal increase of the implant devices include [19,23]

- Electromagnetic fields caused due to wireless communication: The wireless communication among the implant devices produces electromagnetic fields on the human body that could lead to dissipated power in the tissue and thus cause thermal increase. The implant devices usually use the frequency range <1 MHz. For safety reasons, the radiation level being absorbed by the human body needs to be measured. The power dissipation in the human body can be measured in terms of specific absorption rate (SAR), expressed in watts per kilogram, which records the rate at which radiation energy is absorbed by tissue per unit weight. Mathematically, SAR can be expressed as [23]

$$SAR = \frac{\sigma |E|^2}{\rho} \ \text{W/kg}$$

  where E is the induced electric field, ρ is the density of the tissue, and σ is the electrical conductivity of the tissue. Experiments show that exposure to an SAR of 8 W/kg in any gram of tissue in the head or torso for 15 min may have significant risk of tissue damage [24]. Some studies also investigated the heating of an eye by examining the SAR and the temperature of the eye when exposed to a WLAN [25] or infrared radiation [26].
- Thermal increase due to radio frequency (RF) powering: The body sensor nodes might be recharged by an external RF power source. The frequency of the power supply is usually between 2 and 20 MHz [27,28]. The power absorbed by the surrounding tissue of an implant device due to RF powering is one of the sources of thermal increase of the tissue.
- Power dissipated by the implanted electronics: Along with the issue of electromagnetic fields induced in the human body, the induction of current in the circuitry of implant devices could cause significant power dissipation in the circuitry itself, which could in turn cause a thermal increase in the surrounding tissues of the human body [19]. The power consumption of the sensor circuitry depends on its implementation technology and architecture.

5.2.3 Estimating Temperature Increase

The sources of thermal increase, as mentioned above, should be taken into account to measure the thermal increase surrounding the tissue of an implant device. The calculation of SAR is crucial in this regard. SAR can be determined by various techniques or experimental methods. One of the most diffused numerical methods is known as the finite-difference time-domain (FDTD) method [29].

The calculation of SAR is presented in detail in [23]. To measure the SAR due to the wireless communication of the implant device, let us assume that a sensor node has a short dipole antenna consisting of a short conducting wire of length dl with a sinusoidal drive current I. To analyze the effects of radiation on the tissue, the tissue is assumed to be homogeneous with no sharp edges and rough surfaces. The space around the antenna is divided into near field and far field, where the extent of near field is given by $d_0 = \left(\dfrac{\lambda}{2\pi}\right)$. Here, λ is the RF wavelength for wireless communication. According to [7,30], the SAR in the near field ($\leq\lambda/2\pi$) and far field ($>\lambda/2\pi$) can be estimated as follows:

$$SAR_{NF} = \frac{\sigma\mu\omega}{\rho\sqrt{\sigma^2 + \varepsilon^2\omega^2}}\left(\frac{I\ dl\ \sin\theta\ e^{-\alpha R}}{4\pi}\left(\frac{1}{R^2} + \frac{|\gamma|}{R}\right)\right)^2 \tag{5.1}$$

$$SAR_{FF} = \frac{\sigma}{\rho}\left(\frac{\alpha^2 + \beta^2}{\sqrt{\sigma^2 + \omega^2\varepsilon^2}}\frac{I\ dl}{4\pi}\right)^2 \frac{\sin^2\theta\ e^{-2\alpha R}}{R^2} \tag{5.2}$$

where R is the distance from the source to the observation point, θ is the angle between the observation point and the X-Y plane, γ is the propagation constant, σ is the medium conductivity, ε is the relative permittivity, μ is the permeability, ρ is the mass density, and $\sin\theta = 1$.

SAR due to the radiation from RF powering can be measured as [23]

$$SAR_{RF} = \frac{\sigma|E|^2_{RF}}{\rho} = \frac{2\sigma W_d}{\rho\,\mathrm{Re}\left\{\dfrac{1}{\eta}\right\}}\left(\frac{W}{kg}\right) \tag{5.3}$$

where $|E|^2$ is the incident electric field, η is the intrinsic impedance, and W_d is the power density of the incident wave at a depth d.

Combining Equations 5.1 through 5.3, the total SAR is

$$SAR_{all} = \begin{cases} SAR_{RF} + SAR_{NF}, & R \leq \dfrac{\lambda}{2\pi} \\[3mm] SAR_{RF} + SAR_{FF}, & R \geq \dfrac{\lambda}{2\pi} \end{cases} \tag{5.4}$$

Considering all the sources of thermal increase, the rate of the thermal increase can be estimated by the Pennes bioheat equation [31] as follows:

$$\rho C_p \frac{dT}{dt} = K\nabla^2 T - b(T - T_b) + \rho SAR + P_c + Q_m \left(\frac{W}{m^2}\right)$$

where ρ is the mass density, C_p is the specific heat of the tissue, K is the thermal conductivity of the tissue, b is the blood perfusion constant, and T_b is the temperature of the blood and the tissue. On the left side of the equation, dT/dt denotes the rate of thermal increase in the control volume. The right-side terms indicate the heat absorbed inside the tissue. The terms $K\nabla^2 T$ and $b(T - T_b)$ denote the heat transfer due to the thermal conduction and blood perfusion, respectively. P_c refers to the power dissipation density, which is the power consumed by the body sensor circuitry divided by the sensor volume. Thus, the terms ρSAR, P_c, and Q_m indicate heat generation due to radiation of the sensor antenna, power dissipation of the circuitry, and the metabolic heating, respectively.

This bioheat equation can be modeled exploiting the FDTD technique, where the total problem space is discretized into small cells that are marked with a pair of coordinates (x, y). Thus, the temperature of a node at a grid point (x, y) can be estimated as [23]

$$T^{m+1}(x, y) = \left[1 - \frac{\delta_t b}{\rho C_p} - \frac{4\delta_t K}{\rho C_p \delta^2}\right] T^m(x, y) + \frac{\delta_t}{C_p} SAR + \frac{\delta_t b}{\rho C_p} T_b + \frac{\delta}{\rho C_p} P_c$$

$$+ \frac{\delta_t K}{\rho C_p \delta^2} [T^m(x+1, y) + T^m(x, y+1) + T^m(x-1, y) + T^m(x, y-1)]$$

$$(5.5)$$

where $T^{m+1}(x, y)$ is the temperature of the cell (x, y) at time $m + 1$, δ_t is the discretized time step, and δ is the discretized space step. The SAR distribution and temperature increase can be estimated using Equations 5.4 and 5.5.

5.3 Communication Protocols for Minimizing Thermal Effects

Considering thermal effect on the human body as one of the striking criteria, a series of communication protocols have been developed [20,32–38]. The proposed protocols consider node temperature a primary metric for routing decisions. The main objective of the protocols is to maintain the temperature below some threshold and lower the temperature rise rate to avoid significant damage to human body tissue. Table 5.1 shows a list of the thermal-aware communication protocols for BSN.

All the thermal-aware communication protocols employ multihop routing to address high path loss inside the human body. Moreover, to save energy, the transmission power of the implant sensors also needs to be kept low, which necessitates the adoption of multihop routing for resource-constrained BSNs.

Table 5.1   Thermal-Aware Communication Protocols for BSNs

Protocol Name	Author	Publication Year	Type of Communication
TARA [32]	Tang et al.	2005	Multihop routing
LTR [20]	Bag and Bassiouni	2006	Multihop routing
ALTR [20]	Bag and Bassiouni	2006	Multihop routing
LTRT [33]	Takahashi et al.	2007	Multihop routing
HPR [34]	Bag and Bassiouni	2007	Multihop routing
RAIN [35]	Bag and Bassiouni	2008	Multihop routing
TSHR [36]	Tabandeh et al.	2009	Multihop routing
M-ATTEMPT [37]	Javaid et al.	2013	Single-hop and multihop routing
TMQoS [38]	Monowar et al.	2014	Multihop routing

5.3.1 State-of-the-Art Thermal-Aware Communication Protocols

This section provides a brief description of the state-of-the-art thermal-aware protocols.

5.3.1.1 Thermal-Aware Routing Algorithm

The thermal-aware routing algorithm (TARA) [32] is one of the primitive protocols that consider temperature as a routing metric. TARA forwards data packets based on localized temperature information and hop counts to the destination. It measures temperature considering heat generation due to radiation from the antenna, as well as power dissipation due to sensor circuitry. It exploits the FDTD method to estimate the temperature increase of the sensor and its surrounding area, as discussed in Section 5.2.3.

Every node in TARA maintains a neighbor table by exchanging neighborhood information and is aware of the hop-count information to the gateway node. It is assumed that each node knows the gateway location. The forwarder node is selected based on the minimal temperature criteria. A node listens to its neighbor activities and counts the number of packet transmissions and receptions. Based on this, a node evaluates the communication radiation and power dissipation of neighbors and estimates the temperature changes. A node defines the neighbors as hotspots if the estimated temperature exceeds a certain threshold. TARA avoids the hotspots by establishing an alternative route toward the destination using a withdrawal strategy where a packet is sent back to its previous sender if all the neighbors are identified as hotspots. The sender then attempts to select an alternate route to detour the hotspots. After cooling the temperature beneath some threshold, those hotspots can be considered for later routing.

Figure 5.2 presents the protocol operation of TARA. Here, node 0 is the sender that is sending a packet to the gateway node. When the packet reaches node 5 through the path $0 \rightarrow 1 \rightarrow 5$, node 5 evaluates all the possible forwarders (i.e., nodes 4 and 7) as hotspots. Thus, following the withdrawal strategy, the packet is sent back to node 1, which then finds an alternative route for the packet through the path $1 \rightarrow 2 \rightarrow 6 \rightarrow 9 \rightarrow 10 \rightarrow$ gateway.

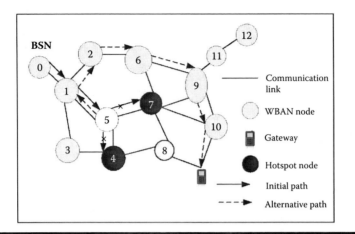

Figure 5.2 Protocol operation of TARA.

5.3.1.2 Least Temperature Routing Protocol

Similar to TARA, Bag and Bassiouni proposed the least temperature routing (LTR) protocol [20]. In LTR, nodes communicate with their neighbors to collect the temperature information in the setup phase. A node chooses the least temperature neighbor or "coolest" neighbor as a forwarder. To prevent a packet traversing unnecessarily with a large number of hops, a threshold parameter MAX_HOPS is defined that depends on the diameter of the BSN. Upon receiving a packet, a node checks the hop count. If it exceeds the MAX_HOPS, then the packet is discarded. To avoid the routing loop, each packet maintains a small list of nodes that it has most recently visited, and if the coolest neighbor is already in the list, it is ignored as a forwarder and the second coolest neighbor is chosen. If the coolest neighbor is a leaf node but not the destination, then the packet is forwarded to a node with lower temperature among the neighboring nodes. The list of recently visited nodes within some past window is included in the packet. Figure 5.3 depicts an example of LTR.

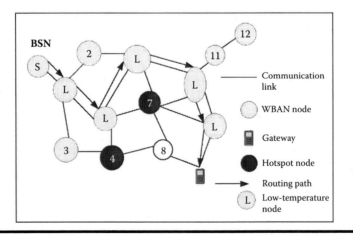

Figure 5.3 Example of LTR. Node S is the source node that is sending a packet to the gateway. The packet reaches the gateway following a path having low-temperature nodes.

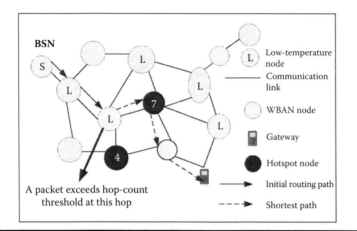

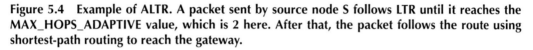

Figure 5.4 Example of ALTR. A packet sent by source node S follows LTR until it reaches the MAX_HOPS_ADAPTIVE value, which is 2 here. After that, the packet follows the route using shortest-path routing to reach the gateway.

5.3.1.3 Adaptive Least Temperature Routing Protocol

The adaptive least temperature routing (ALTR) protocol [20] is a variant of LTR that intends to minimize the packet delivery delay. To minimize the delay, a threshold parameter MAX_HOPS_ADAPTIVE is used. In ALTR, a node having a packet to route checks the hop count of the packet. If the value is less than or equal to MAX_HOPS_ADAPTIVE, the packet follows the route using LTR. However, if the value exceeds the threshold, a shortest-path algorithm is used to route the packet to the destination. To keep the average temperature rise to an acceptable level for a certain topology where the network connectivity is low and the same path is used repeatedly for packet transmission, ALTR employs a "proactive delay" approach. In this case, upon receiving a packet, if a node has no more than two neighbors, and the coolest neighbor has a relatively high temperature, then the node delays the packet by one time unit before forwarding it. Thus, the average temperature rise is kept lower, sacrificing packet delivery delay. Figure 5.4 illustrates an example of ALTR.

5.3.1.4 Least Total-Route Temperature

The least total-route temperature (LTRT) [33] is designed to avoid redundant hops that cause wastage of network bandwidth and minimize total temperature rise. LTRT also uses temperature as a routing metric and chooses a least temperature route from all possible routes from a sender to a destination, instead of choosing the coolest neighbor.

LTRT applies the single-source shortest-path problem in graph theory to the thermal-aware routing problem. First, by collecting the information from the neighboring nodes, it assigns temperature as weight to each sensor node in the network and constitutes a weight graph by transferring the temperature of sensor nodes to the weight of edges ahead.

Then it applies single-source shortest-path algorithms (e.g., Dijkstra's algorithm) on this graph to figure out the route from the sending nodes to the destination nodes. LTRT periodically maintains a route update to avoid excessive temperature rise. Figure 5.5 illustrates an example of LTRT.

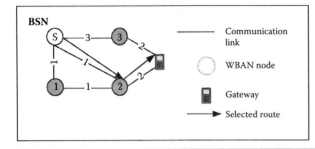

Figure 5.5 Example of LTRT.

5.3.1.5 Hotspot Preventing Routing

Hotspot preventing routing (HPR) [34] was proposed to improve LTR and ALTR. HPR intends to prevent the formation of hotspots in the network, as well as minimize the packet delivery delay exploiting shortest-path routing and tuning a threshold value.

In its setup phase, a node builds a routing table by exchanging information about the shortest path and initial temperature of its neighboring nodes. During routing of a packet, HPR follows shortest-path routing as long as the temperature of the forwarder is less than or equal to the temperature of the sending node plus a threshold. Otherwise, a node predicts the formation of a hotspot and detours the packet through an alternate path. In this case, a node chooses the coolest neighbor that has not been visited before. Similar to LTR, HPR also exploits a MAX_HOPS parameter to prevent a packet from being routed unnecessarily with a large number of hops and drops the packet if it exceeds the parameter value. Figure 5.6 illustrates an example of how HPR works.

HPR estimates the temperature by calculating the number of packets routed by a node over a past window, which actually means the local load. The value of the threshold is dynamically set based on the local load and the temperature of the neighboring nodes.

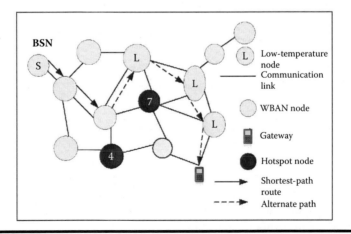

Figure 5.6 Example of HPR. The packet sent by source node S initially follows the shortest path. When it encounters a hotspot, it chooses an alternate path having a low-temperature node.

5.3.1.6 Thermal-Aware Shortest-Hop Routing

Thermal-aware shortest-hop routing (TSHR) [36] is a bit improved version of HPR. It is designed for the application that has a higher priority of delivering a packet to the destination, and if a packet is dropped, it will be retransmitted.

TSHR follows a procedure similar to that of HPR, with the exception of having two thresholds: a fixed threshold and a dynamic threshold. The fixed threshold applies for all nodes and determines the temperature that the nodes cannot exceed. The dynamic threshold is used to identify hotspots and is calculated in a way similar to that of HPR. While routing a packet, if the temperature of a neighbor exceeds the dynamic threshold, TSHR detours the packet through its coolest neighbor that has not been visited before. However, a fixed threshold is used every time during the routing of a packet. A packet will be buffered as long as the temperature of the forwarder node falls below the fixed threshold. This threshold basically signifies the maximum allowable temperature in the network.

5.3.1.7 Routing Algorithm for Network of Homogeneous and ID-Less Biomedical Sensor Nodes

The routing algorithm for network of homogeneous and ID-less biomedical sensor nodes (RAIN) [35] was proposed with the intent to prevent zone formation from having high temperature, reduce energy consumption, and achieve a higher packet delivery rate and lower average packet delivery delay. Another design objective of RAIN is to devise such a routing algorithm that would work in biomedical sensor networks with ID-less sensor motes.

Although RAIN avoids the use of a global ID due to the complexity of global coordination, it uses a local ID, which is temporary, focusing on local coordination. To solve the problem of data routing in a network of ID-less nodes, RAIN assumes that each node has a 16-bit random number generator that generates a random number between 0 and $(2^{16}-1)$ that serves as a node ID during the operational lifetime of the node. The local ID is known only among the neighbor nodes. However, the same node ID might exist in two different localities in the network. The algorithm is designed in such a way that such duplicate network IDs do not affect the performance of the network.

In the setup phase, every node generates a local ID and exchanges it through Hello packets among neighbors. The SINK ID is set to 0. In the routing phase, a packet will be routed to the SINK through multihop routing. A uniquely identifiable packet ID is generated with a format (N, T, R), where N is the local ID of the node, T is the time of packet generation, and R is a random number. Each node maintains a queue containing the packet IDs it has seen in the recent past. At each hop the packet travels, the hop-count value is incremented by 1. Each node estimates the temperature of its neighbors by counting the number of packets transmitted by the neighbors through the overhearing of packets. To prevent duplicate packet transmission, each node checks whether a received packet ID is already in the queue. It drops the packet if it is in the queue. A packet is also dropped if it exceeds a certain hop-count threshold to prevent a packet from moving around in the network unnecessarily. RAIN selects the forwarder node based on some probability that is a function that is inversely proportional to the node temperature. Thus, coolest neighbor is likely to be chosen as a forwarder.

RAIN also employs a status update mechanism to prevent reception of duplicate packets by the SINK and mitigate the energy-hole problem, in which nodes around the sink become depleted of energy very quickly. To mitigate this, the SINK broadcasts a "status update" message to its neighbors on receiving a packet. The status update message contains the packet ID of the received packet and prevents duplicate packet reception by the SINK from its neighbors, thus saving energy of the nodes around the SINK.

5.3.1.8 Mobility-Supporting Adaptive Threshold-Based Thermal-Aware Energy-Efficient Multihop Protocol

The mobility-supporting adaptive threshold-based thermal-aware energy-efficient multihop protocol (M-ATTEMPT) [37] has been proposed with the aim of supporting mobility and reducing energy consumption, along with thermal-awareness, while selecting a route to a sink node. It has three features: (1) single-hop communication for emergency services and on-demand data, (2) multihop communication for regular data delivery, and (3) network lifetime extension for the path with a smaller hop count in multihop communication.

The above features are achieved utilizing four phases. In the initialization phase, each node broadcasts a Hello message among neighbors to determine neighbor information and hop counts to the sink. In the routing phase, routes with fewer hops to the sink are selected from the available routes for less energy consumption. M-ATTEMPT employs direct communication to the sink for emergency data or on-demand data. The single-hop communication is delay-efficient but has high energy consumption. However, for normal data, it employs multihop communication where delay is not the main factor, but rather, saving energy is the primary concern. M-ATTEMPT detects a hotspot node if the temperature of a node exceeds some predefined threshold. In that case, a node breaks its link with all neighbors for a few rounds. It reestablishes the original route when the temperature reaches a normal stage (Figure 5.7).

M-ATTEMPT employs a scheduling phase for communication between the sink and the root nodes. In this phase, the sink creates a time division multiple access (TDMA) schedule and allocates time slots to all nodes. Nodes can communicate with the sink in the assigned time slot for normal data delivery in the data transmission phase.

To support mobility in M-ATTEMPT, nodes with high data rates are placed at less mobile phases on the human body. These high data rate nodes are known as parent nodes and are directly connected to the sink. The first-level child nodes have less energy with lower data rates than the parent nodes. Thus, a hierarchy of nodes is created based on their energy level and data rate.

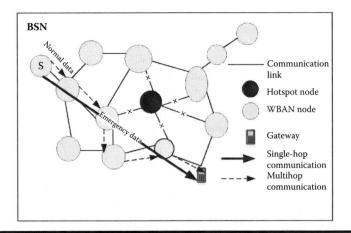

Figure 5.7   Example of M-ATTEMPT routing phase. Source node S sends emergency data to the gateway using single-hop communication. Normal data is transmitted following multihop communication. During multihop communication, when a node identifies itself as a hotspot, it breaks all the communication links with the neighbors and the data is delivered following a route-avoiding hotspot.

Whenever a node is distracted from its original parent node due to mobility and enters into the communication range of another parent node, it sends a joint request to the new parent node. The new parent node accepts the request if the number of its children is less than three.

5.3.1.9 Thermal-Aware Multiconstrained QoS Routing

Thermal-aware multiconstrained QoS routing (TMQoS) [38] is proposed, focusing on QoS provisioning with multiple constraints (delay and reliability), along with keeping the temperature rise of the nodes as low as possible. TMQoS is the first protocol that mainly combines the QoS provisioning issue with thermal awareness in selecting the route of a packet.

TMQoS classifies the traffic into four classes: delay-constrained, reliability-constrained (C1); reliability-constrained, non-delay-constrained (C2); delay-constrained, non-reliability-constrained (C3); and non-delay-constrained, non-reliability-constrained (C4) traffic. A cross-layer proactive routing framework has been developed, as shown in Figure 5.8, which maintains an ongoing

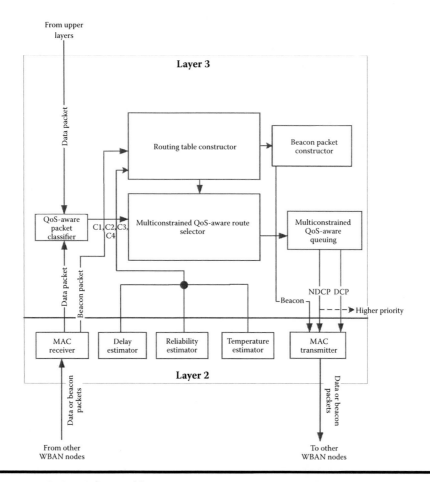

Figure 5.8   TMQoS routing architecture. (From M. M. Monowar, M. M. Hassan, F. Bajaber, M. A. Hamid, and A. Alamri, Thermal-aware multiconstrained intrabody QoS routing for wireless body area networks, *International Journal of Distributed Sensor Networks*, vol. 2014, 2014.)

routing table containing routes with minimum hop count to the gateway or body coordinator (BC) through periodic beacon exchanges. The delay estimator and the reliability estimator modules in layer 2 estimate the hop-to-hop delay and the link reliability for particular neighboring nodes and send the feedback to the routing table constructor module. Also, the temperature estimator module measures the node temperature and gives the feedback to the routing table constructor module to update the temperature value. The routing table constructor module updates the metric values upon either receiving a beacon packet from neighboring nodes or any changes in the value occurring at the local estimation of the values. The routing table constructor module also employs a hotspot avoidance mechanism in which a node identifies itself as a hotspot by getting feedback from the temperature estimator module and immediately notifies its neighboring node of the information through an urgent beacon packet transmission. Eventually, this information is propagated along the network. The multiconstrained QoS-aware route selector module chooses the most suitable route based on the respective QoS requirements. In particular, the minimum temperature path will be selected for C4 traffic, the minimum delay path will be selected for C3 traffic, and the maximum reliable path will be selected for C2 traffic. Since C1 traffic has both delay and reliability constraints, a minimum delay path will be selected first, followed by a maximum reliable path. Multiple copies of the packet on two paths also increase the data reliability. There is also a multiconstrained QoS-aware queuing module that maintains two separate queues for storing delay-constrained packets and non-delay-constrained packets, and delay-constrained packets will be given higher priority for their transmission.

5.3.2 Comparative Study

The primary design consideration of the state-of-the-art thermal-aware protocols is to minimize temperature rise of the implant sensor devices inside the human body. However, there are numerous issues to be taken into account while designing communication protocols for BSN. Table 5.2 presents a comparative study of the state-of-the-art thermal-aware protocols.

The existing communication protocols are evaluated considering several issues that are significant design considerations for BSN:

- Routing approach: The routing approach determines how the routing functionality will be executed and maintained. Generally, routing approaches are of two types: table driven and on demand. In the table-driven approach, every node maintains a routing table containing information about neighbors, and route selection is done based on some routing metrics. For route maintenance, the table needs to be updated periodically. In the on-demand approach, routes are established on demand through flooding of a route request packet by the sending node. Comparatively, table-driven approaches have more overhead due to the periodic routing table maintenance than the on-demand approach. However, table-driven approaches have lower latency in route finding than on-demand approaches.

- Routing decision: Routing decisions can be done by a node on either a hop-by-hop or end-to-end basis. In a hop-by-hop approach, a potential forwarder is selected on a per-hop basis, while in an end-to-end approach, the optimal path from the source to the destination is selected through routing table broadcast. The hop-by-hop approach has much lower protocol overhead, but it may choose a suboptimal path. It also requires more delay, as forwarder selection is done on a per-hop basis after processing. An end-to-end approach possesses high protocol overhead due to the routing table broadcast among neighbors. But this approach is delay efficient and selects an optimal path if the network topology does not change frequently.

Table 5.2 Comparison among Thermal-Aware Routing Protocols

Protocol	Routing Metric	Estimation of Temperature Rise	Routing Approach	Routing Decision	QoS Provisioning	Traffic Heterogeneity	Mobility Support	Energy Efficiency
TARA	Temperature	SAR and heating effects on sensor	Table driven	Hop by hop	No	No	No	No
LTR	Temperature	Number of packet receptions	Table driven	Hop by hop	No	No	No	No
ALTR	Temperature and hop count	Number of packet receptions	Table driven	Hop by hop	No	No	No	No
LTRT	Temperature	Number of packet receptions	Table driven	End to end	No	No	No	No
HPR	Hop count and temperature	Number of packets routed over a past window	Table driven	Hop by hop	No	No	No	No
TSHR	Hop count and temperature	Number of packet transmissions and receptions	Table driven	Hop by hop	No	No	No	No
RAIN	Temperature	Number of packet transmissions by neighbors through overhearing	Table driven	Hop by hop	No	No	No	Yes
M-ATTEMPT	Hop count	SAR and heating effects on sensor	Table driven	Hop by hop	No	Yes	Yes	Yes
TMQoS	Hop count, delay, reliability, temperature	SAR and heating effects on sensor	Table driven	End to end	Yes	Yes	No	No

■ QoS provisioning: Quality of service (QoS) provisioning is one of the significant design criteria for BSN communication protocol. The BSN application requires the meeting of certain QoS parameters, such as delay, reliability, and throughput. If the required criteria are not met, the application objective cannot be preserved.

■ Traffic heterogeneity: BSN can be equipped with diverse types of sensors with various QoS requirements. For instance, EEG, ECG, and EMG generate real-time medical continuous data that must be delivered within a certain deadline with higher reliability. However, respiration monitoring and pH-level monitoring applications have strict reliability requirements, while the traffic belonging to these applications can be processed offline, and hence are non-delay constrained. Thus, addressing traffic heterogeneity satisfying respective QoS requirements is another striking design issue for BSN.

■ Mobility support: The mobility of the human body may disconnect the communication links among the implant devices and even cause changes in topology. An efficient communication protocol for BSNs should address this issue.

■ Energy efficiency: Since the body sensor nodes are battery driven, high energy consumption could reduce the lifetime of the sensor nodes very rapidly. Considering the implanted devices, the lifetime is crucial. It is not desirable or practical to frequently change the node or replace the battery for such implant devices [8]. Thus, consideration of energy efficiency is also a vital metric for communication protocols of BSN.

Considering the above design issues, a comparative study of the protocols is provided below.

TARA is the first protocol that considers temperature a routing metric. TARA achieves significant thermal efficiency in terms of both maximum temperature rise and average temperature rise in the network, compared to a shortest-hop routing (SHR) protocol. However, due to the withdrawal strategy, TARA suffers from high end-to-end delay, lower reliability, and high energy consumption since the packet needs to traverse many hops when it encounters a hotspot and will be detoured arbitrarily.

The problem of TARA, due to the withdrawal strategy, is addressed by both LTR and ALTR. However, LTR is a greedy approach that may be locally optimal, but not globally optimal. A packet in LTR is not always directed to the destination, which significantly increases the hop count, thus resulting in higher delay and low reliability. Although this problem is addressed by ALTR by employing shortest-hop routing after a packet traverses certain hops, it still wastes network bandwidth through unnecessary transmissions until the hop count reaches the threshold value, and also, a packet traverses through hotspots while utilizing shortest-hop routing.

LTRT addresses the problems of TARA, LTR, and ALTR by selecting a least temperature route from all possible routes from a sender node to a destination node exploiting the Dijkstra's algorithm. LTRT, however, employs an end-to-end approach in route selection that incurs much protocol overhead and has high energy consumption. LTRT only chooses the least temperature route, which may not be the least delay path or reliable path or least energy-efficient path.

HPR is bit modified version of ALTR. HPR tries to reduce the maximum temperature rise of a network. However, similar to LTR, it also chooses the coolest neighbor while a hotspot is identified. Thus, the problems of LTR also remain in HPR in terms of energy consumption, packet delivery delay, and reliability.

TSHR, an improvement of HPR, maintains an acceptable temperature rise in the network exploiting a two-temperature threshold. Because of the packet buffering mechanism, TSHR reduces the packet drop rate; however, the incurred delay is much higher than HPR.

RAIN mainly focuses on employing a technique that utilizes a local ID instead of a global ID. Because of the probabilistic forwarder selection, RAIN cannot avoid selection of heated nodes completely, although it is chosen with lower probability. RAIN addresses energy efficiency in the protocol operation by reducing the duplicate packet reception; however, it requires flooding of a status update message by the SINK, which incurs additional energy consumption.

M-ATTEMPT is the only protocol that considers mobility support. It also addresses packet heterogeneity by classifying packets into two classes—emergency packet and normal data packet—although it overlooks the QoS provisioning for diverse traffic types. M-ATTEMPT did some trade-off between delay and energy efficiency for the emergency packet by exploiting single-hop communication. The routing part mainly applies for normal data. M-ATTEMPT achieves energy efficiency for normal data by selecting a route with a smaller hop count, also choosing an energy-efficient path if two paths have the same hop count. M-ATTEMPT, however, cannot reduce the maximum temperature and average temperature rise in the network considerably, as it uses the same shortest path for normal data. Transmitting emergency data with the higher transmission power also might increase node temperature. The mobility support mechanism, as explained in M-ATTEMPT, also has some limitations. A parent node accepts the join request of a new child node that is disconnected from an earlier parent, if it has less than three child nodes. Otherwise, the node will not be connected to any parent.

TMQoS is the only protocol so far that integrates QoS provisioning of diverse types with thermal awareness. Considering multiple QoS metrics such as delay and reliability, along with temperature and hop count, as routing metrics, TMQoS achieves the desired QoS requirements for the respective traffic types. Irrespective of the traffic types, TMQoS chooses route-avoiding hotspots, and thus is able to maintain the temperature rise at an acceptable level. However, because of employing an end-to-end approach, TMQoS incurs more protocol overhead, which would be a costly solution for resource-constrained BSNs.

All of the state-of-the-art approaches presented so far follow a table-driven routing approach. The main intention of this design issue could be avoiding unnecessary delay for route selection and excessive flooding for resource-constrained BSNs. Regarding temperature estimation, TARA, M-ATTEMPT, and TMQoS calculate the SAR and consider both heat generation due to wireless communication and power dissipation of the node circuitry. This is more realistic, as it considers all the sources of temperature rise, but it is computation-intensive with respect to BSNs. However, the other protocols perform temperature estimation by counting the number of packet receptions or transmissions, or both. Although it simplifies the estimation, it may not reflect the actual scenario of temperature rise. QoS provisioning also has been paid little attention so far, with the exception of TMQoS. Although some of the protocols try to reduce packet delivery delay (RAIN, HPR, ALTR) or increase reliability (TSHR), those metrics have not been explicitly considered in selecting the route, and meeting of the QoS parameters has not been significantly observed in those protocols. Being a considerably worthy metric, energy efficiency and mobility support have also not been paid much attention in thermal-aware communication protocols for BSNs.

5.4 Challenges and Open Research Issues

BSNs have tremendous potential in promoting e-healthcare applications. To address the challenge of temperature rise inside the body for implant sensor devices, a good number of solutions have already been offered, as discussed in Section 5.3. However, some research issues still remain to be addressed before intra-BAN communication is widely applied, as summarized below:

■ Although few studies have discussed energy efficiency issues, those protocols achieve energy efficiency only by choosing shortest-hop routes (M-ATTEMPT, HPR) or reducing duplicate packets (RAIN). However, the shortest-hop route may not always be energy-efficient, especially if the available energy of the nodes on the shortest path is very low. Also, due to non-redundancy requirements, the possibility of duplicate packets is also low. Therefore, energy efficiency should be considered a separate metric during route selection for thermal-aware protocols. Moreover, the protocol operation also requires consideration of load balancing issues so that the lifetime of the BSN extends for sufficient periods.

■ Since one of the reasons for a thermal increase is RF power for battery recharging from an external source, an efficient energy scavenging mechanism that reduces thermal effects is also an important research issue that has not been considered so far. Integrating energy efficiency with energy scavenging would provide an optimal solution for intrabody BSNs.

■ The existing thermal-aware protocols employ a table-driven routing approach where a number of parameters are exchanged periodically among the neighbors, increasing protocol overhead for the resource-constrained BSNs. An analytical model is required to determine an optimal frequency of information exchange among neighbors that can reduce protocol overhead. Also, instead of using a table-driven approach, reactive routing approaches can also be applied for some applications by reducing the amount of flooding packets and route finding delay.

■ The mobility issue has also been paid exiguous attention so far. Body movement could cause frequently disconnected links, as well as changes in topology. The only technique (M-ATTEMPT) that addresses mobility also has limitations. Hence, integrating mobility support with a thermal-aware solution is an important research issue.

■ QoS provisioning issues considering traffic heterogeneity in thermal-aware solutions are also worthy of research. Although TMQoS has addressed this issue, the solution has high protocol overhead. Moreover, emergency data handling has not been addressed considerably in thermal-aware solutions. This type of data has hard deadlines and needs to be delivered with 100% reliability. Effective approaches are thus required in this regard.

■ The contemporary thermal-aware solutions mostly focus on temperature issues. No solution is available that considers all the required design issues. Integrating all the design considerations in one solution is quite challenging. However, some trade-offs can be done considering the application requirements, lifetime of BSNs, temperature rise, and so forth. A globally optimal system that unites the required design issues in a single BSN is thus essential.

5.5 Conclusion

BSNs are envisioned to be a very useful technology with high potential to offer a wide range of benefits to patients, healthcare professionals such as doctors and nurses, and society through continuous monitoring and early detection of possible problems. However, if not effectively designed, BSNs could even be a threat to human life. Rise of temperature inside the human body is thus considered a significantly important criterion for designing communication protocols for BSNs.

This chapter reviewed state-of-the-art research on thermal-aware communication protocols for BSNs. In particular, the problem of thermal effects on the human body has been extensively discussed. The contemporary solutions that address the thermal effects have been presented in detail. A comparative analysis among the protocols considering relative advantages and limitations was also performed. To conclude, some challenging open research issues have been discussed. We believe our contribution in this chapter will help a wide range of researchers understand the problem and causes of thermal effects on the human body, and devise communication protocols that could effectively solve the problems of temperature rise, along with meeting the application objectives.

References

1. S. Ullah, H. Higgins, B. Braem, B. Latre, C. Blondia, I. Moerman, S. Saleem, Z. Rahman, and K. S. Kwak, A comprehensive survey of wireless body area networks, *Journal of Medical Systems*, vol. 36, no. 3, pp. 1065–1094, 2012.
2. S. Park and S. Jayaraman, Enhancing the quality of life through wearable technology, *IEEE Engineering in Medicine and Bilogical Magazine*, vol. 22, no. 3, pp. 41–48, 2003.
3. L. Roelens, S. V. d. Bulcke, W. Joseph, G. Vermeeren, and A. L. Martens, Path loss model for wireless narrowband communication above flat phantom, *IEE Electronics Letters*, vol. 42, no. 1, pp. 10–11, 2006.
4. T. Zasowski, F. Althaus, M. Stäger, A. Wittneben, and G. Tröster, UWB for noninvasive wireless body area networks: Channel measurements and results, in *IEEE Conference on Ultra Wideband Systems and Technologies (UWBST 2003)*, Reston, VA, 2003, pp. 285–289.
5. A. D. C. Fort, Ultra wide-band body area channel model, in *IEEE International Conference on Communications (ICC 2005)*, Seoul, 2005, pp. 2840–2844.
6. B. Latré, G. Vermeeren, I. Moerman, L. Martens, and P. Demeester, Networking and propagation issues in body area networks, in *11th Symposium on Communications and Vehicular Technology in the Benelux (SCVT 2004)*, Ghent, Belgium, 2004.
7. Y. Prakash, S. Lalwani, S. K. S. Gupta, E. Elsharawy, and L. Schwiebert, Towards a propagation model for wireless biomedical applications, in *IEEE International Conference on Communications (ICC 2003)*, Anchorage, AK, 2003, pp. 1993–1997.
8. B. Latré, B. Braem, I. Moerman, C. Blondia, and A. P. Demeester, A survey on wireless body area networks, *Wireless Networks*, vol. 17, no. 1, pp. 1–18, 2011.
9. IEEE (Institute of Electrical and Electronics Engineers), IEEE standard for safety levels with respect to human exposure to radio frequency electromagnetic fields, 3 KHz to 300 GHz, C95.1-1999, IEEE, New York, 1999.
10. M. Chen, S. Gonzale, A. Vasilakos, H. Cao, and V. C. M. Leung, Body area networks: A survey, *Mobile Networks and Applications*, vol. 16, no. 2, pp. 171–193, 2011.
11. L. Theogarajan, J. Wyatt, J. Rizzo, B. Drohan, M. Markova, S. Kelly, G. Swider, M. Raj, D. Shire, M. Gingerich, J. Lowenstein, B. Yomtov, Minimally invasive retinal prosthesis, in *IEEE International Solid-State Circuits Conference 2006 (ISSCC) Digest of Technical Papers*, San Francisco, February 6–9, 2006, pp. 99–100.
12. T. Penzel, B. Kemp, G. Klösch, A. Schlögl, J. Hasan, A. Värri, and I. Korhonen, Acquisition of biomedical signals databases, *IEEE Engineering in Medicine and Biology Magazine*, vol. 20, no. 3, pp. 25–32, 2003.
13. S. Arnon, D. Bhastekar, D. Kedar, and A. Tauber, A comparative study of wireless communication network configurations for medical applications, *IEEE Wireless Communications*, vol. 10, no. 1, pp. 56–61, 2003.
14. B. Gyselinckx, J. Penders, and R. Vullers, Potential and challenges of body area networks for cardiac monitoring, *Journal of Electrocardiology*, vol. 40, no. 6, pp. 165–168, 2007.
15. B. Warneke and K. Pister, MEMS for distributed wireless sensor networks, in *9th International Conference on Electronics, Circuits and Systems*, Berkeley, CA, 2002, pp. 291–294.

16. S. D. Haan, NEMS—Emerging products and applications of nano-electromechanical systems, *Nanotechnology Perceptions*, vol. 2, no. 3, pp. 267–275, 2006.

17. L. Schwiebert, S. K. Gupta, and J. Weinmann, Research challenges in wireless networks of biomedical sensors, in *Proceedings of the 7th Annual International Conference on Mobile Computing and Networking*, Rome, Italy, 2001, pp. 151–165.

18. S. Ullah, P. Khan, N. Ullah, S. Saleem, H. Higgins, and K. S. Kwak, A review of wireless body area networks for medical applications, *International Journal of Communications, Network and System Sciences*, vol. 2, no. 8, pp. 797–803, 2009.

19. G. Lazzi, Thermal effects of bioimplants, *IEEE Engineering in Medicine and Biology Magazine*, vol. 24, no. 5, pp. 75–81, 2005.

20. A. Bag and M. Bassiouni, Energy efficient thermal aware routing algorithms for embedded biomedical sensor networks, in *2006 IEEE International Conference on Mobile Adhoc and Sensor Systems (MASS)*, Vancouver, British Columbia, 2006, pp. 604–609.

21. L. Schwiebert, S. Gupta, P. Auner, G. Abrams, R. Iezzi, and P. McAllister, A biomedical smart sensor for the visually impaired, in *Sensors 2002*, Orlando, FL, 2002, pp. 693–698.

22. A. Hirata, G. Ushio, and T. Shiozawa, Calculation of temperature rises in the human eye for exposure to EM waves in the ISM frequency bands, *IEICE Transactions on Communications*, vol. E83-B, no. 3, pp. 541–548, 2000.

23. Q. Tang, N. Tummala, S. Gupta, and L. Schwiebert, Communication scheduling to minimize thermal effects of implanted biosensor networks in homogeneous tissue, *IEEE Transactions on Biomedical Engineering*, vol. 52, no. 7, pp. 1285–1294, 2005.

24. IEC (International Electrotechnical Commission), Medical electrical equipment, Part 2–33: Particular requirement for the safety of magnetic resonance systems for medical diagnosis, 2nd ed., IEC, Geneva, 1995, pp. 60601-2–60601-33.

25. P. Bernardi, M. Cavagnaro, S. Pisa, and E. Piuzzi, SAR distribution and temperature increase in an anatomical model of the human eye exposed to the field radiated by the user antenna in a wireless LAN, *IEEE Transactions on Microwave Theory and Techniques*, vol. 46, no. 12, pp. 2074–2082, 1998.

26. J. A. Scott, The computation of temperature rises in the human eye induced by infrared radiation, *Physics in Medicine and Biology*, vol. 33, no. 2, pp. 243–257, 1988.

27. W. J. Heetderks, RF powering of millimeter and submillimeter-sized neural prosthetic implants, *IEEE Transactions on Biomedical Engineering*, vol. 35, no. 5, pp. 323–327, 1988.

28. W. Mokwa and U. Schnakenberg, Micro-transponder systems for medical applications, *IEEE Transactions on Instrumentation and Measurement*, vol. 50, no. 6, pp. 1551–1555, 2001.

29. E. A. Taflove, *Advances in Computational Electrodynamics: The Finite-Difference Time-Domain Method*, Artech House, Boston, 1998.

30. NCRP (National Council on Radiation and Protection and Measurements), A practical guide to the determination of human exposure to radio frequency fields, Report 119, NCRP, Bethesda, MD, 1993.

31. H. H. Pennes, Analysis of tissue and arterial blood temperature in the resting human forearm, *Journal of Applied Physiology*, vol. 1.1, pp. 93–122, 1948.

32. Q. Tang, N. Tummala, S. K. Gupta, and L. Schwiebert, TARA: Thermal-aware routing algorithm for implanted sensor networks, in *1st IEEE International Conference on Distributed Computing in Sensor Systems (DCOSS '05)*, Los Angeles, CA, 2005, pp. 206–217.

33. D. Takahashi, Y. Xiao, F. Hu, J. Chen, and Y. Sun, Temperature-aware routing for telemedicine applications in embedded biomedical sensor networks, *EURASIP Journal on Wireless Communications and Networking*, vol. 2008, article 572636, 2008.

34. A. Bag and M. Bassiouni, Hotspot preventing routing algorithm for delay-sensitive biomedical sensor networks, in *Proceedings of the IEEE International Conference on Portable Information Devices (PORTABLE)*, Orlando, FL, 2007, pp. 1–5.

35. A. Bag and M. Bassiouni, Routing algorithm for network of homogeneous and ID-less biomedical sensor nodes (RAIN), in *Proceedings of Sensors Applications Symposium (SAS)*, Atlanta, GA, 2008, pp. 68–73.

36. M. Tabandeh, M. Jahed, F. Ahourai, and S. Moradi, A thermal-aware shortest hop routing algorithm for in vivo biomedical sensor networks, in *Proceedings of the 6th International Conference on Information Technology: New Generations (ITNG)*, Las Vegas, NV, 2009, pp. 1612–1613.

37. N. Javaid, Z. Abbas, M. S. Fareed, Z. A. Khan, N. Alrajeh, M-ATTEMPT: A new energy-efficient routing protocol for wireless body area sensor networks, in *4th International Conference on Ambient Systems, Networks and Technologies*, Halifax, Nova Scotia, 2013, pp. 224–231.

38. M. M. Monowar, M. M. Hassan, F. Bajaber, M. A. Hamid, and A. Alamri, Thermal-aware multiconstrained intrabody QoS routing for wireless body area networks, *International Journal of Distributed Sensor Networks*, vol. 2014, article 676312, 2014.

WSNs IN EMERGING COMMUNICATION TECHNOLOGIES

Chapter 6

Electromagnetic Wireless Nanoscale Sensor Networks

Eisa Zarepour, Mahbub Hassan, Chun Tung Chou,
and Adesoji A. Adesina

Contents

Abstract

With recent advances in nanotechnology, researchers are now seriously contemplating the possibility of electromagnetic wireless nanoscale sensor networks (WNSNs). WNSNs open up the possibility to sense and control important physical processes from the very bottom, right at the molecular level. Early indications suggest that such a *bottom-up* approach to sense and control, which has up to now not been possible given the constraints of conventional wireless sensor networks, has the potential to radically improve the performance of many critical medical, industrial, biological, and military applications. In this chapter, we first summarize the recent advancements in developing the required hardware components to realize nanomotes such as nanotransceivers, nanomemories, etc. The channel modelling and applications of WNSNs are then reviewed. The main research trends in designing communication protocols for WNSNs such as designing propagation models and routing protocols are also discussed followed by a brief summary.

6.1 Introduction

Following the success of conventional macroscale wireless sensor networks, researchers are now investigating the viability of nanoscale sensor networks (NSNs), which are formed by establishing communication between devices made from nanomaterials. Technically, a NSN is a network of nanoscale devices (nanomotes) capable of some basic computing, sensing, actuation, and communication tasks. Although nanomotes are not yet available commercially, there have been significant relevant developments in recent years that point to a future when such devices could be produced in bulk. For example, a miniature *hydrogen sensor* consisting of a nanotaper coated with an ultrathin palladium film was reported in [1], where the optical properties of the palladium layer changed when exposed to hydrogen. C. R. Yonzon et al. [2] surveyed many other types of nanosensors that can be used for chemical and biological sensing. Similarly, progress has been recorded in chemical and biological nanoactuators that can be used to accomplish some basic tasks at the molecular level by harnessing interactions between nanoparticles, electromagnetic fields, and heat [3–6].

Finally, work has begun exploring communication possibilities at the nanoscale, which would ultimately connect these nanomotes to form a NSN. The seminal papers by I. F. Akyildiz et al. [7] have captured the imagination of many communication researchers by showing that it is conceptually possible to achieve communication at the nanoscale using different paradigms, such as electromagnetic, optic, acoustic, electromechanical, or some form of molecular-based transceivers. This has sparked a flurry of new research activity to understand the unique properties of nanomaterials that could be used for communication between nanodevices [8–16]. The wireless nanoscale sensor networks (WNSNs) that employ the conventional electromagnetic wave to form a network between nanomotes have been recently investigated as a major class of NSN [17]. The focus of this chapter is on electromagnetic WNSNs. There are two main approaches in designing WNSNs: either scaling down the existing metallic nanoantennas or using the physical properties of novel nanomaterials to design new nanoscale antennas [12]. The first approach is not feasible, as the miniaturization of the conventional metallic nanoantennas imposes an extremely high-frequency band—more than 100 THz. It is not practical for a resource-restricted nanomote to operate in such a high-frequency region due to extremely high path loss and absorption in this band. In the second method, using novel nanomaterials such as carbon nanotubes or graphene, researchers have proposed new nanoantennas that can operate in much lower-frequency bands [8,12,18]. For example, recently a graphene-based nanotransceiver has been proposed that employs the *surface plasmon polariton* to propagate an electromagnetic wave mainly in the frequencies between 0.1 and 10 THz [12].

Since WNSNs can operate at molecular levels, they can be used for totally new kinds of nanotechnology applications that cannot be realized with conventional sensor networks. I. F. Akyildiz and J. M. Jornet [17] have outlined a number of possible interesting new applications of WNSNs in biomedical (health monitoring and drug delivery), environmental (plant monitoring and defeating insect plague), industrial (ultrasensitive touch interfaces), and military (biological and chemical defense) domains. For example, a WNSN can be used to monitor the level of different vital molecules in the blood, such as sodium, glucose, and other ions; to trace cancer biomarkers; and to detect infectious agents in different human bodies [17]. Figure 6.1 illustrates a general schematic architecture for online microscopic environmental monitoring using WNSNs. The data collected by nanomotes will be transmitted to a nearby micro- or macroscale sink, and then it will be transferred to a remote server via a macroscale gateway connected to the Internet.

The aim of this chapter is to investigate the state of the art in developing the required hardware, software, and communication for this emerging technology in wireless sensor networks. The chapter is structured as follows. We first review the recent advancements in developing the required hardware components to realize nanomotes such as nanotransceivers and nanomemories (Section 6.2). The possible WNSN paradigms and their channel modeling and applications are reviewed in Section 6.3. The time-varying WNSN communication channel is introduced in Section 6.4. The main research trends in designing communication protocols for WNSN, such as

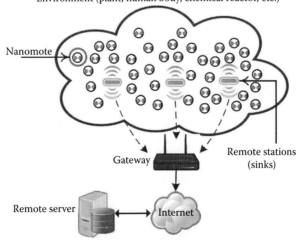

Figure 6.1   A schematic architecture for environmental monitoring via WNSN.

designing coding schemes, medium access control (MAC), and routing protocols, are discussed in Section 6.5, followed by a brief summary in Section 6.6.

6.2 Nanoscale Components for WNSNs

Nanotechnology is enabling the development of devices in a scale ranging from one to a few hundred nanometers. The new characteristics and behavior of nanomaterials that have not been observed at the microscopic level facilitate nanodevices with completely novel functionalities stemming from these unique properties. Like at the macroscale, sensor nodes, processors, memory, batteries, radio transmitters, sensors, and other components are essential required hardware at the nanoscale, in order to have an operational nanomote capable of sensing, processing, and communicating to either the nano- or macroscale. In this section, we review recent advancements in developing the main hardware components required to build a nanomote.

6.2.1 Processing Unit

Nanomaterials such as carbon nanotube (CNT) and graphene nano ribbon (GNR) can be used to build transistors at the nanometer scale [19–21]. For example, a GNR-based nanotransistor consisting of a thin layer of graphene (made of only 10 × 1 carbon atoms) has been experimentally developed [20]. A high-performance nanoscale transistor based on a CNT thin film has been also reported [21].

Miniaturization of silicon-based transistors has been considered another option for nanoscale transistors [22]. For instance, the first silicon-based single-atom transistor was made at the University of New South Wales, Sydney, Australia, in 2012 [23]. The device is fabricated using a combination of scanning tunneling microscopy (STM) and hydrogen-resist lithography.

6.2.2 Data Storage Unit

Several nanomemories based on different technologies and nanomaterials have been reported [24–26], so the capability of saving a single bit in a single atom has been experimentally demonstrated [17]. CNT field-effect transistors are another option for building nanomemories [27,28]. For example, in [28], a single-walled CNT-based transistor was used to construct a nonvolatile charge-storage memory element operating at room temperature.

6.2.3 Sensing and Actuating Unit

Nanotechnology provides the opportunity to design sensors that are much smaller, less power-hungry, and more sensitive than current micro- and macroscale sensors. Several nanosensors reported in [2] take advantage of the unique properties of nanomaterials and nanoparticles to detect and measure new types of events at the nanoscale. In addition, a recently developed graphene-based nanogas sensor can detect a few types of molecules [29].

As another example, a triboelectric nanogenerator (TENG), based on the well-known contact electrification effect, was used as a gas sensor [30]. As the amount of power that can be harvested via a TENG is strongly affected by its surface-adsorbed molecules, different types of molecule can be efficiently detected only by monitoring the output voltage of the TENG. The design and manufacturing of nanoscale actuators have also been well studied [2–6]. For example, the mechanical actuation of an array of polymer nanowires has been employed as a reliable source of mechanical actions at the nanoscale [6].

6.2.4 Communication Unit

There are several options that can be used to establish nanoscale communication. Downscaling existing communication paradigms such as electromagnetic [17], optical [31], acoustic [32,33], or electromechanical communication [34] is one research trend in nanoscale communication. Due to challenges caused by miniaturization of these technologies, defining completely new paradigms such as molecular communication, inspired by natural communication between biological molecules [7,35], has also been considered. The focus of this chapter is on electromagnetic communication, which will be discussed in more detailed in Section 6.3.

6.2.5 Powering Unit

Nanomotes can be powered by nanobatteries or nanoscale energy harvesting interfaces. In this section, we briefly review both options:

1. *Nanobatteries*: Designing nanoscale power supplies with high power density, reasonable lifetime, and contained charge/discharge rates is an active ongoing research topic [36–40]. In [40], a battery was constructed from an array of nanobatteries connected in parallel, each composed of an anode, a cathode, and a liquid electrolyte confined within the nanopores of anodic aluminum oxide as an all-in-one nanosize battery. Nevertheless, at the nanoscale, it is difficult to deploy batteries with significant energy reserve for long network life because the size of the battery is limited. It is also a practical challenge to change or periodically recharge batteries. Energy harvesting is therefore considered the most viable solution for WNSNs. We will review a few options for harvesting energy at the nanoscale in Section 6.3.
2. *Nanoscale energy harvesting*: In recent years, researchers have developed a variety of techniques to harvest energy at the nanoscale using the novel properties of different nanomaterials. In this section, we review some of these techniques. More detailed explanations of these and other techniques can be found in [41].

6.2.5.1 Piezoelectric

The word *piezo* is derived from the Greek word for *pressure*. The piezoelectric effect, which was discovered in 1880 by Jacques and Pierre Curie, can be defined as a property of certain materials to produce electricity when subjected to mechanical stress. Noncentrosymmetricity is a key property required for any material to achieve piezoelectricity, as this effect is only possible with homogenous deformation. Researchers have demonstrated that a single zinc oxide (ZnO) nanowire of 20 nm diameter and 200 nm length can produce power up to 0.5 pW at one cycle of resonance, for example, from mechanical stress, subject to the maximum deformation that can be borne by the nanowire [42]. An array of such nanowires can be assembled to produce higher power in a nanoenergy harvester, albeit at a higher dimension.

6.2.5.2 Flexoelectric

Flexoelectricity is similar to piezoelectricity except it does not require the centrosymmetricity property, making it possible to realize nanoscale energy harvesters with many different types of material [43]. Instead of requiring homogenous deformation in the material, flexoelectricity appears at strain gradients. This effect has been known for decades, but it was not used, due to the small magnitude of generated power. Interestingly, strain gradients at the nanoscale can be huge, creating the potential to harvest energy for nanosensor networks. Indeed, using ferroelectric thin films, researchers have demonstrated that flexoelectric polarization can be bigger than piezoelectricity at the 10 nm scale [44].

6.2.5.3 Thermoelectricity

Temperature difference can be converted into electricity using special materials. The most common form of thermoelectric energy harvesters use the so-called Seebeck effect, which requires a thermal gradient between two sides of the energy harvesting device to drive the diffusion of charge carriers. This type of energy harvester is suitable for environments where thermal gradients exist naturally, such as a wristwatch touching the human skin. However, Seebeck-based thermoelectric harvesters will not work when temperature is uniform in space.

When space is thermally uniform, but the temperature may fluctuate over time, pyroelectricity can be used to harvest thermal energy. Pyroelectricity is based on the spontaneous polarization in certain anisotropic solids due to a time-dependent temperature variation. The amount of power that can be generated is proportional to the rate of temperature change, which makes it directly useful for nanoscale systems. For example, even a small temperature change over a nanosecond can produce a large amount of power. Researchers have already demonstrated that ZnO nanowires can be used to generate pyroelectricity from time-dependent temperature variations [45].

6.2.5.4 Other Technologies

Triboelectric nanogenerators (TENGs), which are based on the well-known contact electrification effect, are another option for harvesting mechanical energy through periodic contact and separation of two nanoscale polymer plates [46]. TENGs can also convert magnetic force variation to electricity [47]. Many other hybrid approaches have been proposed to scavenge energy from different sources at the nanoscale [48–51].

6.3 Electromagnetic Wireless Nanoscale Sensor Networks

Individual nanomotes can expand their capabilities by communication with each other, to be able to perform more complex tasks in wider coverages [7]. Miniaturization of the traditional electromagnetic-based metallic antenna to the nanoscale imposes an extremely high operational frequency (more than 100 THz), which is not a feasible frequency for resource-restricted WNSNs due to extremely high attenuation in the channel. However, the physical property of some novel nanomaterials can be used to design nanoantennas able to work in lower-frequency bands. CNT and GNR are the most promising candidates to form a WNSN, as their wave propagation velocity is up to 0.01 of the speed of light in vacuum. Therefore, the resonant frequency of CNT and GNR nanoantennas can be theoretically up to two orders of magnitude smaller than metallic nanoantennas [17]. Although the CNT- and GNR-based nanoantennas have not been manufactured in bulk, several attempts have been made to characterize their behavior in different conditions [12,52–56]. J. M. Jornet and I. F. Akyildiz have proposed a novel GNR-based nanoradio that exploits the behavior of surface plasmon polariton (SPP) waves in a GNR to generate and propagate an electromagnetic signal [12]. The mathematical modeling and COMSOL simulation shows that such a GNR-based nanoantenna is able to operate at much lower frequencies than its metallic counterparts, ranging from 100 GHz to 10 THz.

The terahertz band is an unlicensed frequency range between 0.1 and 10 THz that is not allocated to specific purposes [18] but has been proposed for high-capacity short-distance macroscale wireless communication [57,58], which can be considered for connecting devices in short ranges, kiosk downloading, board-to-board (B2B) communication, and wireless personal area networks (WPANs) [59]. In a recent case, Japanese researchers developed a small transmitter, *T-ray*, able to transfer information over distances less than 3 m at 3 Gbps (theoretically up to 100 Gbps) in the frequency range from 300 GHz to 3 THz [60].

Nevertheless, both macro- and nanoscale communication in the terahertz band are severely affected by molecular absorption because the terahertz is also the resonant frequency of many molecules. In the next section, we review the effect of molecular absorption on the terahertz channel, based on a model proposed in [11] and [61].

6.3.1 Channel Modeling

As was explained in the opening of Section 6.3, WNSNs operate in the terahertz band; thus, there is a need to characterize the terahertz channel in the nanoscale to design an appropriate propagation model. Although several attempts have been made to characterize wireless communication in the terahertz band [62,63], J. M. Jornet and I. F. Akyildiz [11,64] introduced the first propagation model for terahertz communication at the nanoscale. They proposed a mathematical model based on the radiative transfer theory for path loss and molecular noise, which takes molecular absorption into account. This section reviews the modeling of the terahertz radio channel based on the radiative transfer theory presented in [11]. Radio communication is affected by the chemical composition of the medium (i.e., the existing molecules in the communication channel) in two different ways in the terahertz band. First, the radio signal is attenuated because molecules in the channel absorb energy in certain frequency bands. Second, this absorbed energy is reradiated by the molecules, which creates noise in the channel. This phenomenon, called molecular absorption, will also be affected by the pressure and temperature of the medium.

The effect of each chemical species on the radio signal is characterized by its molecular absorption coefficient at frequency f, $K(f)$. We now explain how the molecular absorption coefficient can be calculated for a given medium.

1. *Molecular absorption coefficient*: As a medium is a mixture of different species, $K(f)$ can be calculated as

$$K(f) = \sum_{i,g} k(f)^{i,g} \tag{6.1}$$

where f is the frequency of electromagnetic wave and $k^{i,g}$ stands for the individual absorption coefficient for the isotopologue* i of gas g. For example, the constituent gases in the normal

---

* Isotopologues are molecules that differ only in their isotopic composition.

air at temperature 15°C and pressure 1 atm are 78.08% nitrogen, 20.94% oxygen, 0.9% argon, 0.03% carbon dioxide, and others less than 0.001%, such as neon and helium. Water vapor varies between 0.01% and 3%.

Each $k(f)^{i,g}$, in m^{-1}, can be obtained from Equation 6.2, where $Q^{i,g}$ is the molecular volumetric density of isotopologue i of gas g in molecules/m^3, p is pressure, and T is the temperature of the medium.

$$k(f)^{i,g} = \frac{p}{p_{STP}} \frac{T_{STP}}{T} Q^{i,g} \sigma^{i,g}(f) \tag{6.2}$$

where $\sigma^{i,g}$ is the *absorption cross section* for the isotopologue i of gas g in m^2/molecule, and T_{STP} and p_{STP} are the standard temperature and pressure (STP), that is, 273.15 K and 1 atm, respectively. For a given gas mixture, the total number of molecules per volume unit, $Q^{i,g}$, of the isotopologue i of gas g in molecules/m^3, at pressure p and temperature T, is obtained from the ideal gas law as

$$Q^{i,g} = \frac{n}{V} q^{i,g} N_A = \frac{p}{RT} q^{i,g} N_A \tag{6.3}$$

where n is the total number of moles of the gas mixture being considered, V stands for the volume, $q^{i,g}$ is the mixing ratio for the isotopologue i of gas g, NA is Avogadro's number, and R is the gas constant. The absorption cross section, $\sigma^{i,g}(f)$, is calculated as

$$\sigma^{i,g}(f) = S^{i,g}(f)^* \ G^{i,g}(f)$$

where $S^{i,g}(f)$ stands for line intensity, which illustrates the strength of the absorption by a specific type of molecules in a frequency f and can be obtained from some public and private databases, such as HITRAN [65], GEISA, and JPL [66]. HITRAN was originally established by the Air Force Cambridge Research Laboratories (AFCRL) in the late 1960s to predict and simulate the transmission and emission of light in the atmosphere; it is the most valuable database for this purpose.

The spectral line shape, $G^{i,g}(f)$, can be calculated via equations mentioned in [61] based on the HITRAN information. The spectral line shape, $G^{i,g}(f)$, is also influenced by the pressure and temperature of the medium.

HITRAN on the web: Recently, an online tool was developed by the Russian Institute of Atmospheric Optics, and the University of Harvard [67] provides all available HITRAN information in a novel manner. This publicly available tool enables researchers to directly obtain all related parameters for molecular absorption, including absorption coefficient, transmittance spectrum, absorption spectrum, and radius spectrum. This information is available for any arbitrary pressure, temperature level, and frequency based on the HITRAN information for 39 species and 10 standard mixtures, such as the U.S. model for air in mean or high altitude. A snapshot of this online tool is presented in Figure 6.2, which shows the absorption coefficient of carbon monoxide over the terahertz band (wave number from 3.3 to 330 cm^{-1}) in the standard temperature and pressure.

Figure 6.3 demonstrates the absorption coefficient extracted from HITRAN on the web for normal air in two seasons (summer and winter), water vapor, and oxide (OH), an

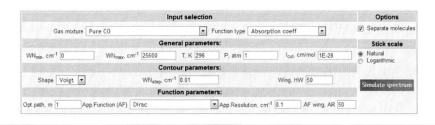

Figure 6.2   Interface of the HITRAN on the web tool. The molecular absorption coefficient can be calculated directly (http://www.hitran.iao.ru/).

intermediate molecule formed in chemical reactors during water formation reactions (O + H→ OH). The average absorption coefficient of water vapor and oxide is two orders of magnitude greater than that of normal air.

2. *Path loss*: The medium absorption coefficient, $K(f)$, determines the attenuation and molecular absorption noise in the radio channel. We first consider attenuation. The attenuation of the radio signal at the terahertz band is due to spreading and molecular absorption. Let $A(f, d)$, $A_{\text{spread}}(f, d)$, and $A_{\text{abs}}(f, d)$ be, respectively, the total attenuation, attenuation due to spreading, and attenuation due to molecular absorption at frequency f and a distance d from the radio source. We have

$$A(f, d) = A_{\text{spread}}(f, d) \times A_{\text{abs}}(f, d) \tag{6.4}$$

$$A_{\text{spread}}(f,d) = \left(\frac{4\pi f d}{c} \right)^2 \tag{6.5}$$

$$A_{\text{abs}}(f, d) = \exp(K(f)d) \tag{6.6}$$

3. *Molecular noise*: There are two main models for molecular absorption noise, which is due to the reradiation of absorbed radiation by the molecules in the channel. The first noise model assumes that the magnitude of the noise is not influenced by the amplitude of the transmitted power, so the molecular noise power spectral density (PSD), $N_{\text{abs}}(f, d)$, is given by [11,61]

$$N_{\text{abs}}(f, d) = k_B T_0 (1 - \exp(-K(f)^* d)) \tag{6.7}$$

where T_0 is the reference temperature 296 K and k_B is the Boltzmann constant. In the second noise model, recently proposed in [68,69], it is assumed that the intensity of the transmitted power affects the noise; therefore, different transmitted symbols create different levels of noise in a given condition. The PSD of molecular absorption noise that affects the transmission of a symbol m, S_{Nm}, is contributed by the atmospheric noise, S_B, and the self-induced noise, S_X:

$$S_{Nm}(f,d) - S_N^B(f,d) + S_{Nm}^X(f,d) \tag{6.8}$$

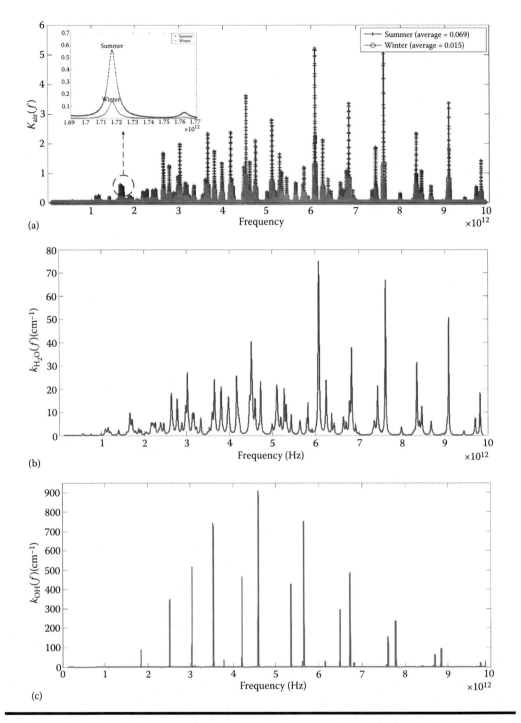

Figure 6.3 Absorption coefficient in cm^{-1} over 0.1–10 THz for standard air, water vapor, and OH, which is an intermediate highly absorbent species in chemical reactors. (a) USA model of the air [67] in mean latitude for both summer and winter. The difference is due to the amount of water vapor in the air in the summer (1.86%) and winter (0.043%). (b) Water vapor, the average is 3.7. (c) Oxide (OH), the average is 3.1.

$$S_N^B(f,d) = \lim_{d \to \infty} (K_B T_0 (1 - e^{-K(f)d})) \left(\frac{c}{\sqrt{4\pi} f_0} \right)^2 \tag{6.9}$$

$$S_{N_m}^X(f,d) = P_m(f)(1 - e^{-K(f)d}) \left(\frac{c}{4\pi d f_0} \right)^2 \tag{6.10}$$

where $P_m(f)$ is the PSD of the transmitted symbol m, and c is the speed of light. In this chapter, we use the first noise model. However, using the second noise model will not affect the contributions and the overall results of this work. Note that neither of these noise models has been experimentally tested due to unavailability of nanoscale transceivers [68].

4. *Channel capacity*: Regarding the achievable capacity of such a propagation model, it has been shown that for normal air composition and using two power allocation schemas, including flat PSD and optimized PSD, for a transmission distance in the order of several tens of millimeters, the terahertz band can be considered a single transmission window (almost 10 THz wide), providing an extremely high-capacity terahertz channel for short distances [11,64]. In this study, the total signal energy is kept equal to 500 pJ, independently of the specific power spectral distribution. As the water molecule is the main absorbent species in normal air, the effect of 1% and 10% water vapor (on the channel) on the capacity has been also investigated, showing, with optimized PSD, the capacity of the channel with 10% is four times lower than the channel with 1% water vapor over the entire terahertz band ranging from 0.1 to 10 THz.

5. *Model's limitation*: The proposed propagation model for WNSN suffers from three main limitations. First, other effective parameters on a wireless signal, such as shadowing, multiple fading, and interference, have not been considered in the model. Second, it seems that the generated molecular noise as a wireless signal can be absorbed by other molecules in the channel; this has not been captured even in the latest noise model. Third, as the nanomaterials are sensitive to environmental conditions, such as pressure and temperature, the effect of these kinds of parameter on the gain of the nanotransmitter should be considered [70]. Authors in [56] and [71] attempted to characterize the effect of temperature on the chemical potential and conductivity of the graphene and its effect on the radiation frequency of the graphene-based plasmonic nanoantennas. However, this model might not be practical for time-varying channels whose pressure and temperature could be dynamic over time. This variation could affect the operating frequency of the nanotransceivers. This should be considered in designing propagation models for time-varying channels.

To sum up, evaluation of the proposed propagation models for WNSNs in practical scenarios, only possible after having the nanoantennas in bulk, is a research gap. Designing more comprehensive propagation models that take all effective parameters of wireless propagations into account is another research issue that should be followed up in the future.

6.3.2 Applications of WNSNs

The microscopic size of nanomotes allows WNSNs to reach and monitor locations that were completely out of reach of conventional macroscale sensor networks. It is expected that nanosensor networks will enable transformational new applications in medical, biological, and chemical fields [10,13,17,72–74]. I. F. Akyildiz and J. M. Jornet have outlined a number of interesting new applications of WNSN in biomedical (health monitoring and drug delivery), environmental (plant monitoring and defeating insect plague), industrial (ultrasensitive touch interfaces), and military (biological and chemical defense) domains [17]. Figure 6.4 presents the schematic of a few proposed applications for WNSNs in the literature [75].

The use of WNSNs in plant monitoring has been proposed in [13] and [76], where the deployment of a chemical sensing nanonetwork in a crop field has been considered to monitor the interaction of the plants with the environment (Figure 6.4b).

Using other nanoscale communication paradigms for nanohealth systems, such as targeted tumor detection and treatment [77,78] and tissue monitoring [33], has also been reported in the literature. Moreover, researchers have shown how communication between synthetic biological and nanotechnological components can be used to amplify in vivo disease targeting [77,78]. It has been demonstrated that such a schema can be used for targeted tumor detection and treatment; it can target more than 40 times higher doses of chemotherapeutics to tumors than noncommunicating schemas. This is a remarkable result for tumor detection and treatment systems [77].

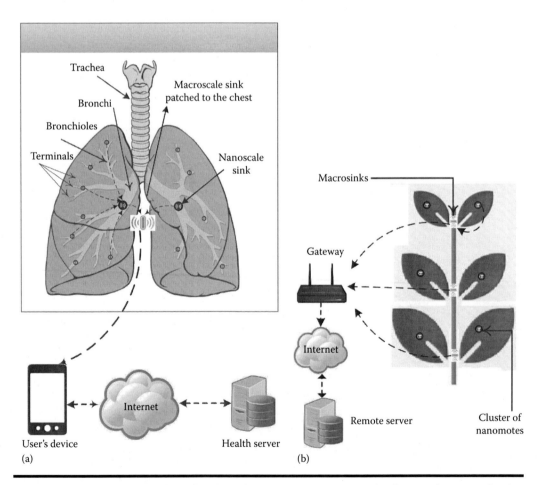

Figure 6.4 A few proposed applications for WNSNs. (a) WNSN for human lung monitoring systems [75]. (b) WNSN for plant monitoring [13]. *(Continued)*

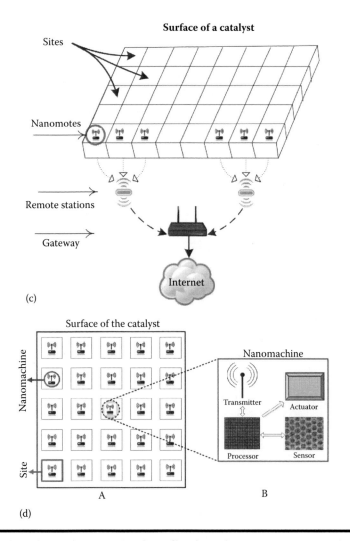

(c)

(d)

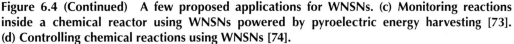

Figure 6.4 (Continued)   A few proposed applications for WNSNs. (c) Monitoring reactions inside a chemical reactor using WNSNs powered by pyroelectric energy harvesting [73]. (d) Controlling chemical reactions using WNSNs [74].

In [73], we have proposed a novel self-powered WNSN architecture for remote detection of chemical reactions. We have shown that *reaction heat* can be used both as a source of energy harvesting for nanosensor nodes and for reaction detection. In principle, remote reaction detection is possible by using the harvested energy to transmit a terahertz pulse of proportional amplitude, because different reactions generate different amounts of energy. In [74] and [79], we have shown how a WNSN could be deployed inside a reactor for bottom-up control of the chemical synthesis, with the ultimate goal of improving the performance of the reactor. Chemical reactors are built to produce some high-value products, but they also generate some low-value materials. The performance of a reactor is measured by its *selectivity*, which refers to the percentage of high-value products in the overall output [80]. By monitoring a reactor at the molecular level and turning off elementary reactions leading to undesired molecular species, a WNSN's enabled catalyst can potentially achieve very high selectivity.

In all aforementioned WNSN applications, distributed reliable communication among nano-motes or between nanomotes and a macroscale sink is envisaged to fulfill the application goal. However, on many occasions, due to variation in the channel condition (its pressure, temperature, or composition), the molecular noise and attenuation would be dynamic over time, and so provid-ing reliable communications would therefore be a challenging task.

6.4 Time-Varying WNSNs

As we discussed in Section 6.3.1, molecular absorption, as a new source of noise and path loss in WNSNs, is highly sensitive to the communication frequency and the channel condition, includ-ing the type of existing molecules in channel, and the pressure and temperature of the medium. In many envisaged applications of the WNSNs, all the channel conditions could be dynamic over time, which leads to a time-varying channel [81]. The aim of this section is to characterize the effect of a WNSN's channel condition variation on the quality of communication. First, we investigate the effect of each individual parameter (composition, pressure, and temperature), and then we introduce a new channel model for time-varying WNSNs followed by a numerical analysis.

6.4.1 Effect of Channel Condition on the Molecular Absorption

Figure 6.5 shows the log-scale representation of the absorption coefficient over the entire terahertz band for three molecules, water vapor (H_2O), nitrogen dioxide (NO_2), and carbon monoxide (CO). The numbers in parentheses are the average absorption coefficient over the terahertz band in cm^{-1}. First, Figure 6.5 shows that molecules might have different absorption spectra. For example, while CO mainly resonates at frequencies less than 5.5 THz, water molecules absorb energy over the entire terahertz band. Second, the amplitude of the absorption is different for different mol-ecules, so the terahertz band is differently affected by various molecules. For example, the average

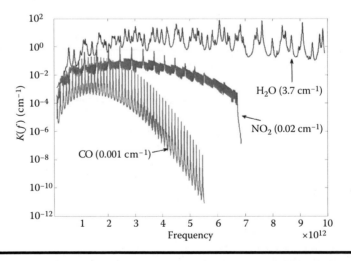

Figure 6.5 **The log-scale spectrum of molecular absorption coefficient (in cm^{-1}) for three chemical species over the terahertz frequency band ranging from 0.1 to 10 THz.**

absorption coefficients of water vapor and carbon monoxide molecules are 3.7 and 0.001 cm⁻¹, which show a difference of four orders of magnitude.

To study the effect of the temperature and pressure of the medium on molecular absorption, we extract the $K(f)$ for water vapor as an example at three different pressures (1, 20, and 50 atm) and temperatures levels (273.15, 400, and 500 K), depicted in Figure 6.6. Figure 6.6a shows that increasing pressure increases absorption at a given temperature. The average absorption coefficients over the entire terahertz band are 3.7, 36, and 168 cm⁻¹ for pressures of 1, 10, and 50 atm, respectively, so the average absorption over the terahertz band increases with the pressure of the medium. On the other hand, Figure 6.6b shows the effect of temperature. While the average absorption of water vapor at a temperature of 273.15 K over the terahertz band is 40 cm⁻¹, it drops

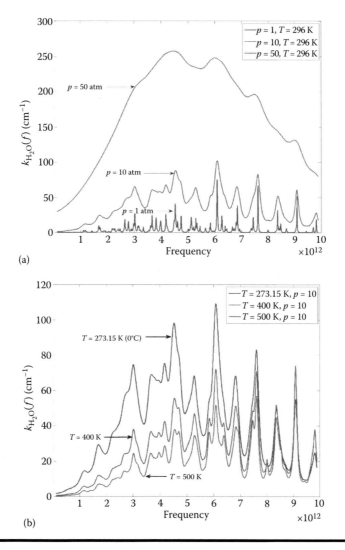

Figure 6.6 Effect of pressure and temperature of the medium on the molecular absorption. (a) Increasing the pressure intensifies molecular absorption at a given temperature. (b) Temperature has a reverse effect on the molecular absorption.

to 18 cm^{-1} for a temperature of 500 K, which means that temperature has an inverse effect on the absorption coefficient.

6.4.2 Composition-Varying Channels

In many applications in which WNSNs are envisaged to be used, the channel condition (composition, pressure, and temperature) could be dynamic over time, which means the molecular absorption, and thus noise and path loss, would be variable over time. We refer to this type of channel as *time-varying WNSN*. For example, a WNSN monitoring a chemical reactor might experience different types of molecules during the synthesis, due to the occurrence of different reactions. Figure 6.7 shows three snapshots of the simulation of a given chemical reactor. The reactor starts with only carbon and hydrogen molecules, but after a few steps, it contains more than 15 different types of molecules. In addition, the temperature of a chemical reactor could also be variable because the exothermic and endothermic reactions release and consume heat, respectively, changing the temperature of the medium.

Table 6.1 highlights the key parameters that lead to a dynamic channel in a few time-varying WNSNs. It also presents the origin of the variation in channel condition and introduces the most dominant molecule that holds the highest average absorption coefficient in each application.

In this chapter, we study composition-varying WNSNs as a subclass of time-varying WNSNs, whose type and concentration of molecules in the channel are dynamic over time. We investigate two case studies, including chemical reactor monitoring and human body health monitoring systems, both using WNSNs.

1. *Monitoring chemical reactor*: Microscopic monitoring of chemical reactors at the molecular level using WNSNs has been recently demonstrated in the literature [74,79]. In order to make our discussion concrete, we consider a WNSN that has been used to monitor the progress of a Fischer–Tropsch (FT) synthesis. FT synthesis is a major process for converting natural gas to liquid hydrocarbons in a batch chemical reactor. The reactor starts with a specific amount of carbon monoxide, CO, and hydrogen, H$_2$. Many chemical species are produced and consumed, via many different chemical reactions, during the synthesis. The synthesis stops when no more new products are produced.

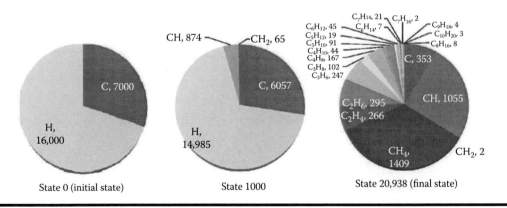

Figure 6.7   Three compositions of a given chemical reactor over time; obtained from our SSA simulation.

Table 6.1   Few Examples for Applications of WNSNs Whose Terahertz Channel Is Time Varying

	Chemical Reactor Monitoring	*Human Cell Monitoring [17,82]*	*Plant Monitoring [13]*
Composition	✓	✓	✓
Pressure	✓	✓	✓
Temperature	✓	–	✓
Timescale	Picosecond	Second	Minute
Source of variation	Occurrence of different reactions	Mainly respiration process and blood pressure	Photosynthesis and respiration, dehydration and rehydration
Dominant molecule	OH	H_2O	H_2O

✓Confirms variation.

The chemical composition within an FT chemical reactor changes over time. We use the stochastic simulation algorithm (SSA) [83], which is a standard algorithm to simulate chemical reactions to study the variation of composition during FT synthesis. Figure 6.8 shows the mole fraction, that is, concentration of six selected species over time during the course of FT catalysis with initial composition consisting of 500 molecules of CO and 1200 molecules of H_2. It shows that the chemical composition in the reactor changes over time.

2. *Nanomote monitoring human body cells*: In this section, we want to demonstrate the existence of the time-varying property of the terahertz channel for intrabody WNSN applications. Human bodies, as the communication channel between nanosensors, could have

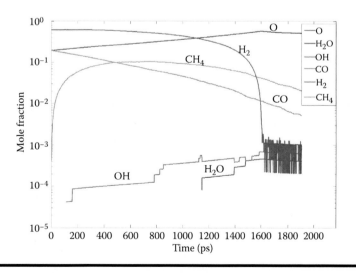

Figure 6.8   Evolution of concentration of different molecules during an FT process. CO and H_2 are consumed while other species are gradually produced.

different compositions depending on the type of body. These compositions could also vary over the time. For example, the amounts of some species (specifically oxygen, carbon dioxide, CO_2, and water molecules) change during the respiration process in all human body cells. The blood circulatory system circulates the blood through the body, carrying oxygen to the cells and collecting the excess water and CO_2 that has been produced during cell respiration, and carries them to the alveolus cells in the lungs (Figure 6.9) [84]. This process affects the composition of most of the living cells. To get more insight into the effect of variations in the human body on the quality of the terahertz communication, we assume a scenario that two nanorobots need to communicate with each other within the blood [85]. The communication channel is therefore mainly affected by the blood composition, which is dynamic over time due to respiration process. Although the chemistry of the blood composition has a complicated form, for the sake of terahertz communication, we consider three main reactions that affect the mole fraction of the most absorbent molecule in the blood, that is, water. Table 6.2 shows these reactions. During the cell respiration (R_1), water will be produced. A fraction of this water then will be used by the blood bicarbonate buffering system $(R_2$ and $R_3)$ to regulate the pH of the blood, and the rest will be carried to the lung to be exhaled [86].

We simulate these reactions via SSA. We assume the initial blood is composed of 7% oxygen (from inhalation), 80% water, 7% glucose (absorption from the intestine), and 6% other molecules. Figure 6.10 shows the evolution of species during one breathing cycle. It shows that the mole fraction of the water increases toward the inhalation and then approaches the initial amount at the end of exhalation because the extra produced water (during the cell respiration) will be exhaled from the lung.

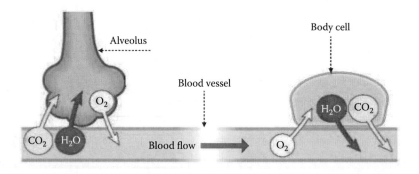

Figure 6.9 Respiration process affects the composition of body cells, the blood itself, and the alveolus cells in the lung. (From The Respiratory System. Available at http://leavingbio.net /respiratory system/the respiratory system.htm.)

Table 6.2 Simple Model for Variation of the Blood Composition

Cell respiration	R_1: $(CH_2O)_6$ (Glucose) + 6 O_2 → 6 CO_2 + 6 H_2O + Energy
Buffering system	R_2: CO_2 + H_2O → H_2CO_3
	R_3: H_2CO_3 + H_2O → H_3O^+ + HCO_3^-

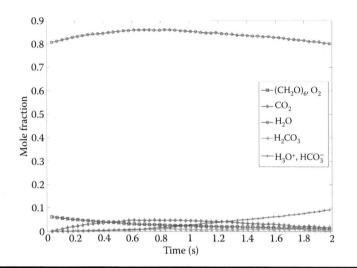

Figure 6.10 Variation of the blood composition over one respiration cycle.

6.4.3 Channel Modeling for Composition-Varying WNSNs

This section introduces the channel modeling of time-varying WNSNs based on the radiative transfer theory presented in [11]. We assume the radio channel is a medium consisting of N chemical species S_1, S_2, ..., S_N. The effect of each chemical species S_i on the radio signal is characterized by its molecular absorption coefficient $K_i(f)$ of species S_i at frequency f, which can be obtained from the HITRAN database [67]. We consider a radio channel in a medium that has time-varying chemical composition. Let $m_i(t)$ be the mole fraction of chemical species S_i in the medium at time t. The medium absorption coefficient $K(t, f)$ at time t and frequency f is a weighted sum of the molecular absorption coefficients in the medium:

$$K(t,f) = \sum_{i=1}^{N} m_i(t) K_i(f) \qquad (6.11)$$

The total attenuation, attenuation due to spreading, and attenuation due to molecular absorption at time t, frequency f, and a distance d from the radio source is [11]

$$A(t,f,d) = e^{K(t,f)d} \times \left(\frac{4\pi f d}{c} \right)^2 \qquad (6.12)$$

The molecular absorption noise, $N_{abs}(t,f,d)$, which is due to the reradiation of absorbed radiation by the molecules in the channel, is given by [11]

$$N_{abs}(t,f,d) = k_B T_0 (1 - e^{-K(t,f)^* d}) \qquad (6.13)$$

where T_0 is the reference temperature 296 K and k_B is the Boltzman constant.

Let $U(t, f)$ be the PSD of the transmitted radio signal at time t and frequency f. The signal-to-noise ratio (SNR) at time t, frequency f, and distance d is

$$\text{SNR}(t, f, d) = \frac{U(t, f)}{A(t, f, d)N_{\text{abs}}(t, f, d)} \tag{6.14}$$

6.4.4 Reliability Analysis

In this section, we study the reliability of a communication for both case studies, that is, terahertz WNSN within human blood and chemical reactors.

1. *Methodology*: In the first case, we use the simulation setup that we used in Section 6.4.2 to calculate the evolution of blood composition over time. For the second one, we consider an FT reactor with initial feeding molecules of 500 carbon monoxide and 1200 hydrogen molecules. The chemical production continues until no more new chemicals can be produced. Molecular absorption coefficients of the chemical species of both the blood and the FT reactor are obtained from the HITRAN database [67]. Then, we follow the procedure in Section 6.4.3 to compute the medium absorption coefficient, attenuation, molecular noise, and SNR. The bit error rate (BER) is calculated using the recently proposed modulation schema for WNSNs, time spread on-off keying (TS-OOK) [87]. The transmitted power has been set to 1 pW (10–12 W), and the distance between two nanomotes is 1 mm.

Results: First, we investigate the reliability of communication within the chemical reactor. Figure 6.11 shows the SNR and BER during the FT synthesis, and as it can be seen, both are fluctuated. SNR starts from 3.2 dB at the beginning of the catalysis, reaches 7.8 dB in $t = 600$ ms, and finally drops to less than –6 dB at the end of the synthesis, which makes BER fluctuate between 0.05 and 0.4.

Figure 6.12 presents the SNR within the blood over one respiration cycle. As can be seen, because the water has the highest molecular absorption coefficient among other species, the SNR follows its reverse pattern, that is, falls when the water increases and vice versa. It starts at 13 dB and gradually decreases to around 4 dB in the middle of the respiration process, and then returns to the initial value at the end of cycle. This pattern is modulated by the respiration process, so we expect that it would be repeated over time. In order to investigate this hypothesis, we calculate the corresponding BER for an adult person with respiration rate of 15 cycles per minute. As can be seen from Figure 6.13, the BER fluctuated over time follows a fixed pattern, which is modulated by the respiration process.

6.4.5 Discussion and Future Work

Given that the terahertz channel is critically dependent on the chemical composition of the medium, how to guarantee reliable communication in environments that exhibit time-varying composition is challenging. Intuitively, WNSNs might be required to implement some form of adaptive communication to guarantee reliability in such environments. However, adaptive communication requires accurate channel estimation, which is challenging at the nanoscale due

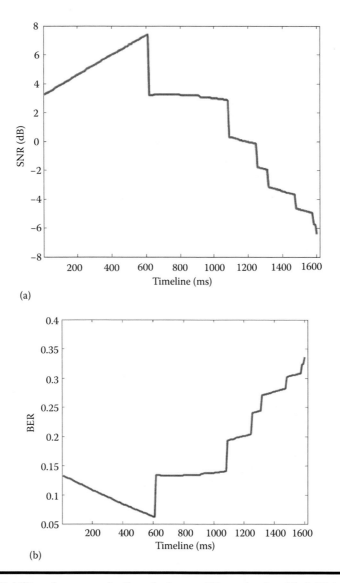

Figure 6.11 **Reliability of communication during an FT synthesis with initial CO = 500, H = 1200. (a) SNR, (b) BER.**

to the extremely limited resources of nanomotes. Recently, some offline adaptation techniques, such as open-loop power adaptation and offline dynamic frequency hopping, have been proposed to address this issue [72,88,89]. These techniques use the prior knowledge of the channel to derive the offline policies. However, in some applications, the prior knowledge of the channel might not be available, or it may be difficult to obtain. In such cases, we need to design a system to perform online estimation of the channel state, but how to do this is a challenge. In addition, designing a mechanism that allows the nanomote to self-synchronize the transmission parameters with the prior offline channel estimation is still an open problem, which is considered a future work.

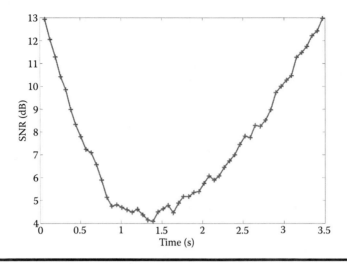

Figure 6.12 **SNR within the blood in one respiration cycle.**

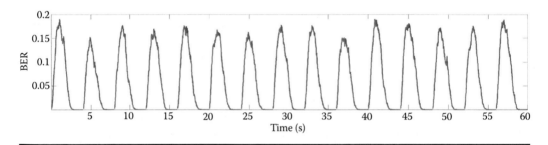

Figure 6.13 **BER within the blood for an adult person (18 years, respiratory rate = 15 per minute).**

6.5 Communication Protocols for WNSNs

A recent insightful article by I. F. Akyildiz and J. M. Jornet [17] has shown how phenomenal progress in nanotechnology has brought us closer to realizing WNSNs. Technically, a WNSN is a network of nanoscale devices capable of some basic computing, sensing, actuation, and communication tasks. Due to the size and energy constraints of nanodevices and also high molecular absorption noise and attenuation in the WNSN channel, designing simple and energy-efficient communication protocols that take molecular absorption into account is an active ongoing research area in WNSNs. Nevertheless, as WNSNs are in an early stage of development and have not yet been experimentally tested, most work is focusing on mathematical modeling and conceptual designing, mostly for the physical and data link layers. In this section, we overview the main efforts in designing communication protocols for WNSNs.

6.5.1 Modulation and Coding

In this section, we investigate the state of the art in designing modulation and coding techniques for WNSNs. We also review the only proposed receiver architecture for nanomotes.

1. *Modulation schema*: Size and energy constraints of nanomotes impose the need to use carrier-less pulse-based modulation techniques for WNSNs, since it is technologically challenging for such resource-restricted nanomotes to generate a continuous high-power carrier frequency in the terahertz band, which is the operating frequency of nanotransceivers [11]. As recently proposed, GNR-based nanotransceivers [12,56] can generate extremely short pulses; an alternative option is to employ well-known pulse based communication (PBC) techniques that have efficiently been used in high-speed short-range communication, such as ultra-wide-band (UWB) systems and tracking and positioning systems, such as *radar*, in the last decade [90,91]. The use of PBC in wireless sensor networks has also been indicated [91,92].

A PBC system relies on exchanging short pulses between transmitters and receivers to demonstrate different symbols in the channel. The main advantages of PBC systems are noise immunity, high data rate, simplicity, low power consumption, and low equipment cost. The promising characteristics of PBC, specifically energy efficiency, simplicity, and high reliability, also make it an appropriate candidate for designing WNSN communication protocols. Several PBC schemas in the literature can be adopted for WNSN, such as OOK, pulse amplitude modulation (PAM), pulse position modulation (PPM), and binary phase shift keying (BPSK). Figure 6.14 demonstrates the pulse train used by each of these

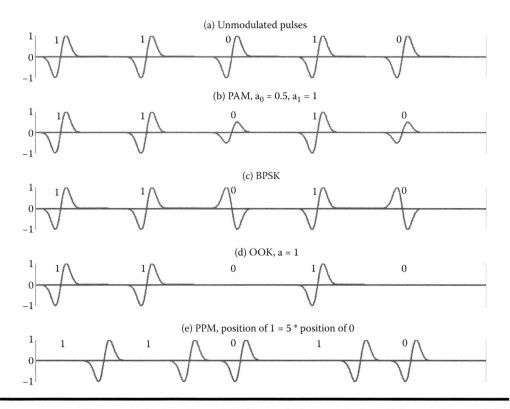

Figure 6.14 Different pulse-based modulation techniques (PBC), which may be used for WNSNs, transmitting a stream of 11010 using a second derivation of Gaussian-shaped pulses. PAM varies the amplitude of the pulses, OOK employs a full pulse and a silence, PPM adds a small random offset to each pulse, and finally, BPSK uses a full pulse and its opposite phase to represent 1 and 0, respectively.

four techniques to transmit a stream of 11010. These modulation schemes have different characteristics and show different performance levels in terms of BER, complexity, energy efficiency, and capacity. For instance, if lower complexity is required, the best option is OOK [90]. On the other hand, if robustness against error and higher power efficiency are the issues to consider, BPSK is the best candidate. In the first attempt to propose modulation schemas for WNSN, OOK was extended—TS-OOK—and operated based on exchanging femtosecond-long pulses between nanomotes [61]. TS-OOK employs a full pulse to represent 1 and no transmission for 0. It has two significant differences with normal OOK. The first difference is that it benefits from initialization preambles and constant-length packets to prevent confusion between transmission of 0 and no transmission and to alleviate the synchronization problem, which improves the performance of OOK in multiaccess scenarios. The second difference is that the time between two consecutive transmissions is significantly larger than the duration of a pulse, allowing a nanomote to process another nanomote's bit streams or transmit its own data. An extended version of this modulation, called RD TS-OOK, introduced in [93], is able to mitigate the collision in multiaccess scenarios more efficiently.

Detecting such short pulses highlights the need for highly sensitive and ultrasimple low-power receivers. While J. Pujol [94] proposed a simple noncoherent receiver for TS-OOK, able to differentiate very short pulses in the terahertz band (Figure 6.15), a more efficient novel receiver architecture for TS- OOK was proposed by R. G. Cid-Fuentes et al. [95], which uses a continuous-time moving average (CTMA) symbol detection scheme. This later receiver bases its symbol decision on the received signal power maximum peak after the CTMA and can be implemented with a single low-pass filter.

Nevertheless, TS-OOK suffers from a few drawbacks. In self-powered WNSNs, in which nanomotes use harvested power from the embedded environment to transmit the data, a silence due to power outage (i.e., there is not enough power in the capacitor to transmit a 1) can be misinterpreted as 0 in the receiver. And although OOK, the original version of TS-OOK, is a simple schema with the lowest contribution in molecular absorption due to using silence to present 0, other PBC systems, such as BPSK, may provide better reliability with higher energy efficiency [96]. In [97], we have investigated the performance of four pulse-based modulation schemas for WNSNs, including PAM, PPM, OOK, and BPSK. Our investigation shows that although OOK is the simplest schema, its reliability performance is lower than that of BPSK. The BPSK has the highest performance and also the highest energy efficiency, but it needs more complex transceivers than the other schemes. The BER performance and complexity of PPM is similar to those of OOK, but its energy efficiency is better than that of OOK.

2. *Interference mitigation and channel coding*: It is expected that an enormous number of nanodevices would be required to accomplish a task at the nanoscale [98,99]. For example, a

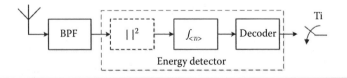

Figure 6.15 **A simple noncoherent scheme, that is, energy detector receiver for TS-OOK. (From J. C. Pujol, Bridging PHY and MAC layers in wireless electromagnetic nanonetworks, Final year project from Technical University of Catalonia, Spain, 2010.)**

network of trillions of nanomotes is required to monitor and potentially control a chemical reactor at the molecular level. On the other hand, the random nature of the nanomotes makes the coordination of these nanomotes unfeasible and creates potential high interface in the channel. This randomness might be originated by the events that are the subject of the WNSNs or might be caused by the modulation schema. For example, using TS-OOK as the modulation schema, nanomotes can start transmitting at any specific time in an uncoordinated manner, creating collisions between symbols that will be translated to the interference and limitation in the capacity of the channel. To alleviate this problem, J. M. Jornet and I. F. Akyildiz proposed a low-weight channel coding as a novel mechanism to reduce interference in WNSNs [100]. This approach aims to mitigate the interference and reduce the number of transmission errors by choosing the appropriate weight of a code, that is, the average number of bits equal to 1 in a codeword, rather than using channel codes to detect and correct transmission errors. The numerical results show that the overall interference can be reduced while the achievable information rate is improved in an interference-limited scenario. In [101], the original framework was extended to calculate the optimal code weight, analyze the codeword error rate (CER), and validate the numerical results via COMSOL simulation.

Although this approach can be used to mitigate the interference in many scenarios, it can only be applied when the type of transmitted data is predetermined, so codewords are definable. However, in the random and unknown environments where WNSNs might be used, this approach may not be efficient. For example, due to the stochastic behavior of chemical reactions, defining codewords for a nanomote that monitors the type of reactions might not be feasible.

6.5.2 Frequency Band Selection Strategies

As we discussed earlier, the noise and attenuation in the terahertz band are frequency sensitive due to sensitivity of molecular absorption to the frequency (see Figure 6.5). That means, first, for a given composition, different frequency regions absorb the energy differently, and second, this absorption pattern over the frequency is composition sensitive; that is, different compositions have different absorption spectra. As a result, selecting an appropriate frequency range for a given channel composition, that is, the subband with the lowest average absorption, can increase SNR, so the communication distance can be improved. However, based on Shannon's well-known capacity formula, selecting a subband of the terahertz region can decrease the capacity of the channel. Therefore, selecting an appropriate frequency region for different environments and investigating the trade-off between the communication distance and the capacity has been investigated [10,13,76,102].

I. T. Javed and I. H. Naqvi [10] proposed some beneficial narrow subchannels for WNSNs in different environments with different compositions, such as various locations in the human body and air. They divide the terahertz band into a number of subchannels, and in each type of medium, they investigate the performance of different subchannels. Finally, they suggest the best subchannel for communication. For example, they have shown that the best subchannel in a given human body is 0.1–0.5 THz, with an average attenuation of 0.02 dB, that is, 400 times smaller than the attenuation caused when the whole terahertz (0.1–10 THz) is used for communication.

In another recent study, deployment of a chemical sensing WNSN in a crop field to monitor the interaction of plants with the environment was examined [13,76]. Four different frequency selection strategies were investigated to maximize available transmission bandwidth while balancing nanomote assignment to a limited number of microdevice sink nodes. Because of the variation in the moisture content of the vegetation, they evaluated the performance of different frequency selection strategies at different levels of moisture ranging from 0.1% to 50%. They have shown that for a medium with less than 10% moisture, the proposed frequency selection strategies can deliver an average successful transmission probability of 100% when 100 nanomotes categorized in five clusters try to communicate with the sinks. However, this falls dramatically to less than 70% when the moisture level is higher than 15%.

In addition, it has been shown that the terahertz communication in normal air is limited to a few millimeters [11]. In [102], the authors attempted to increase this distance by identifying an *absorption-free* subband, that is, the maximum possible frequency range with near-zero absorption. They showed that by using the frequency range of 0.1–0.54 THz, the transmission range can be increased up to 10 m using only 0.1 aJ (10^{-18} J) energy per symbol. However, the capacity drops from 2 Tbps when the entire terahertz band is used with the same energy per symbol (0.1 aJ) to 10 Mbps when the absorption-free region is used.

To sum up, the main focus of these works is the selection of the best terahertz frequency region for a given channel composition. However, any effective solution for time-varying WNSNs, in which the channel conditions change constantly over time, has not been proposed. For example, a WNSN that is monitoring the conditions of a chemical reactor will be exposed to different channel conditions due to the production and consumption of different chemical species over time. Therefore, these static frequency selection approaches cannot efficiently operate for time-varying WNSNs.

6.5.3 *Medium Access Control*

As with conventional wireless sensor networks (WSNs), enabling multiaccess scenarios in WNSNs requires medium access control (MAC) protocols to moderate concurrent transmissions among nanomotes and to efficiently regulate access to the channel. However, due to restriction and specification of the WNSN, the conventional MAC protocols cannot be directly applied to WNSNs, so novel approaches are required. First, very high network density is one of the specifications of WNSNs [98,99]. Although in all-to-all broadcasting the radio range can be regulated for proper distances to reduce the probability of collision, the number of nodes is still a challenge compared to conventional WSNs. In addition, in some applications, such as monitoring chemical reactors at the molecular level, an extremely high message rate is another aspect of the WNSNs. Finally, the MAC design should be more sensitive to the energy of nanomotes because at this scale, power supplies such as nanobatteries and energy harvester interfaces are significantly restricted; a well-defined MAC protocol can preserve the total energy of networks by reducing the control messages and number of collisions and packets lost.

J. M. Jornet et al. attempted to address the multiaccess problem in WNSNs for the first time by suggesting the first MAC protocol for WNSN, called the PHLAME [93]. The proposed framework design is based on a new modulation schema called RD TS-OOK, an extended version of TS-OOK [61]. PHLAME uses the advantages of the low-weight channel coding protocol proposed in [100]. By reducing the code weight, that is, encoding the information using more logical 0's than logical 1's, both molecular absorption noise and interference can be mitigated without affecting the throughput of the WNSNs (information rate). PHLAME provides joint selection of the optimal communication parameters and channel coding scheme for both the sender and the receiver.

Although PHLAME tries to minimize collisions in the network and maximize the probability of successfully decoding the received data packets, it suffers from a few potential drawbacks. First, the network scalability might be an issue for PHLAME, as performance of the proposed approach has only been investigated for a maximum of 10 nanomotes per square millimeter, which might not be a realistic scenario for some applications, such as monitoring chemical reactions at the molecular level, which need thousands of nanomotes per square millimeter. Second, PHLAME relies on the low-weight channel coding, which might not be applicable for WNSNs in some environments where the codeword is hard to define for the transmitted data due to natural randomness in the environment. Finally, the *handshake process* in the PHLAME can exhaust the energy of nanomotes, limiting the throughput of the network, and nanosensors might not have enough computational resources to dynamically find the optimal communication parameters.

To alleviate the last restriction of the PHLAME, P. Wang et al. extended the original PHLAME to increase the throughput and lifetime of WNSNs by jointly optimizing the energy harvesting and consumption processes [103]. In other research, G. Piro et al. attempted to adopt the existing MAC protocols, mainly ALOHA and carrier sense multiple access (CSMA), for WNSNs and investigate their performances [104].

In summary, it seems there is room to design more efficient MAC protocols that can deal with extremely high-density WNSNs.

6.5.4 Higher-Layer Protocols

Some initial attempts to investigate and design other required communication protocols for WNSNs, such as routing schemas, energy models, and security issues, will be discussed in this section.

1. *Routing*: On top of the previous energy-aware MAC protocol [103], M. Pierobon et al. proposed the first routing protocol for WNSNs to optimize the use of harvested energy to guarantee continuous operation of WNSNs while improving the overall network throughput [69]. Due to the resource restriction of the nanomotes, in this work, a hierarchical cluster-based structure is used in which the WNSN is formed from a few clusters, and each cluster is coordinated by a nanoscale controller, a nanodevice with more advanced capabilities than a nanomote. The proposed architecture attempts to offload the network operation complexity from the individual nanomotes to the nanocontrollers.

   When a nanomote wants to transmit data, the protocol employs a metric that is called probability of saving energy (PSE) to either enable multihop communication through other nanomotes or direct transmission to the sink. This mechanism offers a framework to calculate the PSE at the nanoscale controller before the transmission of the data starts. It allows multihop communication only when there is a reasonable PSE, which can finally improve the overall throughput of the network. In multihop communication, the nanocontroller imposes an optimized transmission range to the source nanomote to efficiently determine the distance of the next hop. The performance of this framework has been numerically evaluated in terms of energy, capacity, and delay over many physical distances between nanomotes and the nanocontroller. The results confirmed that using the proposed routing mechanism, multihop communication can significantly save the energy of the network compared to the single-hop communication. For example, at a distance equal to 10 mm, while the single-hop communication requires about 6×10^{-21} J energy per nanomote to successfully deliver a message, the proposed routing protocol can achieve the same throughput with only 2.5×10^{-21} J, which shows around a 55% energy savings.

2. *Energy-efficient protocols*: The limited size of nanomotes imposes a limited capacity on nano-batteries. Manually recharging or replaceing nanobatteries is also impractical, so nanoscale energy harvesting systems are considered the most viable solution for WNSNs. Energy harvesters convert different forms of energy, such as vibration, mechanical stress, and heat, to electricity. However, due to the stochastic behavior of events that generate the required input for an energy harvester, the energy harvesting process is not deterministic. On the other hand, the energy consumption of nanomotes is also stochastic, highlighting the need to jointly optimize the energy production and consumption in WNSN, to maximize network lifetime. To address this issue, J. M. Jornet and I. F. Akyildiz proposed a mathematical energy model for self-powered WNSNs that can capture the correlation between the energy harvesting and energy consumption processes and attempts to jointly optimize them [105,106]. By knowing the distribution of the energy harvesting and consumption process, nanomote energy can be modeled as a nonstationary continuous-time Markov process that captures the evolution of the energy states of nanomotes over time. J. M. Jornet and I. F. Akyildiz mathematically obtained an optimal packet size and number of retransmissions that minimize the end-to-end delay and maximize the network throughput.

J. M. Jornet and I. F. Akyildiz's work can be considered a valuable first attempt to increase the lifetime of self-powered WNSNs. However, it has a few limitations. First, the model assumes a Poisson distribution for the energy harvesting arrival. This might not be the case for all types of energy harvesting, as the amplitude and the arrival time of the harvested power are directly governed by the behavior of the initial source of the energy, such as vibration or heat variation. Second, to employ this model, the distribution of the harvesting process should be known and obtainable, which is not practical in some environments where the events are completely random. Third, the leakage in the nanocapacitor has not been taken into account. Finally, in the energy consumption process, only the consumption of the radio transmitter has been considered; the energy consumption of other components, such as the sensor, memory, and processor, has not been modeled.

3. *Security*: Given the potential future applications of WNSNs in a variety of fields, such as the biomedical, military, and food industries, it is prudent to assume malicious agents will try to adversely affect nanoscale communication. The consequences of not doing so could be serious. For example, compromising nanomotes that try to deliver a drug to a patient's cells might harm or kill the person.

F. Dressler and F. Kargl tried to provide some first insights into security issues in WNSNs and highlighted some of the open research challenges in this area [107,108]. They found that existing security and cryptographic solutions might not be applicable to WNSNs due to their restrictions and specifications.

6.5.5 Efficient Communication Protocols for Time-Varying WNSNs

As we discussed in Section 6.4, the reliability of communication in time-varying WNSNs is challenging due to the dynamic behavior of molecular absorption in their channels. In this section, we overview our recently proposed approaches to overcome highly dynamic molecular absorption to establish highly reliable communication in time-varying WNSNs [72,88,89].

1. *Adaptive power allocation schema*: In time-varying WNSNs, if the nanosensors choose a very high power so that they can overcome the worst possible absorption, this high power creates a high interference when the channel is good. Similarly, the nanosensors cannot choose a

low power that is only suitable for good channel conditions because the nanosensors will not be able to communicate when the channel is bad. For autonomous WNSNs, which are powered by limited-capacity nanobatteries or limited-throughput energy harvesting circuits [17], a better strategy is to adjust the power over time to maximize the selectivity with a minimal amount of power consumption. In [72], we investigate a WNSN that is used to improve the selectivity of Fischer–Tropsch synthesis (FTS) [80] and show that the optimal power allocation strategy can be obtained by formulating the problem as a Markov decision process (MDP). However, the MDP solution requires the nanosensors to know the chemical composition of the reactor at any given time, which is not practical. We therefore propose an alternative for the nanosensors to adjust power following a rule or heuristic computed offline. We have proposed three local policies that have been obtained based on the channel approximation via extensive simulations. By simulating the process details, we demonstrate that the proposed heuristics perform well relative to MDP optimization. Figure 6.16b shows the power–selectivity trade-off for the constant power allocation, MDP, and the three local policies compared to the average selectivity for the uncontrolled FTS. The figure shows that all five power allocation schemes achieve better selectivity than the uncontrolled FTS. As expected, the optimal solution gives the highest selectivity for a given power budget and outperforms all the other schemes.

2. *Frequency hoping*: Different molecules resonate and absorb terahertz at different frequency subbands. Therefore, for a given species composition in the wireless channel, some subchannels are affected less severely than others. This means in time-varying mediums, a subchannel that has low noise at one time may become very noisy at another time, and vice versa. Figure 6.17a plots the molecular absorption coefficient during an FTS for different frequencies as a heat map, where the areas with a high medium absorption coefficient are shown as hot. The hot areas correspond to time–frequency regions where the medium has a high absorption coefficient, and communication between nodes in the WNSN will be difficult. The figure shows that different frequency regions become hot at different times.

In [88], we propose *frequency hopping* as a means to overcome the problem of time-varying noise and attenuation in time-varying channels. We formulate the subchannel selection

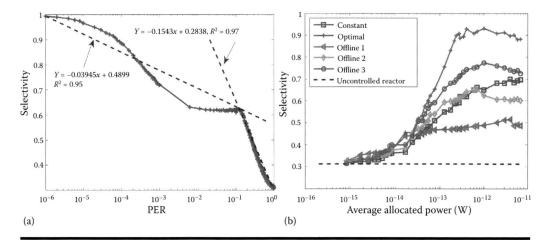

Figure 6.16 (a) Overall selectivity as a function of packet lost probability *p*. (b) Selectivity achieved by different power allocation policies for a given power budget.

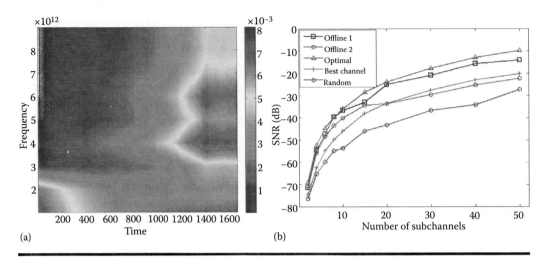

Figure 6.17 **(a) Frequency–time variation of molecular absorption coefficient for an FTS channel. (b) Achievable SNR via different policies versus number of subchannels.**

problem at any given time as an MDP, which allows us not only to optimize the communication performance over the entire reaction process, but also to control the rate of channel switching by imposing different penalties for hopping because constrained WNSN nodes may not be able to switch frequencies rapidly. We show that, compared to conventional non-hopping channel selection, the proposed frequency hopping can significantly improve the signal-to-noise ratio (SNR), capacity, and bit error rate (BER) in time-varying mediums. We propose practically realizable offline policies that obviate the need for observing the channel states, yet perform close to the MDP-based optimal solutions.

Figure 6.17b shows the SNR that can be obtained via different channel selection policies for a power budget of 1 pW and a distance of 1 m between the transmitter and receiver for different available subchannels ranging from 2 to 50. As expected, with an increasing number of subchannels, the SNR increases irrespective of the policy, because the total power is distributed over a smaller range of frequencies, boosting the signal power. The most important result is that the MDP improves SNR by around 10 and 20 dB in comparison with the best channel and random policies, respectively, irrespective of the number of subchannels.

6.6 Conclusion

This chapter surveys the emerging research area of wireless nanoscale sensor networks (WNSNs) to assess the state of the art, its potential, and the outstanding challenges that remain. The survey indicates that WNSNs have enormous potential to improve many existing industrial, biological, and chemical processes. The research area is relatively new, but growing. Researchers have already made some fundamental contributions. In particular, it is now believed with reasonable confidence that WNSNs will communicate in the terahertz band. Researchers have also derived the propagation model for this band, which shows that communication in this band is critically dependent on the chemical (molecular) composition of the channel. Therefore, providing reliable communication in environments that exhibit time-varying composition is a challenging task that has been investigated. These early discoveries have laid the foundation for further research in developing efficient communication protocols suitable for WNSNs.

References

1. J. Villatoro and D. Monzón-Hernández, Fast detection of hydrogen with nano fiber tapers coated with ultra thin palladium layers, *Optics Express*, vol. 13, no. 13, pp. 5087–92, 2005.
2. C. R. Yonzon, D. A. Stuart, X. Zhang, A. D. McFarland, C. L. Haynes, and R. P. Van Duyne, Towards advanced chemical and biological nanosensors—An overview, *Talanta*, vol. 67, pp. 438–48, 2005.
3. L. Novotny, R. X. Bian, and X. S. Xie, Theory of nanometric optical tweezers, *Physical Review Letters*, vol. 79, no. 4, pp. 645–648, 1997.
4. C. Falconi, A. Amico, and Z. Lin Wang, Wireless joule nanoheaters, *Sensors and Actuators*, vol. 127, no. 1, pp. 54–62, 2007.
5. C. Li, E. T. Thostenson, and T.-W. Chou, Sensors and actuators based on carbon nanotubes and their composites: A review, *Composites Science and Technology*, vol. 68, no. 6, pp. 1227–1249, 2008.
6. R. Shankar, T. K. Ghosh, and R. J. Spontak, Mechanical and actuation behavior of electroactive nanostructured polymers, *Sensors and Actuators A: Physical*, vol. 151, no. 1, pp. 46–52, 2009. Available at http://www.sciencedirect.com/science/article/pii/S092442470900003X.
7. I. F. Akyildiz, F. Brunetti, and C. Blázquez, Nanonetworks: A new communication paradigm, *Computer Networks*, vol. 52, no. 12, pp. 2260–2279, 2008.
8. B. Atakan and O. B. Akan, Carbon nanotube-based nanoscale ad hoc networks, *IEEE Communications Magazine*, vol. 48, no. 6, pp. 129–135, 2010.
9. C. T. Chou, Molecular circuits for decoding frequency coded signals in nano-communication networks, *Nano Communication Networks*, vol. 3, no. 1, pp. 46–56, 2012.
10. I. T. Javed and I. H. Naqvi, Frequency band selection and channel modeling for WNSN applications using simplenano, in *Proceeding of the 2013 IEEE International Conference on Communications (ICC)*, Budapest, Hungary, June 2013, pp. 5732–5736.
11. J. M. Jornet and I. F. Akyildiz, Channel modeling and capacity analysis for electromagnetic wireless nanonetworks in the terahertz band, *IEEE Transactions on Wireless Communications*, vol. 10, no. 10, pp. 3211–3221, 2011.
12. J. M. Jornet and I. F. Akyildiz, Graphene-based plasmonic nano-transceiver for terahertz band communication, in *8th European Conference on Antennas and Propagation (EuCAP)*, The Hague, The Netherlands, 2014, pp. 2–6.
13. A. Afsharinejad, A. Davy, and B. Jennings, Frequency selection strategies under varying moisture levels in wireless nano-networks, in *Proceedings of ACM: The First ACM International Conference on Nanoscale Computing and Communication*, Atlanta, GA, 2014, pp. 23–32.
14. C. T. Chou, Molecular communication networks with general molecular circuit receivers, in *Proceedings of ACM International Conference on Nanoscale Computing and Communication*, Atlanta, GA, 2014, pp. 1–9.
15. H. Awan and C. T. Chou, Impact of receiver molecular circuits on the performance of reaction shift keying, in *Proceedings of the Second ACM International Conference on Nanoscale Computing and Communication*, Boston, 2015.
16. C. T. Chou, Impact of receiver reaction mechanisms on the performance of molecular communication networks, *IEEE Transactions on Nanotechnology*, vol. 14, no. 2, pp. 304–317, 2015.
17. I. F. Akyildiz and J. M. Jornet, Electromagnetic wireless nanosensor networks, *Nano Communication Networks*, vol. 1, no. 1, pp. 3–19, 2010.
18. J. M. Jornet and I. F. Akyildiz, Graphene-based nano-antennas for electromagnetic nanocommunications in the terahertz band, in *Proceedings of the Fourth European Conference on on Antennas and Propagation (EuCAP)*, Barcelona, Spain, April 2010, pp. 1–9.
19. S. J. Tans, A. R. M. Verschueren, and C. Dekker, Room-temperature transistor based on a single carbon nanotube, *Nature*, vol. 672, no. 1989, pp. 669–672, 1998.

20. L. Ponomarenko, F. Schedin, and M. Katsnelson, Chaotic Dirac billiard in graphene quantum dots, *Science*, vol. 320, no. 5874, pp. 356–8, 2008.

21. B. K. Sarker, N. Kang, and S. I. Khondaker, High performance semiconducting enriched carbon nanotube thin film transistors using metallic carbon nanotubes as electrodes, *Nanoscale*, vol. 6, no. 9, pp. 4896–902, 2014. Available at http://www.ncbi.nlm.nih.gov/pubmed/24671657.

22. G. P. Lansbergen, Transistors arrive at the atomic limit, *Nature Nanotechnology*, vol. 7, no. 4, pp. 209–10, 2012. Available at http://www.ncbi.nlm.nih.gov/pubmed/22343381.

23. M. Fuechsle, J. A. Miwa, S. Mahapatra, H. Ryu, S. Lee, O. Warschkow, L. C. L. Hollenberg, G. Klimeck, and M. Y. Simmons, A single-atom transistor, *Nature Nanotechnology*, vol. 7, no. 4, pp. 242–6, 2012. Available at http://www.ncbi.nlm.nih.gov/pubmed/22343383.

24. J. S. Juan, M. No´, and C. Schuh, Nanoscale shape-memory alloys for ultrahigh mechanical damping, *Nature Nanotechnology*, vol. 4, pp. 415–419, 2009. Available at http://www.nature.com/nnano/journal/v4/n7/abs/nnano.2009.142.html.

25. A. Chung, J. Deen, J.-S. Lee, and M. Meyyappan, Nanoscale memory devices, *Nanotechnology*, vol. 21, no. 41, p. 412001, 2010. Available at http://www.ncbi.nlm.nih.gov/pubmed/20852352.

26. M.-J. Lee, D. Lee, S.-H. Cho, J.-H. Hur, S.-M. Lee, D. H. Seo, D.-S. Kim et al., A plasma-treated chalcogenide switch device for stackable scalable 3D nanoscale memory, *Nature Communications*, vol. 4, p. 2629, 2013. Available at http://www.ncbi.nlm.nih.gov/pubmed/24129660.

27. M. S. Fuhrer, B. M. Kim, T. Drkop, and T. Brintlinger, High-mobility nanotube transistor memory, *Nano Letters*, vol. 2, no. 7, pp. 755–759, 2002. Available at http://dx.doi.org/10.1021/nl025577o.

28. M. Radosavljevi, M. Freitag, K. V. Thadani, and A. T. Johnson, Nonvolatile molecular memory elements based on ambipolar nanotube field effect transistors, *Nano Letters*, vol. 2, no. 7, pp. 761–764, 2002. Available at http://dx.doi.org/10.1021/nl025584c.

29. J. Kim, S. D. Oh, J. H. Kim, D. H. Shin, S. Kim, and S.-H. Choi, Graphene/Si-nanowire heterostructure molecular sensors, *Scientific Reports*, vol. 4, p. 5384, 2014. Available at http://www.pubmedcentral.nih.gov/articlerender.fcgi?artid=4064328&tool=pmcentrez&rendertype=abstract.

30. H. Zhang, Y. Yang, Y. Su, J. Chen, C. Hu, Z. Wu, Y. Liu, C. Ping Wong, Y. Bando, and Z. L. Wang, Triboelectric nanogenerator as self-powered active sensors for detecting liquid/gaseous water/ethanol, *Nano Energy*, vol. 2, no. 5, pp. 693–701, 2013. Available at http://linkinghub.elsevier.com/retrieve/pii/S2211285513001444.

31. A. Alu` and N. Engheta, Wireless at the nanoscale: Optical interconnects using matched nanoantennas, *Physical Review Letters*, vol. 104, no. 21, 2010. Available at http://link.aps.org/doi/10.1103/PhysRevLett.104.213902.

32. V. Loscrí, V. Mannara, E. Natalizio, and G. Aloi, Efficient acoustic communication techniques for nanobots, in *Proceedings of the 7th International Conference on Body Area Networks (BodyNets '12)*, Institute for Computer Sciences, Social-Informatics and Telecommunications Engineering, Brussels, Belgium, 2012, pp. 36–39. Available at http://dl.acm.org/citation.cfm?id=2442691.2442702.

33. G. Santagati and T. Melodia, Opto-ultrasonic communications in wireless body area nanonetworks, in *2013 Asilomar Conference on Signals, Systems and Computers*, Pacific Grove, CA, November 2013, pp. 1066–1070.

34. H. Kai, S. Nara, K. Kinbara, and T. Aida, Toward long-distance mechanical communication: Studies on a ternary complex interconnected by a bridging rotary module, *Journal of the American Chemical Society*, vol. 130, no. 21, pp. 6725–6727, 2008 [pMID: 18447353]. Available at http://dx.doi.org/10.1021/ja801646b.

35. S. F. Bush, *Nanoscale Communication Networks*, Artech House, Norwood, MA, 2010.

36. F. Vullum and D. Teeters, Investigation of lithium battery nanoelectrode arrays and their component nanobatteries, *Journal of Power Sources*, vol. 146, no. 1–2, pp. 804–808, 2005.

37. F. Vullum, D. Teeters, A. Nyten, and J. Thomas, Characterization of lithium nanobatteries and lithium battery nanoelectrode arrays that benefit from nanostructure and molecular self-assembly, *Solid State Ionics*, vol. 177, no. 26–32, pp. 2833–2838, 2006.

38. C. Chan, High capacity Li ion battery anodes using Ge nanowires, *Nano Letters*, vol. 8, no. 1, pp. 307–309, Jan. 2008. Available at http://www.ncbi.nlm.nih.gov/pubmed/21891841 and http://pubs.acs.org/doi/abs/10.1021/nl0727157.

39. C. Chan, Silicon nanotube battery anodes, *Nano Letters*, vol. 9, no. 11, pp. 3844–3847, 2009. Available at http://www.ncbi.nlm.nih.gov/pubmed/22447161.

40. C. Liu, E. I. Gillette, X. Chen, A. J. Pearse, A. C. Kozen, M. A. Schroeder, K. E. Gregorczyk, S. B. Lee, and G. W. Rubloff, An all-in-one nanopore battery array, *Nature Nanotechnology*, vol. 9, no. 12, pp. 1031–1039, 2014. Available at http://www.nature.com/doifinder/10.1038/nnano.2014.247.

41. Z. L. Wang and W. Wu, Nanotechnology-enabled energy harvesting for self-powered micro-/nano-systems, *Angewandte Chemie (International ed. in English)*, vol. 51, no. 47, pp. 11700–11721, 2012.

42. Z. L. Wang and J. Song, Piezoelectric nanogenerators based on zinc oxide nanowire arrays, *Science (New York, N.Y.)*, vol. 312, no. 5771, pp. 242–246, 2006. Available at http://www.ncbi.nlm.nih.gov/pubmed/16614215.

43. X. Jiang, W. Huang, and S. Zhang, Flexoelectric nano-generator: Materials, structures and devices, *Nano Energy*, vol. 2, no. 6, pp. 1079–1092, 2013. Available at http://linkinghub.elsevier.com/retrieve/pii/S2211285513001493.

44. Q. Deng, M. Kammoun, A. Erturk, and P. Sharma, Nanoscale flexoelectric energy harvesting, *International Journal of Solids and Structures*, vol. 51, no. 18, pp. 3218–3225, 2014. Available at http://linkinghub.elsevier.com/retrieve/pii/S0020768314002121.

45. Y. Yang, W. Guo, K. C. Pradel, G. Zhu, Y. Zhou, Y. Zhang, Y. Hu, L. Lin, and Z. L. Wang, Pyroelectric nanogenerators for harvesting thermoelectric energy, *Nano Letters*, vol. 12, no. 6, pp. 2833–2838, 2012. Available at http://www.ncbi.nlm.nih.gov/pubmed/22545631.

46. S. Wang, L. Lin, and Z. Wang, Nanoscale triboelectric-effect-enabled energy conversion for sustainably powering portable electronics, *Nano Letters*, vol. 12, pp. 6339–6346, 2012. Available at http://pubs.acs.org/doi/abs/10.1021/nl303573d.

47. M. Han, X.-S. Zhang, X. Sun, B. Meng, W. Liu, and H. Zhang, Magnetic-assisted triboelectric nano-generators as self-powered visualized omnidirectional tilt sensing system, *Scientific Reports*, vol. 4, p. 4811, 2014. Available at http://www.pubmedcentral.nih.gov/articlerender.fcgi?artid=4001096&tool=pmcentrez&rendertype=abstract.

48. M. Han, X. Zhang, B. Meng, W. Liu, and W. Tang, r-Shaped hybrid nanogenerator with enhanced piezoelectricity, *ACS Nano*, vol. 7, no. 10, pp. 8554–8560, 2013. Available at http://pubs.acs.org/doi/abs/10.1021/nn404023v.

49. B. J. Hansen, Y. Liu, R. Yang, and Z. L. Wang, Hybrid nanogenerator for concurrently harvesting biomechanical and biochemical energy, *ACS Nano*, vol. 4, pp. 3647–3652, 2010.

50. S. Lee, S. H. Bae, L. Lin, S. Ahn, C. Park, S. W. Kim, S. N. Cha, Y. J. Park, and Z. L. Wang, Flexible hybrid cell for simultaneously harvesting thermal and mechanical energies, *Nano Energy*, vol. 2, pp. 817–825, 2013.

51. J. Yin, Z. Zhang, X. Li, J. Zhou, and W. Guo, Harvesting energy from water flow over graphene, *Nano Letters*, vol. 12, pp. 1736–1741, 2012.

52. G. W. Hanson, Fundamental transmitting properties of carbon nanotube antennas, *IEEE Transactions on Antennas and Propagation*, vol. 53, no. 11, pp. 3426–3435, 2005.

53. P. Burke, Quantitative theory of nanowire and nanotube antenna performance, *IEEE Transactions on Nanotechnology*, vol. 5, no. 4, pp. 314–334, 2006. Available at http://ieeexplore.ieee.org/lpdocs/epic03/wrapper.htm?arnumber=1652847.

54. J. Weldon, K. Jensen, and A. Zettl, Nanomechanical radio transmitter, *Physica Status Solidi (B)*, vol. 245, no. 10, pp. 2323–2325, 2008.

55. I. Llatser, C. Kremers, A. Cabellos-Aparicio, J. M. Jornet, E. Alarcón, and D. N. Chigrin, Graphene-based nano-patch antenna for terahertz radiation, *Photonics and Nanostructures—Fundamentals and Applications*, vol. 10, no. 4, pp. 353–358, 2012. Available at http://linkinghub.elsevier.com/retrieve/pii/S1569441012000727.

56. J. M. Jornet and I. F. Akyildiz, Graphene-based plasmonic nano-antenna for terahertz band communication in nanonetworks, *IEEE Journal on Selected Areas in Communications*, vol. 31, no. 12, pp. 685–694, 2013. Available at http://ieeexplore.ieee.org/lpdocs/epic03/wrapper.htm?arnumber=6708549.

57. H.-J. Song and T. Nagatsuma, Present and future of terahertz communications, *IEEE Transactions on Tera-Hertz Science and Technology*, vol. 1, no. 1, pp. 256–263, 2011.

58. T. Kleine-Ostmann and T. Nagatsuma, A review on terahertz communications research, *Journal of Infrared, Millimeter, and Terahertz Waves*, vol. 32, no. 2, pp. 143–171, 2011.

59. T. Kürner, Towards future THz communications systems, *Terahertz Science and Technology*, vol. 5, no. 1, pp. 11–17, 2012.

60. K. Ishigaki, M. Shiraishi, S. Suzuki, M. Asada, N. Nishiyama, and S. Arai, Direct intensity modulation and wireless data transmission characteristics of terahertz-oscillating resonant tunnelling diodes, *Electronics Letters*, vol. 48, no. 10, p. 582, 2012.

61. J. M. Jornet and I. F. Akyildiz, Information capacity of pulse-based wireless nanosensor networks, in *Proceeding of the 2011 8th Annual IEEE Communications Society Conference on Sensor, Mesh and Ad Hoc Communications and Networks*, Salt Lake, UT, June 2011, pp. 80–88.

62. R. Piesiewicz, T. Kleine-Ostmann, N. Krumbholz, D. Mittleman, M. Koch, J. Schoebei, and T. Kürner, Short-range ultra-broadband terahertz communications: Concepts and perspectives, *IEEE Antennas and Propagation Magazine*, vol. 49, pp. 24–39, 2007.

63. F. Rana, Graphene terahertz plasmon oscillators, *IEEE Transactions on Nanotechnology*, vol. 7, no. 1, 2008.

64. J. M. Jornet and I. F. Akyildiz, Channel capacity of electromagnetic nanonetworks in the terahertz band, in *2010 IEEE International Conference on Communications*, Cape Town, South Africa, May 2010, pp. 1–6.

65. L. Rothman, I. Gordon, A. Barbe, D. Benner, P. Bernath, M. Birk, V. Boudon et al., The HITRAN 2008 molecular spectroscopic database, *Journal of Quantitative Spectroscopy and Radiative Transfer*, vol. 110, no. 9–10, pp. 533–572, 2009.

66. R. L. Poynter and H. M. Pickett, Submillimeter, millimeter, and microwave spectral line catalog, *Applied Optics*, vol. 24, p. 2235, 1985.

67. Y. L. Babikov, I. E. Gordon, and S. N. Mikhailenko, "HITRAN on the web," a new tool for HITRAN spectroscopic data manipulation, in *Proceedings of the ASA-HITRAN Conference*, Reims, France, August 29–31, 2012, pp. 1–9. Available at http://hitran.iao.ru/.

68. J. M. Jornet, Fundamentals of electromagnetic nanonetworks in the terahertz band, PhD dissertation, Georgia Institute of Technology, Atlanta, 2013.

69. M. Pierobon, J. M. Jornet, N. Akkari, S. Almasri, and I. F. Akyildiz, A routing framework for energy harvesting wireless nanosensor networks in the terahertz band 1, *Wireless Networks (Springer)*, vol. 20, no. 5, pp. 1169–1183, 2014. Available at http://link.springer.com/10.1007/s11276-013-0665-y.

70. E. M. Hajaj, O. Shtempluk, V. Kochetkov, A. Razin, and Y. E. Yaish, Chemical potential of inhomogeneous single-layer graphene, *Physical Review B*, vol. 88, no. 4, p. 045128, 2013. Available at http://link.aps.org/doi/10.1103/PhysRevB.88.045128.

71. I. Llatser, C. Kremers, and D. Chigrin, Radiation characteristics of tunable graphennas in the terahertz band, *Radioengineering*, vol. 21, no. 4, pp. 946–953, 2012. Available at http://www.radioeng.cz/fulltexts/2012/12_04_0946_0953.pdf.

72. E. Zarepour, M. Hassan, C. T. Chou, and A. A. Adesina, Power optimization in nano sensor networks for chemical reactors, in *Proceedings of ACM International Conference on Nanoscale Computing and Communication*, Atlanta, GA, 2014, pp. 1–9.

73. E. Zarepour, M. Hassan, C. T. Chou, and A. A. Adesina, Remote detection of chemical reactions using nanoscale terahertz communication powered by pyroelectric energy harvesting, in *Proceedings of the Second ACM International Conference on Nanoscale Computing and Communication*, Boston, 2015, pp. 1–6.

74. E. Zarepour, A. A. Adesina, M. Hassan, and C. T. Chou, Innovative approach to improving gas-to-liquid fuel catalysis via nanosensor network modulation, *Industrial and Engineering Chemistry Research*, vol. 53, no. 14, pp. 5728–5736, 2014.

75. E. Zarepour, N. Hassan, M. Hassan, C. T. Chou, and M. Ebrahimi Warkiani, Design and analysis of a wireless nanosensor network for monitoring human lung cells, Technical Report 201510, School of Computer Science and Engineering, University of New South Wales, Sydney, Australia, July 2015.

76. A. Afsharinejad, A. Davy, B. Jennings, and S. Balasubramaniam, GA-based frequency selection strategies for graphene-based nano-communication networks, in *2014 IEEE International Conference on Communications (ICC)*, Sydney, Australia, June 2014, pp. 3642–3647. Available at http://ieeexplore.ieee.org/lpdocs/epic03/wrapper.htm?arnumber=6883887.

77. G. von Maltzahn, J.-H. Park, K. Y. Lin, N. Singh, C. Schwöppe, R. Mesters, W. E. Berdel, E. Ruoslahti, M. J. Sailor, and S. N. Bhatia, Nanoparticles that communicate in vivo to amplify tumour targeting, *Nature Materials*, vol. 10, no. 7, pp. 545–52, 2011. Available at http://www.pubmedcentral .nih.gov/articlerender.fcgi?artid=3361766&tool=pmcentrez&rendertype=abstract.

78. L. Felicetti, M. Femminella, G. Reali, and P. Liò, A molecular communication system in blood vessels for tumor detection, in *Proceedings of ACM NANOCOM*, Atlanta, GA, 2014, pp. 21–30. Available at http://dl.acm.org/citation.cfm?id=2619978.

79. E. Zarepour, A. A. Adesina, M. Hassan, and C. T. Chou, Nano sensor networks for tailored operation of highly efficient gas-to-liquid fuels catalysts, in *Proceedings of Australasian Chemical Engineering Conference (Chemeca 2013)*, Brisbane, Australia, October 2013, pp. 1–9.

80. A. Adesina, Hydrocarbon synthesis via Fischer-Tropsch reaction: Travails and triumphs, *Applied Catalysis A: General*, vol. 138, no. 2, pp. 345–367, 1996.

81. E. Zarepour, M. Hassan, C. T. Chou, A. A. A. Adesina, and M. Ebrahimi Warkiani, Reliability analysis of time-varying wireless nanoscale sensor networks, in *Proceedings of the 15th IEEE International Conference on Nanotechnology*, Rome, Italy, July 2015.

82. K. Yang, A. Alomainy, and Y. Hao, In-vivo characterisation and numerical analysis of the THz radio channel for nanoscale body-centric wireless networks, in *Radio Science Meeting (Joint with AP-S Symposium), 2013 USNC-URSI*, Lake Buena Vista, FL, July 2013, pp. 218–219.

83. D. Gillespie, Exact stochastic simulation of coupled chemical reactions, *Journal of Physical Chemistry*, vol. 81, no. 25, pp. 2340–2361, 1977.

84. The respiratory system. Available at http://leavingbio.net/respiratory system/the respiratory system.htm.

85. V. Loscri and A. Vegni, An acoustic communication technique of nanorobot swarms for nanomedicine applications, *IEEE Transactions on NanoBioscience*, vol. 14, no. 6, pp. 598–607, 2015.

86. W. J. P. D. Marshall and S. K. Bangert, *Clinical Chemistry*, 5th ed., New York: Mosby, 2004 [includes index].

87. J. M. Jornet and I. F. Akyildiz, Femtosecond-long pulse-based modulation for terahertz band communication in nanonetworks, *IEEE Transactions on Communications*, vol. 62, no. 5, pp. 1742–1754, 2014.

88. E. Zarepour, M. Hassan, C. T. Chou, and A. A. Adesina, Frequency hopping strategies for improving terahertz sensor network performance over composition varying channels, in *IEEE International Symposium on a World of Wireless, Mobile and Multimedia Network (WoWMoM)*, Sydney, Australia, June 2014, pp. 1–9.

89. E. Zarepour, Adaptive protocols for nano-scale sensor networks over composition varying channels, in *2014 IEEE 15th International Symposium on a World of Wireless, Mobile and Multimedia Networks (WoWMoM)*, Sydney, NSW, Australia, June 2014, pp. 1–2.

90. M. Ghavami, L. Michael, and R. Kohno, *Ultra Wideband Signals and Systems in Communication Engineering*, West Sussex, England: John Wiley & Sons, 2004.

91. B. Otis and J. Rabaey, *Ultra-Low Power Wireless Technologies for Sensor Networks*, New York: Springer, 2007.

92. J. Zhang, P. V. Orlik, Z. Sahinoglu, A. F. Molisch, and P. Kinney, UWB systems for wireless sensor networks, *Proceedings of the IEEE*, vol. 97, no. 2, pp. 313–331, 2009.

93. J. M. Jornet, J. Capdevila, P. Joan, and J. Solé Pareta, PHLAME: A physical layer aware MAC protocol for electromagnetic nanonetworks in the terahertz band, *Nano Communication Networks*, vol. 3, no. 1, pp. 74–81, 2012.

94. J. C. Pujol, Bridging PHY and MAC layers in wireless electromagnetic nanonetworks, Final year project from Technical University of Catalonia, Spain, 2010.

95. R. Cid-Fuentes, J. M. Jornet, I. Akyildiz, and E. Alarcón, A receiver architecture for pulse-based electromagnetic nanonetworks in the terahertz band, in *International Workshop on Molecular and Nano Scale Communication (MoNaCom)*, Ottawa, Canada, 2012, pp. 1–6.

96. A. Goldsmith, *Wireless Communications*, Cambridge: Cambridge University Press, 2005.

97. E. Zarepour, M. Hassan, C. T. Chou, and S. Bayat, Performance analysis of carrier-less modulation schemes for wireless nanosensor networks, in *Proceedings of the 15th IEEE International Conference on Nanotechnology*, Rome, Italy, July 2015.

98. I. F. Akyildiz and J. M. Jornet, The Internet of nano-things, *IEEE Wireless Communications*, vol. 17, pp. 58–63, 2010.

99. J. M. Jornet and I. F. Akyildiz, The Internet of multimedia nano-things, *Nano Communication Networks*, vol. 3, pp. 242–251, 2012.

100. J. M. Jornet and I. F. Akyildiz, Low-weight channel coding for interference mitigation in electromagnetic nanonetworks in the terahertz band, in *2011 IEEE International Conference on Communications (ICC)*, June 2011, pp. 1–6. Available at http://ieeexplore.ieee.org/lpdocs/epic03/wrapper.htm?arnumber=5962987.

101. J. M. Jornet, Low-weight error-prevention codes for electromagnetic nanonetworks in the terahertz band, *Nano Communication Networks*, vol. 5, no. 1–2, pp. 35–44, 2014. Available at http://linkinghub.elsevier.com/retrieve/pii/S1878778914000039.

102. P. Boronin, V. Petrov, D. Moltchanov, Y. Koucheryavy, and J. M. Jornet, Capacity and throughput analysis of nanoscale machine communication through transparency windows in the terahertz band, *Nano Communication Networks*, vol. 5, no. 3, pp. 72–82, 2014. Available at http://linkinghub.elsevier.com/retrieve/pii/S1878778914000222.

103. P. Wang, J. M. Jornet, M. Abbas Malik, N. Akkari, and I. F. Akyildiz, Energy and spectrum-aware MAC protocol for perpetual wireless nanosensor networks in the terahertz band, *Ad Hoc Networks*, vol. 11, no. 8, pp. 2541–2555, 2013. Available at http://linkinghub.elsevier.com/retrieve/pii/S157087051300139X.

104. G. Piro, S. Abadal, A. Mestres, E. Alarcón, J. Solé-Pareta, L. A. Grieco, and G. Boggia, Initial MAC exploration for graphene-enabled wireless networks-on-chip, in *Proceedings of ACM: The First Annual International Conference on Nanoscale Computing and Communication (NANOCOM' 14)*, Atlanta, GA, 2014, pp. 63–70.

105. J. M. Jornet, A joint energy harvesting and consumption model for self-powered nano-devices in nanonetworks, in *2012 IEEE International Conference on Communications (ICC)*, Ottawa, Ontario, Canada, June 2012, pp. 6151–6156.

106. J. M. Jornet and I. F. Akyildiz, Joint energy harvesting and communication analysis for perpetual wireless nanosensor networks in the terahertz band, *IEEE Transactions on Nanotechnology*, vol. 11, no. 3, pp. 570–580, 2012. Available at http://ieeexplore.ieee.org/xpls/abs all.jsp?arnumber=6144047.

107. F. Dressler and F. Kargl, Security in nano communication: Challenges and open research issues, in *2012 IEEE International Conference on Communications (ICC)*, Ottawa, Ontario, Canada, June 2012, pp. 6183–6187.

108. F. Dressler and F. Kargl, Towards security in nano-communication: Challenges and opportunities, *Nano Communication Networks*, vol. 3, no. 3, pp. 151–160, 2012. Available at http://linkinghub.elsevier.com/retrieve/pii/S1878778912000294.

Chapter 7

Evolution of Wireless Sensor Networks toward Internet of Things

Zeeshan Ali Khan and Ubaid Abbasi

Contents

Abstract

Wireless sensor networks (WSNs) are playing a key role in several application domains, including transportation, smart grid, healthcare, and environmental monitoring. This wide range of technologies and applications has created new deployment and integration scenarios using wireless sensor networks. The current perspective is to avoid proprietary standards and embrace Internet Protocol (IP)–based sensor networks with nonproprietary and open standards. This allows the connectivity of wireless sensor networks with Internet and enables different objects to be involved in the Internet of Things (IoT). Wireless sensor networks are providing a virtual layer in which information about the physical world can be accessed by any computational system. Thus, WSNs are considered a valuable asset for bringing the vision of IoT into reality. This chapter highlights the key factors for the evolution of wireless sensor networks toward the IoT. Initially, the background information about WSNs and IoT, highlighting the important application areas, is provided. In a later part, the role of wireless sensor networks in the context of IoT is discussed. Finally, the future research challenges for visualizing the concept of IoT are proposed.

7.1 Introduction

The concept of the Internet of Things (IoT) may be implemented in the near future, as heterogeneous communication technologies will be integrated in the future Internet [1]. The IoT can be described as a network of interconnected things using standardized communication protocols. Owing to the low cost of wireless sensor network (WSN) deployment and operation, a number of environmental monitoring, healthcare, agriculture, multimedia, and smart home applications can be realized by WSNs. However, there is a need to interconnect these widely spread WSN implementations in order to obtain a wide-scale sensor network. Thus, the ability of heterogeneous sensing systems to operate together is an important challenge [2].

The sensor network is normally a private network with limited communication to external devices. There is a need to have specific devices or gateways in WSNs that are able to exchange information generated by the sensor network with the external world. The most recent trend is to use the Internet Protocol (IP) for connectivity between the Internet and a particular WSN [3].

In this way, smart sensor devices having their own IP address are connected to the external world to form an IoT. This will allow us to realize important applications, and it will give the future Internet several more features and much more flexibility [4].

In this chapter, a state of the art related to WSNs and IoT is discussed. Furthermore, this chapter depicts the problems that will be faced while WSNs evolve toward the IoT. Moreover, various applications have been discussed that can be realized by integrating the concept of WSNs and IoT.

The chapter is organized as follows. Sections 7.2 and 7.3 discuss the idea of IoT and WSNs, respectively, while Section 7.4 depicts the integration of these two technologies. Related research projects have been discussed in Section 7.5, while Section 7.6 describes the potential applications. Finally, Section 7.7 concludes the chapter.

7.2 Internet of Things

The era of computing is changing quickly, and the future will be outside the realm of the conventional desktop. The Internet of Things (IoT) is appearing as a new paradigm that is getting huge attention in the context of modern wireless communication. In this paradigm, the surrounding objects will be part of the network. As a result, the information and communication system are invisibly embedded in the environment around us. The basic concept behind the IoT is the pervasive presence of different objects around us. These objects include radio frequency identification (RFID) tags, sensors, actuators, and mobile phones. These objects will cooperate with each other to achieve a common objective [5]. Although The IoT paradigm has its own concept and set of characteristics, it also shares cohesion with other areas of computer science. The IoT combines different technologies that include sensors, semantics, the cloud, data modeling, and storing and communication technologies. Due to its huge significance, IoT has been declared one of the disruptive technologies for the future by the U.S. National Intelligence Council [6].

RFID is considered the underpinning technology of IoT, which permits the microchips to transfer the identification information through wireless channels. RFID is extensively used in logistics, pharmaceuticals, retailing, and so forth, for identifying, tracking, and monitoring the objects attached with tags [7]. Apart from RFID, WSNs are also considered the base technology for IoT. The WSNs use interconnected intelligent sensors to monitor traffic-, industrial-, and healthcare-related applications. Different devices that can be part of IoT are shown in Figure 7.1 [8].

7.2.1 Definition

From the perspective of IoT, "things" could be defined as a real/physical or digital/virtual objects that have an identity and exist and move in space and time. Things are usually identified by assigned identification numbers, names, addresses, and so forth. The concept of things is the essential part in the construction of IoT because things are responsible for interacting and communicating with themselves and with the environment while reacting to the physical world events and running the processes that trigger the action.

The concept of IoT has started to become an integral part of the future Internet having a dynamic network infrastructure with self-configuring capabilities that uses standardized communication protocols, while the things have unique identities that use intelligent interfaces to seamlessly integrate into the information network. The research in the area of IoT is still in the

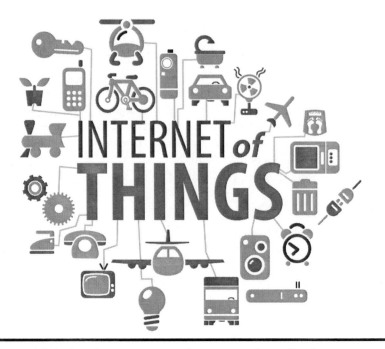

Figure 7.1 Potential devices for an IoT. (From http://smartdatacollective.com, accessed March 14, 2015.)

embryonic stage; therefore, it is difficult to extract any standard definition. However, the following definitions have been proposed by a few researchers.

> **Definition 1:** "The semantic origin of the expression is composed by two words and concepts: Internet and Thing, where Internet can be defined as the world-wide network of interconnected computer networks, based on a standard communication protocol, the Internet suite (TCP/IP), while Thing is an object not precisely identifiable. Therefore, semantically, Internet of Things means a world-wide network of interconnected objects uniquely addressable, based on standard communication protocols" [9].
>
> **Definition 2:** "Things have identities and virtual personalities operating in smart spaces using intelligent interfaces to connect and communicate within social, environment, and user contexts" [10].
>
> **Definition 3:** "The Internet of Things allows people and things to be connected Anytime, Anyplace, with Anything and Anyone, ideally using Any path/network and Any service" [11].

7.2.2 Vision and Technology

The communication and processing capabilities of a processing device are becoming more and more versatile and accessible due to advancements in technology. Thus, there are opportunities for further interconnectivity between these devices to efficiently use these precious resources.

In the earlier sections, we discussed the unique identification of the things. The aim of this unique identification is to assign a unique identifier. This identifier can be globally unique (such as public IP addresses) or unique within the context of that particular environment. In the case of

IoT, there might be a need for multiple unique identifiers for a single object. As an example, a single object can have multiple sensors or RFID tags associated with it. Each of these sensors might be individually addressable, and the identity of the object can be constructed from the individual identities of those sensors. Thus, the large-scale deployments of IoT need the development of new mechanisms and schemes to address the objects globally.

The IoT environment is highly dynamic and heterogeneous due to the presence of different objects at different times. In this environment, scalability and interoperability seem to be a huge concern. Thus, it requires having a layered architecture that eliminates the need for any specific language, protocol, operating system, or specific technology dependency. Moreover, the IoT infrastructure must utilize the ownership of the data in a distributed manner. As a consequence, things can control the sharing of information to other object and entities. The IoT architecture should also provide the standards for collecting the fragments of different information from different sources (even from unknown sources), in order to achieve comprehensive end-to-end traceability as far as it is permitted.

The vision and concept of IoT has a variety of dimensions. These dimensions include the mobility, heterogeneity, network topology, network size, and infrastructure. In the IoT, the network will dynamically change and continuously evolve due to the joining and exiting of things. The automated discovery of the things is very essential to scale the network management capabilities.

7.2.3 Characteristics of IoT

The IoT architecture will allow the creation of a dynamic network of physical world objects within virtual space by combining the characteristics of RFIDs, ubiquitous sensor networks, and other identifiable things. Some of the important characteristics of IoT are highlighted below.

- IoT consists of objects or things that are real-world entities or virtual entities having the ability to identify each other.
- These entities are able to exchange information with each other using the communication protocols and infrastructure.
- IoT can use services that act as an interface to things. These things may have the sensors that can interact with different environments.
- Things in IoT can communicate and collaborate with each other and with different computing devices.
- Things in IoT can adapt to the environment by extracting the patterns from the environment and learn from other things. These entities of IoT have the ability to makes decisions, evolve, and propagate information.

7.3 Wireless Sensor Networks

Recent advancements in microelectronics, sensors, and communication have led to the development of tiny devices that sense, analyze, and communicate the data generated from some sensors. Such a network is called a wireless sensor network, as it can monitor and communicate measured physical parameters. These devices have limited battery and processing capabilities, and they are installed in locations where it may be difficult to replace their battery [12]. There are a number of applications where these networks can be deployed, such as environmental monitoring [13], healthcare applications [14], smart grid deployment [15], and disaster management [16]. Therefore, the solutions prefer to reduce the overall energy consumption of the wireless sensor networks; this is a hot area of research in this domain.

In such a network, a large number of sensor nodes are deployed to monitor a physical phenomenon, and the position of the sensor does not need to be predetermined. The placement of these nodes can be random for any application. However, the networks need to have the ability of self-organization in case of a random deployment, and the protocols need to be developed by considering this fact.

In order to realize such a network, wireless ad hoc networking techniques need to be integrated in the devices. The protocols need to be proposed based on the targeted applications. For example, the sensor network deployed for medical healthcare needs to track doctors, monitor vital signs of patients, and transmit these important measured parameters to the doctors [14]. However, for disaster management applications, it is required to track and guide the first responders inside a building [16].

Some characteristics of sensor networks are listed below:

- The number of sensor nodes in a sensor network can be much larger than the number nodes in an ad hoc network.
- The deployment of sensor nodes is dense.
- The failure can occur in sensor nodes.
- The sensor network topology may change frequently.
- Most of the sensor devices have limited energy and computation capabilities.

The sensor network may comprise a large number of nodes that are close to each other. For energy consumption minimization, multihop communication can have better results in comparison with single-hop communication [17].

7.3.1 Sensor Node Composition

In order to achieve the functionalities described in Section 7.2, the wireless sensor node consists of the following components:

- Sensors: These are used for sensing a physical phenomenon.
- A/D converter: This is used to convert the analog data received from the sensor.
- Transceivers: In order to send and receive the sensed data, transceivers are used.
- Processor: This is used for processing, forwarding, and transmitting the sensed data.
- Battery: The energy is supplied to various components of the sensor node by a battery.

7.3.2 Classification of Wireless Sensor Networks

There are four main types of sensor networks that have been discussed in this section.

7.3.2.1 Classical WSNs

The classical applications of wireless sensor networks include measurement of physical environmental parameters such as temperature, humidity, pressure, and noise levels that can be used to realize an environmental monitoring application. These devices are used for continuous monitoring and reporting of these parameters, and in the case of any unusual situation, the remedial activity can be started by the central base station that is receiving this data [18].

7.3.2.2 Wireless Multimedia Sensor Networks

The development of wireless multimedia sensor networks (WMSNs) is due to the availability of cheaper cameras and microphones [19,20]. It has led to the interconnection of these devices that can gather multimedia content, such as images, video, and audio streams that can be used to extract particular information. These networks have gathered the attention of the research community owing to the challenges related to the implementation of such networks [21]. A number of different applications can be realized in various domains, including the Internet of Things (IoT). The main difference between WMSNs and classical WSNs is the bandwidth requirement, as large volumes of data need to be sent on the network. The other characteristic is that traffic is delay intolerant.

One example of a WMSN node is the Cyclops image capturing and inference module [22], which is designed for imaging applications and can be interfaced with a sensor node such as MICA2 [23] or MICAz [24]. Moreover, WMSNs can store and process the multimedia data that is gathered from different sources.

The applications for realizing a WMSN may consist of surveillance networks that have the capability of processing, capturing, sending, and receiving the multimedia data. Law enforcement agencies can use such a network for monitoring different events and localities [25]. Another example is that of a traffic monitoring system that also is used for enforcing the maximum speed limit and congestion avoidance systems in cities and highways [26].

The research directions include the development of flexible architectures for providing specific services to the end user by considering the data delivery throughput and energy consumption constraints.

7.3.2.3 Wireless Underground Sensor Networks

Wireless underground sensor networks can be used to monitor various physical parameters, such as water and mineral contents [27] for agriculture applications. The measured parameters can be transferred to a central device called the sink by using a wireless connection. This makes the solution very attractive, as it is not needed to deploy a wired connection for data delivery. However, these devices need to be energy-efficient, as the battery replacement is difficult and costly in such a network. The other issue is the communication channel quality and the error rate, which can be improved by using special antennas and signal processing techniques. The other applications include the monitoring of underground plumbing infrastructure [28] and landslide or earthquake monitoring through buried sensors [29].

The research challenges are due to the location of these devices, as the underground wireless channel has a much higher error rate than its terrestrial counterpart. The hardware and communication protocols need to be devised by considering these facts. The major design challenges are

■ The bandwidth is limited and the channel quality is lower.
■ The battery power is limited, and it cannot be recharged easily by solar power.
■ The nodes may fail and the network has to reorganize itself accordingly.

7.3.2.4 Underwater Sensor Networks

These networks have underwater monitoring applications, such as disaster prevention, surveillance applications, and pollution monitoring. This network consists of sensor devices that can collaborate to realize the mentioned applications. The network should possess a capability of self-organization, and each device has to coordinate its operation by sharing location, configuration, and movement information [30]. Thus, the nodes adapt inside the network based on their position in the network.

The major applications include environmental monitoring where the network can monitor the chemical level in streams, rivers, lakes, and oceans [31]. Monitoring of ocean currents and winds may lead to the application of such a network in climate prediction. Another example of application is given in [32], where an underwater sensor network is used to detect extreme temperature gradients that can be used as a breeding ground of some microorganisms. Underwater sensor networks have other applications in undersea exploration missions, disaster prevention in coastal areas [33], and distributed surveillance of an area [34].

The design of underwater sensor networks should be able to transmit the data in real time for surveillance and climate prediction applications. To achieve these objectives, new network protocols need to be developed that are specific to the underwater sensor networks [35]. These protocols should be able to timely send the data to the destination node. Major challenges for the deployment of such networks are

- The underwater network has limited bandwidth.
- The channel quality is lowered due to the multipath and fading phenomenon.
- Propagation delay is nearly five times higher than in the case of terrestrial networks.
- The battery power is limited and solar power cannot be used to recharge it.
- Due to the nature of the environment, the nodes are prone to failure and the network has to reorganize itself in real time.

7.4 Evolution of WSN toward IoT

The IoT is a global network of interconnected things or objects having a unique IP address [36,37]. Heterogeneous communication technologies will be integrated in the future Internet to realize an IoT application. A number of useful applications, such as building automation, healthcare applications, environmental and water monitoring, agriculture applications, and transportation applications, can be realized by using the WSNs if they are able to interconnect with the outside world, as classical WSNs consist of a private network with minimal connection to the worldwide devices. However, integrating the IoT technologies with WSNs will lead to improvement of the implementation of existing applications, in addition to already installed ones. Such a smart sensor network will consist of smart objects that can communicate with the outside world. The block diagram of a proposed network integrating IoT with WSN is described in Figure 7.2. Using this model, IP-enabled sensor nodes can communicate the information to a central node that will be responsible for representing, storing, searching, and organizing the generated information by using the semantic technologies. There is also a need to use application specific gateways to export WSN data to the devices connected to the Internet (Figure 7.3) [4].

However, there are a number of research challenges that will be faced while finding a solution for the evolution of WSN toward IoT. The main ones are security, hardware, and software [38], and these are discussed in detail.

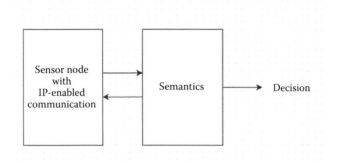

Figure 7.2   Block diagram of IoT-enabled WSN node.

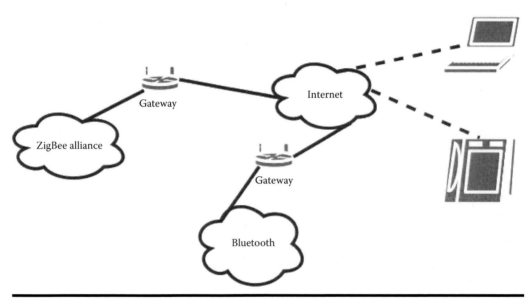

Figure 7.3   Internetworking among heterogeneous WSNs. (From L. Mainetti, L. Patrono, and A. Vilei, Evolution of wireless sensor networks towards the Internet of Things: A survey, in ***2011 19th International Conference on Software, Telecommunications and Computer Networks (SoftCOM),*** **September 15–17, 2011, pp. 1–6.)**

7.4.1 Network Security

There are a lot of security concerns that are associated with the WSN that need to be addressed. It consists of node compromise, unauthorized access, and denial-of-service attacks. These issues need to be addressed before integration of the WSN into IoT.

1. Node compromise: The WSN consists of thousands of nodes working together to gather specific information from the surrounding area. Due to the dynamic dispersion of these nodes, it is impractical to monitor each node to see whether it is compromised [39]. Another problem is the inclusion of false nodes in the network that will corrupt the sensed data by adding false information to it. As the sensor devices have strict energy constraints, the

solutions to minimize such attacks would be expensive, as they require more energy [40]. Moreover, if these devices are connected to the Internet through IoT, new threats will be added through Internet connection that may corrupt the sensed data [41]. Therefore, the research for integration of WSNs into IoT needs to look into these issues.

2. Unauthorized data access: The other issue with the WSNs is that data can be read by an intruder if it is sent in an unencrypted format. The standard way is to use the encryption algorithms [42] for transmitting each packet over the network. However, it is an issue for energy-constrained devices, as they cannot spend most of their energy on data encryption; thus, new schemes need to be developed. Moreover, a network may receive several unauthorized data requests, and the network should be able to check the authenticity of each request [43]. Therefore, this is one of the research directions that have to be looked into while integrating WSN with IoT.

3. Denial of service: A malicious outsider may also launch an attack so that the entire network is unable to perform its tasks. The types of attacks may include sending useless attacks so that the device's energy is depleted quickly. There are a number of solutions to minimize such attacks [39]; however, the solution combining WSN and IoT needs to look into this issue as well.

7.4.2 Hardware

The other issue that will be faced while integrating WSNs with IoT is the nature of hardware devices. The sensor devices consist of transceivers, batteries, processing devices, and sensors. The research issues consist of energy consumption minimization, maximizing the node's processing capability, and having the hardware security of the device.

1. Energy: The job of a sensor device is to have minimal energy consumption while performing the sensing, transmission, and analyzing job. This is so because the batteries of these devices cannot be changed easily due to their geographical location. Thus, there is a need to have research for developing small-sized batteries having a huge amount of energy [44].

2. Processing: The sensor devices have to implement a range of applications, starting from simple environmental parameter measurements to the capture of multimedia data. Based on application requirements, the sensor devices need to have an ample amount of processing resources to meet the overall objective. Thus, another area of research is development of energy-efficient devices capable of the timely processing of data [45].

3. Sensor device's security: WSN devices are realized using an electronic device that has a microprocessor to perform its tasks. These autonomous devices may be exposed to security attacks; thus, there is a need to design them so that they are less susceptible to attacks based on the targeted applications. These attacks can be done by tampering the device to make it malfunction or by reading the information leaked by the device. Such types of attacks are called side-channel attacks.

   If a sensor node is physically accessible to the end user, it is vulnerable to side-channel attacks. The side channel can be accessed by equipment that can interact with it. One example is the case of simple power analysis (SPA), where a block of data can be correlated with its instantaneous power consumption. This helps in determining the sequence of instructions being executed and thus is useful in planning an attack.

There are two types of side-channel attacks [46]:

1. Active attacks: The normal operation of a device is altered so that it moves into a faulty state, resulting in information leak. Sometimes, the chip can be destroyed as a result of such an attack. In other cases, the state of the device is reversible, as is the case of a glitch attack [47]. However, these attacks are difficult to carry out and require good knowledge of the circuit.
2. Passive attacks: In passive attacks, the computation information is extracted to deduce the actions that it is performing. These attacks are easy to realize using inexpensive devices.

There are some methods that can be used to secure the sensor devices and they include

- Physical shielding: In this process, the sensor device is shielded so that tampering is not possible [48,49]. However, with powerful tools, a similar attack can be carried out. Thus, an active attack can be minimized by guarding the physical shielding of the sensor device.
- Autodestruction: This type of measure can be taken in the cases where an attack has been detected by the sensor device, so it should be autodestructed or reset. But the constraint is to detect the attack and not the false alarm, as the noise in the system is not always the cause of a glitch attack.
- Removal of information leakage: If the attacker knows the key, the side channel leaks useful information. The last solution is to ensure that no information is leaked. However, it is quite an expensive technique, as regular chips normally leak a lot of information. The goal is to minimize the leakage of such information, and it includes computation time detection (timing attacks [50]), electromagnetic emissions [51], and instantaneous power consumption (SPA [52]). This can be reduced by using a computer-aided design that increases the chip security, by decorrelating time and data.

With the integration of WSN with IoT, the hardware security is a research area that needs to be considered for a few critical applications.

7.4.3 Software

The coordination of thousands of sensor devices under limited energy constraints is also a challenge. The developed algorithms have to consider these energy constraints in order to realize an efficient network. There is a need to have necessary data processing (e.g., compression and aggregation) before sending it to the next node so that the transmission energy consumption of a node can be minimized [53]. The network should be self-organizing with minimal human interaction. These challenges are unavoidable in any network, and robust solutions need to be presented. Finally, there is a need to develop smart applications for meeting a desired sensing goal [54]. Some of the important software challenges that might be faced while integrating WSN into IoT are described below.

7.4.3.1 Heterogeneity and Integration

One of the important characteristics of an IoT-based network is the integration of heterogeneous devices for achieving a certain goal. These diverse heterogeneous devices make the integration process complicated. There are different IoT platforms that devise their own mechanisms for incorporating the data streams for other nonproprietary sources [55,56]. The other approach, used by Sensinode [57] and SmartThings [58], is to use their own proprietary hardware, which is only compatible with their respective platforms. In order to create a large network of devices and attract developers (in sensor and actuator technologies), there must be a mechanism to integrate the proprietary and open-source devices and platforms. This can be achieved to some extent by abstracting the technical details of the platform to the developer.

7.4.3.2 Scalability

The term *scalability* refers to the adaptability to changes with respect to the load, increasing number of devices, or users. In the case of IoT, the platform deployed at a smaller scale, such as smart gateways, can allow access to any necessary devices. These gateways are limited in resources, especially hardware resources. As a result, the devices may not be able to connect when a saturation stage is achieved. Therefore, we need to find a trade-off for small- and medium-scale deployments in order to optimize the performance with an increasing number of devices. The other way could be to use cloud services to manage the large number of devices per user.

7.4.3.3 Data Mining

One of the important concepts in IoT phenomena is semantics. For managing a number of things that are generating a large amount of data, it is necessary to implement the techniques for extracting useful data and then making a decision. Simple filtering mechanisms are typically defined by means of first-order rule engines or data source combinations (e.g., aggregation or fusion of data sources) by applying functions to aggregate them. Data filtering techniques are very often related to the detection of events. Other approaches take advantage of semantic web techniques in order to disseminate data by performing complex inferences [59].

7.5 Current Research Projects

The convergence of IoT and WSNs presents different challenges when integrated into a single network having uniform middleware architecture. The challenges are tackled by different research projects. In this section, we discuss the current ongoing research in this domain.

7.5.1 Current and Future Research

The IoT has huge potential to develop an information-based economy and society. There are a large number of ongoing research projects in different application fields. In this subsection, we highlight the key objectives and achievements for these projects.

The AmI-4-SME [60] focused on converting the key solutions into a suitable price range. This project has three building blocks for realizing innovative human-centered solutions. The first building block is the RFID-based sensor system, mobile readers, and middleware that are compatible for integration with small and medium enterprises (SME) infrastructures. The second block is the speech recognition system for implementing configurable human interaction on mobile devices, and the third is to enable a flexible, secure, and efficient configuration, mapping, and interfacing of legacy systems, services, and mobile devices.

Table 7.1   Comparative Analysis of IoT Projects

Project/Research Area	Security	Hardware	Software
AmI-4-SME	Yes	Yes	Yes
ASPIRE	–	Yes	Yes
CASCADAS	Yes	–	Yes

ASPIRE [61] aims to focus on nonproprietary platforms and standards to resolve the issues concerning heterogeneity and integration, as discussed in Section 7.4. The open-source objective of ASPIRE requires flexibility of the hardware and standards that support the RFID solutions being built based on the ASPIRE middleware platform. The objective of ASPIRE is to provide a vendor-free middleware that can support different tag formats. Thus, this platform provides the flexibility of being programmable and configurable.

The implementation of RFID and EPCglobal standard solutions is hindered by a number of technical, social, and educational challenges. The objective of the BRIDGE project was to research, develop, and implement tools to enable the deployment of EPCglobal applications in Europe. Seven business work packages were set up to identify the opportunities, establish the business cases, and perform trials and implementations in various sectors, including anticounterfeiting, pharmaceuticals, textiles, manufacturing, reusable assets, products in service, and retail nonfood items.

The goal of CASCADAS [62] is to provide an automatic component-based framework that can support the deployment of a novel set of services, via distributed applications, which can cope with dynamic and uncertain environments, that is, having self-configuration, self-healing, self-optimization, and self-protection (self-CHOP) capabilities.

Wireless sensor networks are seen as one of the most promising technologies that will bridge the physical and virtual worlds, enabling them to interact. Expectations go beyond the research visions, toward their deployment in real-world applications that would empower business processes and future business cases. Advances in networked embedded systems were applied to embed business logic in the physical entities to create so-called collaborative business items. The business processes can be further improved by efficient utilization of these devices.

The classification of the targeted research area by each of the mentioned projects is listed in Table 7.1.

7.6 Applications of IoT

The term *things* can have different perceptions depending on the domain in which it is used. In industrial applications, the thing may typically be anything that participates in the product life cycle. It may refer to the trees, a building, condition measurement devices, and so forth, in an environmental application. The thing may also be related to devices within public spaces or devices for ambient assisted living and so on in society. Although IoT has huge potential for developing numerous applications, only a few of them have actually been deployed. In the following subsections, we discuss the important applications of IoT.

7.6.1 Aerospace and Aviation

IoT can assist in enhancing the safety and security issues in the aerospace and aviation industry. The important threat to the aviation industry is the use of suspected unapproved parts (SUPs). A SUP is an aircraft part that does not meet the requirements of an original and approved aircraft part. Thus, the SUP not only breaches the quality constraint of the aviation industry, but also violates the security standards of an aircraft. In the United States, 28 incidents of aircraft accidents have been reported due to SUP issues [63].

The IoT can resolve the issue of SUP to a large extent by using RFID tags. The data of the origin and safety-critical events of the material could be stored electronically along with that part. This data can be placed in a database, as well as on RFID tags. Later on, this data could be compared before installing it on the aircraft. Moreover, the existing sensors attached to the aircraft for different purposes, such as air pressure and temperature, provide useful data that can be used for maintenance, planning, troubleshooting, and reducing the energy consumption of the aircraft.

7.6.2 Smart Cities, Homes, and Buildings

Smart cities refers to the ecosystems that have emerged by deploying advanced communication infrastructure in the cities. The IoT can contribute in different ways by optimizing the traffic control systems, power grids, vehicle parking, and so forth. The IoT technologies can assist in the detection of congestion on a certain route. Nowadays, the majority of travelers carry smart phones equipped with different sensors. These sensors can transmit information related to congestions, accidents, and climatic issues such as fog and sand storms. Thus, it is possible to monitor the traffic and provide appropriate routing advice [64]. The concept of smart homes and buildings using IoT can be extremely beneficial in terms of operational and societal advantages. The operational perspective can be the optimization of resources such as electricity and water, while the societal perspective can be the reduction of carbon. The sensors can be a major contributor to this, which actually monitor the current consumption and then also detect the user need at a specific interval of time. On the basis of this monitoring and detection system, the decision could be made to control the resources.

7.6.3 Automotive Industry

The automotive industry utilizes the concept of smart things for monitoring and reporting different issues related to vehicles. RFID technology can enhance customer service by improving the logistics, streamlining the production of vehicles, and increasing quality control. The originality of the part is one of the important problems in the automotive industry. Sometimes it is difficult to differentiate between genuine parts and fake parts. This problem can be resolved by providing information on the manufacturer, the origin and history of the product, the serial number associated with the product, and the exact location of the product.

RFID technology can be extremely useful in the manufacturing and maintenance process (putting RFID tags on different parts for future reference) and provides an effective method for managing recalls. The use of RFID or any other identifiable technology provides insight about every part and its location, which helps in accelerating the assembly process and locating components in a fraction of time. Wireless technology is a solution to enable real-time locating systems (RTLSs) and to connect with other IoT-based subnetworks. This will improve vehicle tracking and support automotive manufacturers to manage the process of testing and verifying vehicles.

The higher bit rates and lower interference is possible with the use of dedicated short-range communication (DSRC). Intelligent transportation system (ITS) applications, such as vehicle safety services and traffic management, will be significantly advanced by the use of vehicle-to-vehicle (V2V) and vehicle-to-infrastructure (V2I) communications and will be fully integrated in the IoT infrastructure.

7.6.4 Environmental Monitoring

The environment conservation and related applications can utilize IoT technologies. The new standards aim for a data rate of up to 1 Mbps for WSNs and RFIDs that can help to realize applications such as disastrous situation handling and healthcare applications. The smart environment domain targets the prevention of critical events in unpopulated areas where environmental risks are present. The critical events consist of forest fire detection [65], avalanches, landslides, and so forth. The challenge is to monitor these events without installing fixed communication infrastructures and by using minimal energy and processing resources. The other applications may include the collection of remote seismic data [66] by using microphones or specialized sensors. The main challenge for such applications is the need for data processing in hard conditions. Pollution monitoring in urban and remote areas can also be realized by using wireless sensor networks [67].

Classical and underground WSNs incorporating IoT technologies and appropriate sensors can be used to monitor environmental monitoring parameters, explained briefly in this section.

7.6.5 Logistics Applications

In this domain, the system supports the storage and transportation of goods. The application should support goods monitoring [68], improper storage detection for a particular item, and tracking the current location of goods [69,70]. The challenges include system trust and reliability for goods monitoring, large-scale operations, and standard tagging solutions.

These transportation applications can be well realized using IoT integrating the sensor devices. Utilizing IoT technologies in conjunction with classical WSNs integrating appropriate sensors may allow the transportation of cargo in an efficient manner, as a huge amount of cargo is transported by using air, sea, and ground transportation mechanisms.

7.6.6 Food Traceability

IoT offers promising solutions for recalling a certain product when a certain problem arises. The IoT devices are attached to certain items so that based on the requirement, an alert can be generated, and this will minimize the chance of spreading a disease. Classical WSNs incorporating IoT technologies and appropriate sensors can be used to monitor a certain product.

7.6.7 Agricultural Applications

IoT can be used for various applications in agriculture, such as smart agriculture and smart animal farming. Smart agriculture focuses on soil and environmental condition monitoring for maximizing the yield and ensuring the product quality [71]. The main challenges of this domain are operations in harsh environments and lack of proper infrastructure. These smart greenhouses can be efficiently controlled by using classical WSNs [72]. Single farmers can also benefit from this technology to deliver crops to a certain region by direct marketing, and this will lead to efficient supply chain management.

Another application is smart animal farming for enhancing the productivity and quality of animal meat and associated products by periodically monitoring the animal health. It also takes care of animal behavior for detecting any signs of sickness or stress in the animals [73]. The main challenge is the integration of data obtained from a number of sensors deployed on the animals that are moving about the farm. Moreover, the traceability of animals requires the use of IoT for detecting them. The identification of animals is necessary for the application of disease control and eradication, as necessary. This can lead to accurate identification of the health status of herds, which will eventually increase the overall gain. Another application is the meat traceability from farms to markets during various operation phases [74].

Classical and underground WSNs can be used to realize most of these applications by appropriately integrating IoT technologies.

7.6.8 Media Coverage

Media news is envisaged to be gathered through an IoT infrastructure that has the capability to capture certain multimedia footage. The devices present near any such location can be asked to gather particular information related to an event. This will revolutionize the media. Wireless multimedia sensor networks in conjunction with IoT can be used to realize these applications.

7.6.9 Smart Water Monitoring

These applications include monitoring of water quality (level of various chemicals) and water distribution statistics (water leakage, level of water in reservoirs, etc.) by means of pressure and flow sensors [75]. Similarly, level monitoring of lakes, rivers, and dams may help to prevent floods in a certain region [76]. The challenges are continuous monitoring, processing, and communication of sensed data under real-time constraints. These challenges can be realized using underwater sensor networks in conjunction with IoT technologies.

7.6.10 Vehicle Insurance

Insurance premiums can be reduced by using IoT technologies supported by insurance companies. If clients are willing to install electronic recorders that can record acceleration, speed, and other relevant parameters of the vehicle, then they may be offered an attractive rate [77]. Insurance companies can remotely monitor various parameters through installed devices by triggering an economic action at the time of an accident. The same may be applied to other applications in similar categories, such as equipment and building. Classical WSNs incorporating IoT technologies and appropriate sensors can be used to monitor certain vehicles for realizing such applications.

7.6.11 Material Recycling

IoT and WSNs can be used to have a greener environment by ensuring an optimal recycling process. The collection of recyclable material and the disposal of e-waste can be achieved by using RFID to identify a certain electronic product. The recycled material can be processed in an efficient manner by using the installed IoT infrastructure. Classical WSNs incorporating IoT technologies and appropriate sensors can be used to realize these applications.

7.6.12 Home Automation

These applications are centered on the domestic and commercial buildings for the improvement of building safety and sustainability by remotely controlling the appliances and by implementing theft prevention mechanisms. Building automation optimizes and monitors the subsystems installed in buildings, such as heating, air conditioning, and ventilation [78]. Moreover, appliance control is needed for energy usage optimization in smart grid environments [79]. The classical WSNs can be used to realize these targeted applications in combination with IoT technologies.

7.7 Conclusion

The concept of IoT provides a method of communication between uniquely identifiable objects. These objects can be anything having the capability of communication. The vision of IoT allows participating objects to transport themselves, to optimize their performance and energy, and to reconfigure themselves in a new environment. Therefore, the technologies, such as sensors, smart phones, nanoelectronics and other wireless identifiable devices, are the essential part of an IoT architecture. These wireless devices combine the characteristics of sensor networks and other wireless technologies to react autonomously to real-world situation without human intervention.

From the perspective of technology, there are several areas that should be the focus of research activities for realizing the vision of IoT. The development of efficient energy sources and energy generation devices will be the essential factor for the effective and long-lasting usage of IOT objects. Moreover, the intelligence of IoT devices from the perspective of context awareness and interdevice communication is extremely important. On the basis of this communication and interpretation, IoT devices will reach a logical conclusion for performing an action. Apart from the above-mentioned issues, integration, interoperability, and security are major concerns for implementing the concept of IoT.

References

1. G. Kortuem, F. Kawsar, V. Sundramoorthy, and D. Fitton, Smart objects as building blocks for the Internet of Things, *IEEE Internet Computing*, vol. 14, no. 1, pp. 44–51, 2009.
2. M. Zorzi, A. Gluhak, S. Lange, and A. Bassi, From today's INTRAnet of things to a future INTERnet of things: A wireless- and mobility-related view, *IEEE Wireless Communications*, vol. 17, no. 6, pp. 44–51, 2010.
3. J. Vasseur and A. Dunkels, *Interconnecting Smart Objects with IP—The Next Internet*, Morgan Kaufmann, Burlington, MA, 2010.
4. L. Mainetti, L. Patrono, and A. Vilei, Evolution of wireless sensor networks towards the Internet of Things: A survey, in *2011 19th International Conference on Software, Telecommunications and Computer Networks (SoftCOM)*, Dalmatia, Croatia, September 15–17, 2011, pp. 1–6.
5. D. Giusto, A. Iera, G. Morabito, and L. Atzori (eds.), *The Internet of Things*, Springer, Berlin, 2010.
6. National Intelligence Council, Disruptive civil technologies—Six technologies with potential impacts on US interests out to 2025, Conference Report CR 2008-07, National Intelligence Council, Washington, DC, April 2008, http://www.dni.gov/nic/NIC_home.html.
7. R. van Kranenburg, E. Anzelmo, A. Bassi, D. Caprio, S. Dodson, and M. Ratto, The Internet of Things, in *Proceedings of 1st Berlin Symposium of the Internet Society*, Berlin, Germany, 2011, pp. 25–27.
8. http://smartdatacollective.com/ (accessed March 14, 2015).

9. European Commission, Internet of Things in 2020: Roadmap for the future, Technical report, Working Group RFID of the ETP EPOSS, May 2008. Available at http://ec.europa.eu/information society/policy/rfid/documents/iotprague2009.pdf (accessed on December 12, 2014).
10. T. Lu and W. Neng, Future Internet: The Internet of Things, in *3rd International Conference on Advanced Computer Theory and Engineering (ICACTE)*, Chengdu, Sichuan Province, China, August 2010, vol. 5, pp. V5-376–V5-380. Available at http://dx.doi.org/10.1109/ICACTE.2010.5579543.
11. P. Guillemin and P. Friess, Internet of Things strategic research roadmap, Technical report, Cluster of European Research Projects, September 2009. Available at http://www.internet-of-things-research .eu/pdf/IoT Cluster Strategic Research Agenda 2009.pdf (accessed December 25, 2014).
12. A. Boonsongsrikul, S. Kocijancic, and S. Suppharangsan, Effective energy consumption on wireless sensor networks: Survey and challenges, in *2013 36th International Convention on Information and Communication Technology Electronics and Microelectronics (MIPRO)*, Opatija, Primorje-Gorski Katar County, Croatia, May 20–24, 2013, pp. 469–473.
13. N. Barroca, L. M. Borges, F. J. Velez, F. Monteiro, M. Górski, and J. Castro-Gomes, Wireless sensor networks for temperature and humidity monitoring within concrete structures, *Construction and Building Materials*, vol. 40, pp. 1156–1166, 2013.
14. R. Cavallari, F. Martelli, R. Rosini, C. Buratti, and R. Verdone, A survey on wireless body area networks: Technologies and design challenges, *IEEE Communications Surveys and Tutorials*, vol. 16, no. 3, pp.1635–1657, 2014.
15. Z. A. Khan and Y. Faheem, Cognitive radio sensor networks: Smart communication for smart grids— A case study of Pakistan, *Renewable and Sustainable Energy Reviews*, vol. 40, pp. 463–474, 2014.
16. R. Cavallari, F. Martelli, R. Rosini, C. Buratti, and R. Verdone, A survey on wireless body area networks: Technologies and design challenges, *IEEE Communications Surveys and Tutorials*, vol. 16, no. 3, pp. 1635–1657, 2014.
17. F. Fabbri, C. Buratti, and R. Verdone, A multi-sink multi-hop wireless sensor network over a square region: Connectivity and energy consumption issues, in *2008 IEEE GLOBECOM Workshops*, New Orleans, Southeastern, LA, November 30–December 4, 2008, pp. 1–6.
18. I. F. Akyildiz, W. Su, Y. Sankarasubramaniam, and E. Cayirci, Wireless sensor networks: A survey, *Computer Networks*, vol. 38, no. 4, pp. 393–422, 2002.
19. E. Gurses and O. B. Akan, Multimedia communication in wireless sensor networks, *Annals of Telecommunications*, vol. 60, no. 7–8, pp. 799–827, 2005.
20. S. Misra, M. Reisslein, and X. Guoliang, A survey of multimedia streaming in wireless sensor networks, IEEE *Communications Surveys and Tutorials*, vol. 10, no. 4, pp. 18–39, 2008.
21. I. F. Akyildiz, W. Su, Y. Sankarasubramaniam, and E. Cayirci, Wireless sensor networks: A survey, *Computer Networks*, vol. 38, no. 4, pp. 393–422, 2002.
22. M. Rahimi, R. Baer, O. Iroezi, J. Garcia, J. Warrior, D. Estrin, and M. Srivastava, Cyclops: In situ image sensing and interpretation in wireless sensor networks, in *Proceedings of the ACM Conference on Embedded Networked Sensor Systems (SenSys)*, San Diego, CA, November 2005.
23. MICA2 mote specifications, Moog Crossbow, Milpitas, CA. Available at http://www.xbow.com.
24. MICAz mote specifications, Moog Crossbow, Milpitas, CA. Available at http://www.xbow.com.
25. I. F. Akyildiz, T. Melodia, and K. R. Chowdhury, A survey on wireless multimedia sensor networks, *Computer Networks*, vol. 51, no. 4, pp. 921–960, 2007.
26. M. Cesana, A. Redondi, N. Tiglao, A. Grilo, J. M. Barcelo-Ordinas, M. Alaei, and P. Todorova, Real-time multimedia monitoring in large-scale wireless multimedia sensor networks: Research challenges, in *2012 8th EURO-NGI Conference on Next Generation Internet (NGI)*, Karls Krona, Blckinge County, Sweden, June 25–27, 2012, pp. 79–86.
27. Advanced Aeration Systems, Rz-aer tech sheet. Available at http://www.advancedaer.com/.
28. U.S. Water News, Street spies detect water leakage, December 1998. Available at http://www.uswater news.com.
29. K. Imanishi, W. Ellsworth, and S. G. Prejean, Earthquake source parameters determined by the SAFOD pilot hole seismic array, *Geophysical Research Letters*, vol. 31, no. 12, 2004.
30. Underwater acoustic sensor networks, BWN Laboratory, Georgia Institute of Technology, Atlanta. Available at http://www.ece.gatech.edu/research/labs/bwn/UWASN/.

31. X. Yang, K. G. Ong, W. R. Dreschel, K. Zeng, C. S. Mungle, and C. A. Grimes, Design of a wireless sensor network for long-term, in-situ monitoring of an aqueous environment, *Sensors*, vol. 2, pp. 455–472, 2002.
32. B. Zhang, G. S. Sukhatme, and A. A. Requicha, Adaptive sampling for marine microorganism monitoring, in *IEEE/RSJ International Conference on Intelligent Robots and Systems*, Sendai, Tohoku Region, Japan, 2004, pp. 1115–1122.
33. N. N. Soreide, C. E. Woody, and S. M. Holt, Overview of ocean based buoys and drifters: Present applications and future needs, in *16th International Conference on Interactive Information and Processing Systems (IIPS) for Meteorology, Oceanography, and Hydrology*, Honolulu, HI, January 2004, pp. 2470–2472.
34. E. Cayirci, H. Tezcan, Y. Dogan, and V. Coskun, Wireless sensor networks for underwater surveillance systems, *Ad Hoc Networks*, vol. 4, pp. 431–446, 2004.
35. J. G. Proakis, E. M. Sozer, J. A. Rice, and M. Stojanovic, Shallow water acoustic networks, *IEEE Communications Magazine*, vol. 39, no. 11, pp. 114–119, 2001.
36. L. Mainetti, L. Patrono, and A. Vilei, Evolution of wireless sensor networks towards the Internet of Things: A survey, in *19th International Conference on Software, Telecommunications and Computer Networks (SoftCOM)*, 2011.
37. D. Miorandi, S. Sicari, F. De Pellegrini, and I. Chlamtac, Internet of Things: Vision, applications and research challenges, *Ad Hoc Networks*, vol. 10, no. 7, pp. 1497–1516, 2012.
38. D. Partynski and S. G. M. Koo, Integration of smart sensor networks into Internet of Things: Challenges and applications, in *IEEE International Conference on Green Computing and Communications (GreenCom), 2013 IEEE and Internet of Things (iThings/CPSCom), and IEEE Cyber, Physical and Social Computing*, Beijing, China, August 20–23, 2013, pp. 1162–1167.
39. H. Chan and A. Perrig, Security and privacy in sensor networks, *Computer*, vol. 36, no. 10, pp. 103–105, 2003.
40. J. Undercoffer, S. Avancha, A. Joshi, and J. Pinkston, Security for sensor networks, in *CADIP Research Symposium*, Baltimore, 2002, pp. 1–11.
41. D. Christin, A. Reinhardt, P. Mogre, and R. Steinmetz, Wireless sensor networks and the Internet of Things: Selected challenges, in *Proceedings of the 8th Fachgespräch Drahtlose Sensornetze*, Hamburg, Germany, 2009, pp. 54–57.
42. A. Perrig, J. Stankovic, and D. Wagner, Security in wireless sensor networks, *Communications of the ACM*, vol. 47, no. 6, pp. 53–57, 2004.
43. R. Roman and J. Lopez, Integrating wireless sensor networks and the Internet: A security analysis, *Internet Research*, vol. 19, no. 2, pp. 246–259, 2009.
44. D. Puccinelli and M. Haenggi, Wireless sensor networks: Applications and challenges of ubiquitous sensing, *IEEE Circuits and Systems Magazine*, vol. 5, no. 3, pp. 19–31, 2005.
45. I. F. Akyildiz, W. Su, Y. Sankarasubramaniam, and E. Cayirci, Wireless sensor networks: A survey, *Computer Networks*, vol. 38, no. 4, pp. 393–422, 2002.
46. S. Guilley and R. Pacalet, SoCs security: A war against side-channels, *Annals of Telecommunications*, vol. 59, no. 7–8, pp. 998–1009, 2004.
47. G. Canivet, P. Maistri, R. Leveugle, J. Clediere, F. Valette, and M. Renaudin, Glitch and laser fault attacks onto a secure AES implementation on a SRAM-based FPGA, *Journal of Cryptology*, vol. 24, no. 2, pp. 247–268, 2011.
48. R. Anderson and M. Kuhn, Tamper resistance—A cautionary note, in *Proceedings of the Second Usenix Workshop on Electronic Commerce*, Oakland, CA, November 1996, pp. 1–11.
49. R. Anderson and M. Kuhn, Low cost attacks on tamper resistant devices, in *Proceedings of 5th International Workshop of Security Protocols*, Paris, France, April 7–9, 1997, vol. 1361, pp. 125–136.
50. P. Kocher, J. Jaffe, and B. Jun, Timing attacks on implementations of Diffie-Hellman, RSA, DSS, and other systems, in *Proceedings of CRYPTO '96*, Santa Barbara, CA, 1996, vol. 1109, pp. 104–113.
51. K. Gandolfi, C. Mourtel, and F. Olivier, Electromagnetic analysis: Concrete results, in *Proceedings of CHES '01*, Paris, France, 2001, vol. 2162, pp. 251–261.
52. P. Kocher, J. Jaffe, and B. Jun, Differential power analysis: Leaking secrets, in *Proceedings of CRYPTO '99*, Santa Barbara, CA, 1999, vol. 1666, pp. 388–397.

53. N. Kimura and S. Latifi, A survey on data compression in wireless sensor networks, in *International Conference on Information Technology: Coding and Computing (ITCC) 2005*, Las Vegas, NV, 2005, vol. 2, pp. 8–13.

54. J. Blumenthal, M. Handy, F. Golatowski, M. Haase, and D. Timmermann, Wireless sensor networks—New challenges in software engineering, in *Proceedings of IEEE Conference on Emerging Technologies and Factory Automation (ETFA) 2003*, 2003, vol. 1, pp. 551–556.

55. R. Bhattacharya, C. Florkemeier, and S. Sarma, Towards tag antenna based sensing—An RFID displacement sensor, in *Proceedings of 2009 International Conference on RFID*, Orlando, FL, 2009, pp. 95–102.

56. G. Marrocco, C. Occhiuzzi, and F. Amato, Sensor-oriented passive RFID, in *The Internet of Things*, Springer, Berlin, Germany, 2010, Part 4, pp. 273–282.

57. R. Bhattacharyya, C. Floerkemeier, S. Sarma, and D. Deavours, RFID tag antenna based temperature sensing in the frequency domain, in *Proceedings of 2011 IEEE International Conference on RFID*, Orlando, FL, 2011, pp. 70–77.

58. J. Gao, J. Siden, and H. E. Nilsson, Printed electromagnetic coupler with an embedded moisture sensor for ordinary passive RFID tags, *IEEE Electron Device Letters*, vol. 32, no. 12, pp. 1767–1769, 2011.

59. K. Chang, Y. H. Kim, Y. J. Kim, and Y. J. Yoon, Functional antenna integrated with relative humidity sensor using synthesized polyimide for passive RFID sensing, *Electronics Letters*, vol. 47, no. 5, pp. 7–8, 2007.

60. AmI-4-SME project. Available at http://www.ami4sme.org/ (accessed February 5, 2015).

61. ASPIRE project. Available at http://www.fp7-aspire.eu/ (accessed February 5, 2015).

62. CASCADAS project. Available at http://acetoolkit.sourceforge.net/cascadas/ (accessed February 5, 2015).

63. D. Bandyopadhyay and J. Sen, Internet of Things: Applications and challenges in technology and standardization, *Wireless Personal Communications*, vol. 58, pp. 49–59, 2011.

64. D. Miorandi, S. Sicari, F. D. Pellegrini, and I. Chlamtac, Internet of Things: Vision, applications and research challenges, *Ad Hoc Networks*, vol. 10, no. 7, pp. 1497–1516, 2012.

65. L. Yu, N. Wang, and X. Meng, Real-time forest fire detection with wireless sensor networks, in *Proceedings of International Conference on Wireless Communications, Networking and Mobile Computing 2005*, Wuhan, Hubei Province, China, 2005, vol. 2, pp. 1214–1217.

66. G. Werner-Allen, K. Lorincz, M. Ruiz, O. Marcillo, J. Johnson, J. Lees, and M. Welsh, Deploying a wireless sensor network on an active volcano, *IEEE Internet Computing*, vol. 10, no. 2, pp. 18–25, 2006.

67. L. E. Cordova-Lopez, A. Mason, J. D. Cullen, A. Shaw, and A. I. Al-Shamma'a, Online vehicle and atmospheric pollution monitoring using GIS and wireless sensor networks, *Journal of Physics*, vol. 76, p. 012019, 2007.

68. M. Forcolin, E. Fracasso, F. Tumanischvili, and P. Lupieri, EURIDICE—IoT applied to logistics using the intelligent cargo concept, in *2011 17th International Conference on Concurrent Enterprising (ICE)*, Aachen, North Rhine-Westphalia, Germany, June 20–22, 2011, pp. 1–9.

69. B. Anderseck, A. Hille, S. Baumgarten, T. Hemm, G. Ullmann, P. Nyhuis, J.-M. Potthast et al., smaRTI: Deploying the Internet of Things in retail supply chains, *Logistics Journal*, pp. 9–14, 2013.

70. A. Carullo, S. Corbellini, M. Parvis, and A. Vallan, A wireless sensor network for cold-chain monitoring, *IEEE Transactions on Instrumentation and Measurement*, vol. 58, no. 5, pp. 1405–1411, 2009.

71. T. Wark, P. Corke, P. Sikka, L. Klingbeil, Y. Guo, C. Crossman, P. Valencia, D. Swain, and G. Bishop-Hurley, Transforming agriculture through pervasive wireless sensor networks. *IEEE Pervasive Computing*, vol. 6, no. 2, pp. 50–57, 2007.

72. D. D. Chaudhary, S. P. Nayse, and L. M. Waghmare, Application of wireless sensor networks for greenhouse parameter control in precision agriculture, *International Journal of Wireless and Mobile Networking*, vol. 3, no. 1, pp. 140–149, 2011.

73. A. Scalera, P. Brizzi, R. Tomasi, T. Gregersen, K. Mertens, J. Maselyne, A. Van Nuffel, E. Hessel, and H. Van den Weghe, The pigwise project: A novel approach in livestock farming through synergistic performances monitoring at individual level, in *Proceedings of Conference on Sustainable Agriculture through ICT Innovation (EFITA 2013)*, Torino, Piedmont, Italy, June 2013, pp. 1–8.

74. P. Brizzi, D. Conzon, F. Pramudianto, M. Paralic, M. Jacobsen, C. Pastrone, R. Tomasi, and M. A. Spirito, Bringing the Internet of Things along the manufacturing line: A case study in controlling industrial robot and monitoring energy consumption remotely, in *IEEE International Conference on Emerging Technologies and Factory Automation (ETFA 2013)*, Cagliari, Sardinia, Italy, 2013, pp. 1–8.

75. D. Trinchero, A. Galardini, R. Stefanelli, and B. Fiorelli, Microwave acoustic sensors as an efficient means to monitor water infrastructures, in *Proceedings of IEEE MTT-S International Microwave Symposium Digest 2009 (MTT '09)*, Boston, 2009, pp. 1169–1172.

76. M. Castillo-Effer, D. H. Quintela, W. Moreno, R. Jordan, and W. Westhoff, Wireless sensor networks for flash-flood alerting, in *Proceedings of the Fifth IEEE International Caracas Conference on Devices, Circuits and Systems 2004*, Punta Cana, Dominican Republic, 2004, vol. 1, pp. 142–146.

77. V. Coroama, The smart tachograph—Individual accounting of traffic costs and its implications, in *Proceedings of Pervasive 2006*, Dublin, Ireland, May 7–10, 2006, pp. 135–152.

78. A. Wheeler, Commercial applications of wireless sensor networks using ZigBee, *IEEE Communications Magazine*, vol. 45, no. 4, pp. 70–77, 2007.

79. T. Hubert and S. Grijalva, Realizing smart grid benefits requires energy optimization algorithms at residential level, in *IEEE Conference* on Innovative Smart Grid Technologies (ISGT), Anaheim, CA, January 17–19, 2011, pp. 1–8.

Chapter 8

Wireless Sensor Network Management Using Satellite Communication Technologies

Marios I. Poulakis, Stavroula Vassaki, Georgios T. Pitsiladis, Charilaos Kourogiorgas, Athanasios D. Panagopoulos, Georgios Gardikis, and Socrates Costicoglou

Contents

Abstract

Modern satellite communications constitute an inevitably promising solution for the management of remote area wireless sensor networks (WSNs). Toward this direction, this chapter addresses the exploitation of satellite communication technologies in WSNs, proposing possible implementation architectures. Satellite-based WSNs can be used in various scenarios, such as environmental monitoring, critical and emergency infrastructure monitoring, and surveillance/monitoring, while they constitute an appropriate designing solution for M2M systems. In this chapter, first, the state-of-the-art technologies for WSNs and satellite systems are briefly discussed. Next, two efficient architecture scenarios are proposed for the practical implementation of satellite-based WSNs. The first scenario addresses the direct communication between sensors and satellite (direct access architecture), and the latter addresses the indirect communication of sensors with the satellite through a gateway (gateway-based access architecture). In the direct-access scenario, the high-capability sensors individually communicate directly with the satellite, while the low-capability sensors employ the collaborative beam-forming technique to increase transmission range and reach the faraway located satellite. Regarding the second architecture, a fixed or mobile gateway gathers the total data of the sensors and forwards them to the satellite. A general model of link budget analysis for the gateway–satellite communication is also presented. Finally, some major issues appearing in those systems, including the synchronization, the QoS, and the connectivity concepts in addition to other challenges and future applications, are discussed.

8.1 Introduction

Wireless sensor networks (WSNs) constitute an essential component for modern communication technologies (Karl and Willig, 2007). They are amenable to supporting several real-world applications due to the flexibility they provide. Some examples of applications are the surveillance and monitoring of remote areas, emergency communications, support for supervisory control and data acquisition (SCADA) systems, critical infrastructures, and environmental monitoring. Moreover, the concept of machine-to-machine (M2M) communications implies end-to-end communication among a large number of devices (such as meter readers and monitoring sensors) without any human intervention (Chen and Lien, 2013). M2M communication systems are characterized by an enormous number of devices, low data rates per device, energy efficiency requirements, and large coverage areas. An appropriate example of wireless technology that can be employed in those systems is WSNs.

Moreover, many applications require the installation of sensors in areas where there are no terrestrial infrastructures or broadband wireless terrestrial communications, such as forests, open seas, and islands. In these common applications, the employment of satellite communication technologies is the only efficient and feasible solution to achieve the desired communication. Additionally, the current advances of satellite communication networks (Maini and Agrawal, 2011) (e.g., the use of Ka-band frequencies for capacity and high communication data rates) make them an inevitably promising solution in remote area WSN applications (Celandroni et al., 2013). However, the coexistence of these two types of communication networks has led to the necessity of the efficient design and implementation of unified satellite-based WSN architectures.

The objective of this chapter is twofold. First, a brief survey of the existing technologies for WSNs and satellite systems is presented to familiarize the readers with these communication systems. Specifically, the most common standards of WSNs are surveyed, reviewing the functionalities of a pure WSN environment and highlighting the issues that might arise in the integration with other network elements, such as a satellite network. Furthermore, modern satellite technologies, including well-known standards, systems, and modern satellite constellations, are briefly discussed.

Afterwards, efficient architecture scenarios for the practical implementation of satellite-based WSNs are proposed. Particularly, two architecture scenarios are presented and discussed. The first scenario addresses the direct communication between sensors and a satellite (*direct access architecture*). In this scenario, two different approaches are investigated: the approach where the sensor nodes have high capabilities and can individually communicate directly with the satellite, and the case of sensors with low capabilities where the collaborative beamforming (CB) technique is employed to reach faraway satellites. CB is a signal transmission technique that can improve the transmission range and energy efficiency of wireless networks. It constitutes the only way to succeed in direct communication between low-capability sensors and the satellite without a gateway. In the direct access scenario, it is more appropriate to use lower operating frequencies (e.g., S-band), due to smaller path loss and the communication and energy constraints of wireless sensor nodes. In addition, regarding the second architecture scenario, the indirect communication of sensors with a satellite through a gateway (*gateway-based access architecture*) is considered. Specifically, the fixed or mobile gateway gathers the total data of the sensors and forwards it to the satellite. Since gateways have more complex hardware, they can use higher operating frequencies (e.g., Ka-band), where the performance degradation due to the atmospheric phenomena can be compensated by employing diversity techniques, to increase capacity.

Finally, several major issues that should be taken into consideration in satellite-based WSN communication systems are discussed. In particular, the discussion focuses on synchronization issues among the wireless sensors, satisfaction of certain quality of service (QoS) requirements, as well as the sustainment of connectivity among wireless sensors and between the gateway nodes and the satellite. Applications and challenges of satellite-based WSNs are also discussed.

8.2 Survey on WSNs and Satellite Technologies

This section presents a brief survey of the WSNs and satellite technologies that are candidates to be employed in a unified satellite-based WSN.

8.2.1 WSN Technologies: Standardization Activities

In the case of WSNs, the standards define the functions and protocols that are required for the communication among sensor nodes and also their interface with other types of networks. Given that each application of WSNs has different requirements (e.g., battery life and number of nodes), there have been various technologies and standards developed for WSNs that focus on the characteristics of each application. Some of the most commonly employed standards include IEEE 802.15.4, ZigBee, 6LoWPAN, and WirelessHART. In this section, the specific WSN standards are described and compared in terms of their main performance characteristics.

8.2.1.1 IEEE 802.15.4

The basic requirement for low-power consumption led to the release of the well-known IEEE 802.15.4 standard in 2003. The specific standard was defined by the IEEE 802.15.4 Working Group, and it was the first standard for low-rate wireless personal area networks (LR-WPANs). It has been designed to provide low power consumption, low data rates, low cost, and low-complexity wireless sensor communication (Zheng and Lee, 2004). Thus, it targets wireless sensor applications that require short-range communication and have limited resources of power, with the objective to maximize battery life. This standard specifies the two lowest layers of the protocol stack: the physical (PHY) layer and the medium access control (MAC) layer. The upper layer is independently defined by other standards, such as ZigBee, 6LoWPAN, and WirelessHART, which are subsequently described.

According to the IEEE 802.15.4 standard, two types of network nodes are defined: the full-function device (FFD) and the reduced-function device (RFD). The first type refers to nodes that can fully implement the standard, as they are equipped with all the network functionalities. Hence, they can act as either a network coordinator or a network-end device. On the other hand, RFDs constitute basic nodes that can support only a specific set of functionalities due to limited processing and memory capabilities, and they can act only as network-end devices.

Furthermore, the specific standard allows the formation of two network topology types for communication among the network devices: the star topology and peer-to-peer topology. According to the first topology, all the devices communicate with a central controller using a master–slave network model. In particular, an FFD has the role of the network coordinator, whereas all the other nodes (RFDs or FFDs) can communicate only with the coordinator. The peer-to-peer topology allows more complex network formations, such as mesh or hierarchical networks, as the devices are free to communicate directly with each other. Even in this case, devices should at first communicate with the coordinator before their participation in a peer-to-peer communication.

8.2.1.2 ZigBee

The ZigBee specification was defined in 2004 by the ZigBee Alliance as a stack profile that defines the layers above the PHY and MAC layer defined by the IEEE 802.15.4 standard (ZigBee Alliance, 2008) for LR-WPANs. ZigBee defines network, security, and application layers, as depicted in Figure 8.1. In particular, ZigBee is divided into the following main sections: the ZigBee section that consists of the network layer, the application support sublayer, the security service provider, and the ZigBee device object, as well as the ZigBee application section where the developers can develop their own application profiles. The network layer is responsible for routing over the network, whereas the application layer provides an adequate framework for distributed application development (ZigBee Alliance, 2005).

The ZigBee standard constitutes a simple low-data-rate, low-cost, low-power-consumption standard that is employed in embedded applications (Baronti et al., 2007). Its main contribution

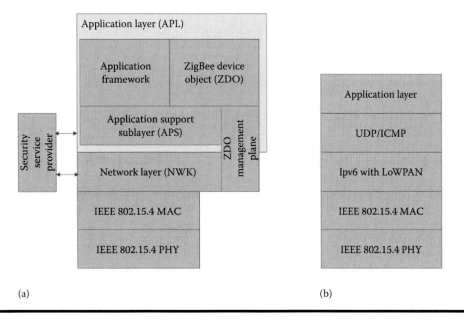

Figure 8.1 (a) ZigBee and (b) 6LoWPAN protocol stacks.

is that it adds mesh networking capabilities by extending the use of peer-to-peer topology. Thus, hundreds to thousands of ZigBee devices can be connected, forming a mesh network. The ZigBee standard consists of three types of devices: ZigBee coordinator, ZigBee router, and ZigBee end device. The first type of device is capable of initiating the formation of the network and "bridging" networks together. ZigBee routers are responsible for the association of device groups and the provision of multihop communication across them. Finally, ZigBee end devices refer to the sensors that collect data and can communicate only with the router or the coordinator.

8.2.1.3 6LoWPAN

The 6LoWPAN standard was developed by the Internet Engineering Task Force (IETF) in 2007, and it refers to IPv6 communication over IEEE 802.15.4 for LR-WPANs. The specific standard enables IPv6 communication directly over low-rate and low-power IEEE 802.15.4 sensor networks (Montenegro et al., 2007) (Figure 8.1). Hence, the wireless sensors in a LoWPAN are accessible from the Internet without using a gateway, and they can communicate directly with other Internet Protocol (IP) devices. To enable the low-power devices to have all the benefits of IP communication, 6LoWPAN provides an adaptation layer, new packet format, and address management. The adaptation layer is employed for the functionalities of header compression, fragmentation, and reassembly, as the IPv6 packet sizes are larger than the frame size of the IEEE 802.15.4, whereas the address management mechanism refers to the formation of device addresses for communication. The most common transport protocol that is employed with 6LoWPAN is the User Datagram Protocol (UDP), which can also be compressed using the LoWPAN format. Moreover, it should be noted that even though 6LoWPAN was originally targeted at IEEE 802.15.4 radio standards, it was later generalized for all similar link technologies, with additional support for IP routing (Hui and Thubert, 2010; Shelby and Bormann, 2011; Shelby et al., 2011).

8.2.1.4 WirelessHART

WirelessHART is an open-standard wireless networking technology proposed by the HART Communication Foundation that is also based on IEEE 802.15.4 for LR-WPANs. The specific standard is more suitable for industrial applications, such as process measurement and control applications (Chen et al., 2010; Song et al., 2008). Similar to ZigBee, WirelessHART is a stand-alone standard, meaning it does not support communication with other networks without the use of a specific gateway device. Regarding the protocol stack that is specified by WirelessHART, the physical layer is based on the IEEE 802.15.4 standard. The network layer is the core of the WirelessHART standard, including responsibilities such as routing, topology control, end-to-end security, and session management. WirelessHART is designed to support mesh, star, and combined network topologies. A WirelessHART network consists of the following nodes: wireless field devices, gateways, process automation controller, host applications, and network manager. Field devices are connected to process or plant equipment. The gateways permit communication among the wireless field devices and host applications, whereas the process automation controller serves as a single controller for a continuous process. The network manager deals with the configuration of the network and the communication's scheduling between the devices. Additionally, it manages the routing and network traffic and monitors network health. Each network contains only one network manager, and it may be incorporated in the gateway, host application, or process automation controller.

Finally, Table 8.1 presents the basic characteristics of the aforementioned WSN standards for comparison purposes.

8.2.2 Satellite Technologies

In this section, the most common satellite communication standards, systems, and satellite constellations are described.

Table 8.1 Characteristics of WSN Standards

Characteristic	ZigBee	6LoWPAN	WirelessHART
Frequency band	868 MHz, 915 MHz, 2.4 GHz	2.4 GHz	2.4 GHz
Maximum throughput (kbps)	250	200	250
Range (m)	10–100	1–100	1–100
Maximum number of nodes	65.536	~100	~30.000
Battery life (days)	100–1000	100–365	760+
Interoperability	High	Low	High
Reliability	Low	Low	High
Application	Industrial/environment monitoring, healthcare, energy-efficient buildings	Monitoring applications, healthcare	Manufacturing/industrial environments

8.2.2.1 Satellite Standards

8.2.2.1.1 DVB Standards

The family of Digital Video Broadcasting (DVB) (2005) standards consists of different versions of the DVB standard, depending on the properties of the transmission channel, such as DVB-C for cable, DVB-S/S2 for satellite, DVB-RCS/RCS2 for the return channel, and DVB-SH for satellite handheld terminals. Subsequently, some of the basic versions of the DVB standard are briefly described to highlight their main characteristics and applications.

8.2.2.1.1.1 DVB-S/S2   DVB-S is the first DVB standard that transports baseband TV signals in the format of an MPEG-2 transport stream over the satellite channel (Maral and Bousquet, 2009). It employs quadrature phase shift keying (QPSK) modulation with direct mapping, convolutional forward error correction (FEC), and concatenated Reed–Solomon channel coding. The specific standard has been adopted by the majority of satellite operators for data and television broadcasting services, as it does not require local access to the fixed telecommunication network.

The significant evolution of satellite transmission technology and the demand increase, from both the operators and the consumers, have led to the development of the second-generation system for satellite broadband services, DVB-S2 (Morello and Mignone, 2006). DVB-S2 benefits from the developments in technology and specifically the advances in channel coding and modulation. In particular, the use of the adaptive coding and modulation (ACM) technique leads to optimization of the transmission parameters and different levels of error protection for different services. Furthermore, the new channel coding schemes that are employed increase the spectrum utilization efficiency, resulting in a capacity gain of around 30% compared to DVB-S.

DVB-S2 is compatible with both MPEG-2 and MPEG-4 coded TV services, and it has been designed for various types of satellite applications, including the following:

- Broadcast services: In particular, standard-definition and high-definition TV (HDTV) for distribution in the fixed satellite service (FSS) and broadcast satellite service (BSS) bands
- Interactive services: Referring to data services that include Internet access for consumer applications, such as integrated receiver decoders and personal computers
- Digital TV contribution and satellite news gathering
- Professional applications

8.2.2.1.1.2 DVB-RCS/DVB-RCS2   DVB-RCS constitutes the first standard for the return link that was released in 2001 by the European Telecommunications Standards Institute (ETSI). The core of DVB-RCS is the use of a multifrequency time division multiple access (MF-TDMA) scheme for the return link that guarantees high spectral efficiency for multiple users. It provides detailed specification of the waveforms for networks that employ MF-TDMA with fast frequency hopping, the IP encapsulation methods, and furthermore, all the necessary control messages for TDMA return channel operation. Regarding the forward links, DVB-RCS specifies the use of either DVB-S or DVB-S2. DVB-RCS supports various access schemes leading to a more responsive and efficient system than the demand-assigned satellite systems. The sum of these characteristics resulted in the adaptation of DVB-RCS systems to a wide range of market segments (small and large networks, fixed and mobile terminals). However, the rapid evolution of the communication technology could be followed only with the definition of a second-generation system. Hence, in 2011, the DVB-RCS2 specification was approved by the DVB project (DVB, 2010). Similar to the previous generation, DVB-RCS2 specifies the physical and MAC layer protocols of the air interface that is employed between the satellite operator hub and the interactive user terminal. Furthermore, the network layer and the necessary functions of the management and control planes of the terminal are defined by the standard. Thus, depending on the system's design parameters and the satellite link characteristics, DVB-RCS2 can provide tens of megabits per second down to the terminals and up to 10 Mbps from each terminal.

After the introduction of DVB-RCS2, its mobility extensions were approved in 2012 (DVB-RCS2+M), supporting mobile terminals and direct connectivity (terminal to terminal). The specific standard also supports live handovers between satellite spot beams, spread-spectrum features to satisfy the regulatory constraints for mobile terminals, and link-layer FEC.

8.2.2.1.2 Inmarsat BGAN

BGAN refers to the Broadband Global Area Network of Inmarsat that offers land, maritime, and aeronautical high-speed voice and data services using geostationary (GEO) satellites (Franchi et al., 2000). In particular, the BGAN system provides various services to fixed and mobile users employing three Inmarsat-4 satellites (Chini et al., 2009). The first two satellites were initially located in the Atlantic and Indian Oceans so as to provide service over Europe, the United States, the Middle East, and Asia. However, after the launch of the third satellite, Inmarsat repositioned its satellites to ameliorate the performance of BGAN users, providing high-quality services on a global basis.

BGAN offers an IP data service with speeds up to 492 kbps, using a transportable terminal and a flat-plate directional antenna. Specifically, BGAN satellites are bent-pipe: the feeder link uses the C-band, having a global coverage beam, whereas the user link employs the L-band, using a deployable antenna (up to 256 beams). The system provides communication from 4.5 to 492 kbps to three classes of portable user terminals, where class 1 terminals reach the maximum throughput of 492 kbps, class 2 can achieve 464 kbps in downstream and 448 kbps in upstream, and class 3 can reach 384 kbps and 240 kbps in downlink and uplink, respectively. Furthermore, BGAN allows the adaptation of transmission power, bandwidth, coding rate, and modulation scheme, depending on the terminal capabilities and the channel conditions, so as to achieve high transmission efficiency.

In 2012, Inmarsat launched its new service, BGAN M2M, that provides a solution to support real-time M2M applications. The specific service satisfies the requirements of low latency and small, low-cost terminals that have been set for SCADA, asset monitoring, and smart metering applications.

8.2.2.2 Satellite Systems and Constellations

8.2.2.2.1 O3b Satellite Constellation

Other three billion (O3b) constitutes a satellite constellation of medium altitude for telecommunications and data backhaul (Cochetti, 2014). The eight O3b satellites were launched during 2013–2014, and it is planned to launch 12 more satellites over the next few years. Each O3b satellite consists of 12 steerable antennas that operate at the Ka-band (downlink, 17.70–20.20 GHz; uplink, 27.50–30.00 GHz), 10 beams of which are employed for links with mobile and fixed terminals and 2 beams that are used for continuous links with a gateway earth station. The specific frequency band is employed for both the links from the satellite to the mobile terminals and the links from the satellite to the gateway. Some of the advantages of the O3b constellation are low latency levels (~120 ms compared to the 500 ms of GEO), high capacity (1.2 Gbps), and quickly deployable infrastructure and global reach.

8.2.2.2.2 Iridium

Iridium constitutes a satellite constellation that provides data and voice services over the entire surface of the earth. It consists of 66 satellites in low earth orbit (LEO) at a height of 781 km (Richharia and Westbrook, 2010). The satellites communicate through intersatellite links at the Ka-band with the neighbor satellites. During recent years, Iridium has established new services targeted at M2M communications and high-speed data and voice and data communications. Furthermore, the second generation of Iridium satellites (Iridium NEXT) has been scheduled around 2015, maintaining the existing constellation architecture and providing numerous benefits, such as higher data rates, enhanced voice quality, and global coverage.

8.2.2.2.3 OrbComm

The OrbComm system originates from the 1980s concept of microsatellites that were created for providing the U.S. military with easily deployed data communications (Cochetti, 2014). Nowadays, OrbComm provides low-cost, two-way data services using LEO satellites focusing on M2M communications. In particular, the constellation consists of 31 LEO satellites, which operate at the VHF band (137–138 and 148–149.9 MHz) and at the UHF band (400.1 MHz), for communications and administration, respectively.

8.3 Satellite-Based WSN Architectures

Satellite-based WSNs can be employed in various application scenarios, such as monitoring for environment protection or agricultural applications; critical and emergency infrastructure monitoring, including industrial areas and energy supply parks; and finally, surveillance and monitoring systems of remote areas, including sea and land border surveillance. In addition, M2M communications imply end-to-end communication among a large number of devices without any human intervention; thus, an appropriate design example for M2M systems, especially in areas where there are no terrestrial infrastructures or broadband wireless terrestrial communications, such as forests, open seas, or islands, is the paradigm of satellite-based WSNs.

However, appropriate system architectures must be proposed, so that the satellite-based WSNs are feasible for implementation. This section presents two different satellite-based WSN architecture scenarios. The first scenario addresses direct communication of the sensors with the satellite (*direct access architecture*), where the high-capability sensors individually communicate directly with the satellite, while the low-capability sensors employ the CB technique to reach the satellite. With regard to the second architecture scenario, indirect communication of sensors with the satellite through a gateway (*gateway-based access architecture*) is considered, where a fixed or mobile gateway gathers the total data of the sensor nodes and forwards it to the satellite.

8.3.1 Direct Access Architecture

In the direct access architecture scenario, the wireless sensors communicate directly with the satellite without need of a gateway. According to the communication and energy capabilities of the wireless sensor nodes, the implementation of this architecture can be separated into the following two models, which are described below: the single-sensor-to-satellite model for high-capability sensors and the multisensors-to-satellite model for low-capability sensors. Figure 8.2 illustrates the direct access architecture for both models.

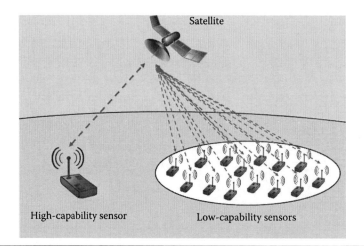

Figure 8.2 **Direct access architecture.**

8.3.1.1 Single-Sensor-to-Satellite Model

In the case where the wireless sensors are equipped with the required satellite transceivers (high-capability sensors), the communication to the satellite can be implemented individually and directly. The sensor nodes can be either placed at specific positions (fixed sensors) or located on moving cars, trains, ships, and aircraft (mobile sensors), which are currently described as earth stations on moving platforms (ESOMPs), using some of the standards that were described previously (e.g., BGAN, DVB-RCS2, and DVB-S2).

8.3.1.1.1 Link Budget Analysis and Outage Probability

The first step in designing a satellite-based system is to perform a link budget analysis. Link budget forms the cornerstone of the space system design, and it is performed to analyze the critical factors in the transmission chain and optimize various link parameters, such as transmission power, bit rate, and link availability, to get the desired performance. Below, the link budget analysis is briefly presented for the communication of sensors (or gateways) with the satellite. The power balance equation that describes the link budget is given in decibel watts by (Maini and Agrawal, 2011)

$$P_r(dBW) = G_r(dBi) + G_t(dBi) + P_t(dBW) - L_{tot}(dB) \qquad (8.1)$$

where P_r and P_t are the received and the transmitted power, respectively, G_r and G_t denote the antenna gains of the receiver and the transmitter, respectively, and L_{tot} represents the total satellite link losses. A key metric in the satellite link budget calculation is the equivalent isotropically radiated power (EIRP) of the transmitter, which can be defined as follows:

$$EIRP_t(dBW) = G_t(dBi) + P_t(dBW) \qquad (8.2)$$

Moreover, regarding the total losses that the transmitted signal experiences, the major components are the propagation losses consisting of the free space losses (given in decibels by $20 \cdot \log_{10}(4\pi d/\lambda)$, where d is the distance to the satellite and λ is the wavelength of the signal), the shadowing (Loo, 1985; Abdi et al., 2003), the attenuation due to vegetation (ITU-R P.833-7), the atmospheric losses (e.g., rain, clouds, and fog) essential at frequencies above 10 GHz (Panagopoulos et al., 2004), and the waveguide losses at the receiver or transmitter.

In satellite systems, there is typically a target minimum received signal-to-noise ratio (SNR) threshold, SNR_{th}, below which the system performance becomes unacceptable. At lower operating frequencies, where the direct access architecture links operate, the dominant propagation impairments, which significantly affect the direct component of the received signal and can lead to inadequate link quality, are composite multipath and shadowed fading. However, considering shadowing effects, the received SNR is distributed according to a specific distribution (e.g., composition of Rice or Rayleigh and lognormal or gamma distributions), with some probability of falling below an SNR_{th}. Consequently, another important metric in satellite communications is the link's outage probability, which represents the probability the satellite service will be unavailable. This refers to the probability that the received SNR (SNR_r) does not exceed an SNR_{th}, and it is given by

$$P_{outage} = P_r\{SNR_r < SNR_{th}\} \tag{8.3}$$

As a result, the outage probability can be expressed as the cumulative distribution function (CDF) of the received SNR at SNR_{th}. A reliable and commonly used shadowing model for land (mobile) satellite links is the Rician shadowed model (Abdi et al., 2003). This model consists of a Rician channel in which the direct line-of-sight component is either completely or partially blocked by buildings, trees, hills, and other objects and, as such, is subject to random shadowing modeled by a gamma distribution. The probability density function (PDF) and the CDF of a Rician shadowed model can be found in Abdi et al. (2003) and Paris (2010), respectively.

Figure 8.3 depicts the outage probability versus the satellite SNR threshold for different shadowing conditions (light, average, and heavy shadowing) according to the Abdi model (Abdi et al., 2003). It can be observed that as the required received SNR value at the satellite increases or the shadowing conditions become heavier, the probability increases that the satellite communication link faces an outage.

8.3.1.1.2 Satellite Random Access

Access methods allow various nodes to share a common transmission medium. Particularly, random access (RA) techniques are very robust to large populations of nodes and bursty traffic. Thus,

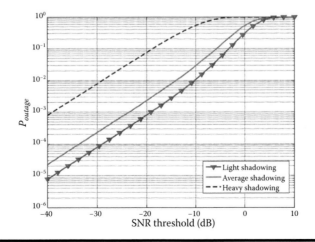

Figure 8.3 Outage probability versus satellite SNR threshold for various shadowing conditions.

efficient RA schemes have to be utilized in satellite-based WSNs, since multiple sensors want to access the satellite. The classical RA schemes are known to not perform very well in the satellite environment. Particularly, schemes based on collision detection and avoidance are hard to apply due to large propagation delays, while for transmissions with high collision probability, a large number of retransmissions are required, enlarging the already high latencies. Therefore, traditional Slotted ALOHA (SA) (Roberts, 1975) and Diversity Slotted ALOHA (DSA) (Choudhury and Rappaport, 1983) are used in satellite communications only for initial access to the satellite band or for short packet transmissions.

However, recent advances in RA schemes based on interference cancellation enhance the performance of RA techniques, making them more appropriate for satellite communications. Thus, satellite networks are now able to efficiently cope with bursty traffic generated by a very large population of nodes. First, Casini et al. (2007) presented the Contention Resolution Diversity Slotted ALOHA (CRDSA) protocol. CRDSA generates two replicas of the same physical layer packet (called burst) at random times within a frame, as in DSA, but with a little extra signaling to point to the "twin" packet location. CRDSA is designed in a way to resolve burst collisions through a simple, yet effective iterative interference cancellation approach that uses frame composition information from the replica bursts. The advantages of CRDSA lie in the improved packet loss ratio and reduced packet delivery delay performance versus channel load jointly with a much higher operational throughput than that of SA and DSA. It is noted that the DVB-RCS2 standard optionally supports CRDSA on the return link for both data and signaling traffic.

Furthermore, an enhanced version of the CRDSA protocol, called CRDSA++, was presented in del Río Herrero and DeGaudenzi (2009). CRDSA++ provides two main enhancements compared to the original CRDSA: (1) an increased (optimized) number of packet repetitions (three to five) and (2) exploitation of the received packet power unbalance to further boost the RA performance. Additionally, a generalized scheme of CRDSA was introduced in Liva (2011), the Irregular Repetition Slotted ALOHA (IRSA), which allows a variable repetition rate for each burst to provide a higher-throughput gain over CRDSA. Afterwards, Paolini et al. (2011) proposed the Coded Slotted ALOHA (CSA) scheme as a further generalization of the IRSA scheme. CSA splits the bursts into segments and then encodes those segments using local codes before the transmission. At the receiver, iterative interference cancellation combined with decoding of the local code is performed to recover collisions.

Another CRDSA-like scheme has been presented recently. The Multi-Slots Coded ALOHA (MuSCA) scheme was proposed in Bui et al. (2012) as an improvement of the CRDSA, IRSA, and CSA protocols. Instead of transmitting replicas, this system replaces them by several parts of a single word of an error correcting code. It is also different from CSA because collided data is used in the iterative decoding of the frame. In MuSCA, the entity in charge of the decoding mechanism collects all bursts of the same user (including the interfered slots) before decoding and implements a successive interference cancellation process to remove successfully decoded signals.

The aforementioned RA mechanisms are synchronous schemes. This excludes partial collisions (e.g., every collision arises from complete superposition of packets), although the signaling overhead in order to keep slot synchronization among all transmitters may become a major drawback. Toward this destination, the Contention Resolution ALOHA (CRA) was proposed (Kissling, 2011), relaxing the synchronization accuracy for slotted RA. CRA removes the notion of slots inside the CRDSA or IRSA frames, allowing the replica packets from individual transmitters to be sent with a random delay (and possibly different duration) within the frame. CRA represents an interesting evolution of the original CRDSA; however, it still requires the transmitter to remain synchronized at the frame level. Furthermore, an extension of the frame-based CRA protocol is the Enhanced Contention Resolution ALOHA (ECRA) (Clazzer and Kissling, 2013), which makes a further attempt to decode those packets that were detected but not successfully decoded due to the collisions.

Finally, truly asynchronous RA schemes have been proposed for satellite systems. An enhancement of packet-based Spread-Spectrum ALOHA (SSA) (Abramson, 1996) was proposed for mobile satellite systems, dubbed Enhanced Spread-Spectrum ALOHA (E-SSA) (del Río Herrero and DeGaudenzi, 2012). E-SSA exploits a recursive sliding-window interference cancellation algorithm. Finally, the Asynchronous Contention Resolution Diversity ALOHA (ACRDA) was introduced in DeGaudenzi et al. (2014) as the evolution of the CRDSA. ACRDA provides better throughput performance with reduced demodulator complexity and lower transmission latency than its predecessor, while allowing truly asynchronous access to the shared medium. ACRDA reduces the gap between the CRDSA and E-SSA schemes for systems that do not adopt spread-spectrum techniques performing better than CRDSA.

8.3.1.2 Multisensors-to-Satellite Model

In contrast with the previous model, a typical sensor node may not be equipped with a communication module capable of extremely long-distance communication directly with a satellite (e.g., ZigBee). For this case of wireless sensors (low-capability sensors), the only possible way to succeed in direct communication between a WSN and a satellite without using a gateway is by employing a modern signal transmission technique, the CB (Poulakis et al., 2013). Particularly, according to this technique, a set of distributed wireless nodes organize themselves into groups and collaboratively send their shared messages to the same destination as a virtual antenna array. The idea was launched in order for the battery-limited devices equipped with a single antenna to be able to use beamforming (Ochiai et al., 2005). CB can improve transmission range, energy efficiency of wireless networks, and data security by eliminating signals in undesired directions (Ochiai et al., 2005; Feng et al., 2010). Nevertheless, there are some main technical challenges when implementing CB. The most important are the feasibility of precise phase and time synchronization between the collaborative nodes to produce the optimal output, accurate channel estimation and localization of nodes, and efficient data dissemination among them. Two possible scenarios can be used, depending on the node synchronization procedure (Ochiai et al., 2005): the closed-loop and open-loop scenarios.

As noted above, synchronization among sensors is critical. Synchronization can be achieved using reference signals (e.g., GPS signals) or a scalable protocol such as the Timing-Sync Protocol for Sensor Networks (TPSN) (Ganeriwal et al., 2003), which provides synchronization for a whole network. Moreover, the concept of clustering in WSNs (Boyinbode et al., 2011; Vassaki et al., 2014) can be employed in order to organize the sensor nodes into groups (clusters) that form individual virtual antenna arrays, while cluster heads contribute to the synchronization of the group and data dissemination, playing the role of coordinators.

Afterwards, the antenna characteristics derived by CB, and specifically the antenna directivity with respect to the number of nodes, are presented. A reasonable assumption when one deals with CB in WSNs is to consider that the distributed nodes are randomly located by nature, while other possible geometrical locations of sensors can be found in Ochiai and Imai (2009). Specifically, the following analysis refers to the geometrical configuration of N sensor nodes located within a disk of radius R, according to a uniform distribution. All nodes are located on the x-y plane and form a planar array, and particularly, the kth node's location in polar coordinates is denoted by (r_k, ψ_k), where $r_k \in [0, R]$ is the radial coordinate and $\psi_k \in [0, 2\pi]$ is the azimuth. Furthermore, the locations of reference points are expressed in spherical coordinates as (A, ϕ, θ), where A is the radial coordinate, the angle $\theta \in [0, \pi/2]$ denotes the elevation direction, and the angle $\phi \in [0, 2\pi]$ is the azimuthal direction. Moreover, the destination that the signal has to reach is located in (A, ϕ_0, θ_0). Other assumptions that are considered for the sensor nodes are the following: all the nodes transmit with the same energy, and they are equipped with ideal isotropic antennas; the path losses are identical for all nodes, and there is no reflection or scattering of the signal; and finally, the nodes are perfectly synchronized and sufficiently separated such that mutual coupling effects are negligible.

A key metric for the CB system is the derived directivity of the virtual antenna, defined as how much radiated energy is concentrated in the desired direction relative to a single isotropic antenna. Given the statistical properties of the achievable beampattern of the random sensor array, it is reasonable to compute the average directivity (Ochiai et al., 2005; Poulakis et al., 2013); however, due to its complexity, an alternative measure can be used, representing a tight lower bound of average directivity (from now on called directivity for the sake of simplicity) that is defined as

$$\tilde{D}_{av} \triangleq \frac{\int_0^{\pi/2} \int_0^{2\pi} P_{\max} \sin\theta \, d\theta \, d\phi}{\int_0^{\pi/2} \int_0^{2\pi} P_{av}(\phi, \theta) \sin\theta \, d\theta \, d\phi} \tag{8.4}$$

It is noted that unlike well-designed deterministic linear arrays, the directivity of a given realization is very likely to be less than N, whereas the limit N can be approached only by increasing R/λ, where λ denotes the wavelength of the signal. For example, numerical calculations showed that the directivity $\tilde{D}_{av}$ increases as R increases, and it is equal to N for R greater than 10–15 m, considering an operating frequency of 2.4 GHz for several numbers of nodes. Furthermore, degradation in the achieved directivity can be caused due to phase errors (imperfect phase knowledge) and node localization errors (imperfect location knowledge) (Poulakis et al., 2013). However, if R is carefully chosen and tolerable phase and localization errors are considered, the directivity of CB can be approximated by N, a value that is used in the following analysis. Moreover, the antenna gain of CB (G_{CB}) can be defined as

$$G_{CB} = \varepsilon \cdot \tilde{D}_{av} \cdot G_s \tag{8.5}$$

where ε is the antenna's efficiency (considered equal to 1 for the sake of simplicity) and G_s is each individual sensor antenna gain. Considering that P_s is the transmitted power of each individual sensor, the total transmitted power of all sensors is $P_{tot} = N \cdot P_s$. Applying G_{CB} to the total transmitted power, the EIRP of the total WSN produced by CB can be defined as

$$EIRP_{CB} = G_{CB} \cdot P_{tot} = G_{EIRP} \cdot EIRP_s \tag{8.6}$$

where $G_{EIRP} = \tilde{D}_{av}N$ is the EIRP gain that CB can achieve in comparison with the EIRP of a single sensor $EIRP_s = G_s \cdot P_s$. The $EIRP_{CB}$ can approach N^2 for reasonable localization and phase errors.

Similar to the single-sensor-to-satellite model, designing a WSN CB link to the satellite, the link budget analysis gives the received power at the satellite in decibel watts:

$$P_{sat}(dB_W) = G_{sat}(dB_i) + EIRP_{CB}(dB_W) - L_{tot}(dB) \tag{8.7}$$

where G_{sat} expresses the antenna gain of the satellite. Moreover, if error control coding is applied, the required received power at the satellite in order to achieve the same performance (e.g., specific bit error rate [BER]) will be reduced (in decibel units) by the factor of the coding gain (CG) (Howard et al., 2006). Thus, given the performance requirements of the communication link, a threshold value of $EIRP_{CB}$ can be calculated:

$$EIRP_{CB,th}(dB_W) = P_{sat,u}(dB_W) - G_{sat}(dB_i) - CG(dB) + L_{tot}(dB) \qquad (8.8)$$

where $P_{sat,u}$ is the required received power for the uncoded signal, as well as a corresponding $EIRP$ gain threshold:

$$G_{EIRP,th}(dB) = EIRP_{CB,th}(dB_W) - EIRP_s(dB_W) \qquad (8.9)$$

Consequently, the required number of collaborative sensor nodes (N_{th}) that fulfill the desired link's performance requirements can be computed as

$$N_{th} = \arg_N \{G_{EIRP}(N) = G_{EIRP,th}\} = \arg_N \{\tilde{D}_{av}(N) \cdot N = G_{EIRP,th}\} \qquad (8.10)$$

From Equation 8.10, it can be observed that the value of a single-sensor node's EIRP ($EIRP_s$) is very significant. Particularly if the characteristics of sensor nodes are slightly improved, the required number of nodes will be significantly reduced.

Finally, numerical calculations regarding the required number of sensor nodes for several performance requirements are presented in Figures 8.4 and 8.5. GEO and LEO satellites are considered, while the sensor nodes operate at the 2.4 GHz frequency band (S-band). Furthermore, we assume realistic technical characteristics for the communication link (e.g., free space and vegetation losses, binary phase shift keying (BPSK)/QPSK modulation, and Reed–Solomon [255,239] coding). In particular, Figure 8.4 presents the values of N_{th} for a GEO satellite versus the desired BER values considering 30° and 60° satellite elevation angles. As can be seen, the required number of sensor nodes decreases as BER increases (i.e., less strict quality requirements) and as elevation angle increases (i.e., reduced path losses).

Figure 8.5 illustrates the corresponding results for the LEO satellite. The behavior of N_{th} is similar to that of the GEO satellite, although the values of N_{th} are significantly lower than before

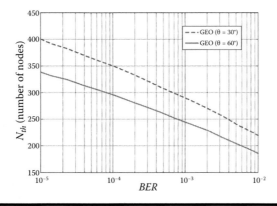

Figure 8.4   Sensor node requirements versus BER for several elevation angles (GEO satellite).

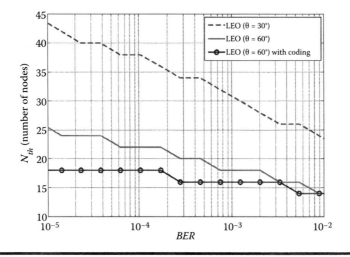

Figure 8.5 Sensor node requirements versus BER for several elevation angles (LEO satellite).

due to the smaller altitude of LEO compared to that of GEO satellites (i.e., reduced path losses). Furthermore, an additional curve is presented where channel coding is employed that significantly reduces the required number of nodes.

8.3.2 Gateway-Based Access Architecture

Apart from the direct access of the sensors to the satellite, gateways may be used for the transmission of the data obtained by sensors to the satellite. In this case, a gateway could be mobile or fixed. In the former, the mobile gateway can be an unmanned aerial vehicle (UAV), ship, or truck, while in the latter case, it is a fixed ground station. In particular, the gateway-based architecture is a multihop link with at least two hops. The sensor nodes form groups (clusters) and send their information toward a gateway directly or via a sink or a cluster head node using a WSN protocol (e.g., ZigBee). Afterwards, the gateway forwards the data to the satellite using some of the previously described standards (e.g., DVB-RCS2 and DVB-S2). In Figure 8.6, the gateway-based architecture is shown for a fixed and a mobile gateway.

More specifically, since the gateways are connected to a great number of sensors or sensor clusters and the hardware of gateways can be more complex for communication purposes than that of sensors, the gateways may use higher operating frequency bands, such as the Ka-band, for transmission to the satellite. The employment of higher-frequency bands has the advantage of no congestion, in contrast with frequencies below 5 GHz; while using higher frequencies for uplink transmission, higher bandwidth is used, and therefore higher data rates. Consequently, more sensors can be supported through the gateways as well as sensors with high capabilities for high-data-rate transmission and real-time monitoring of data.

However, in operating frequencies above 10 GHz, atmospheric phenomena affect the electromagnetic waves and cause the signal's attenuation. In particular, clouds, water vapor, and precipitation cause the attenuation of the power of the signal, while tropospheric turbulence causes scintillation of the signal amplitude (Panagopoulos et al., 2004). Atmospheric attenuation can take values of several decibels in a time percentage that is crucial for the system's availability. The introduction of a static attenuation margin leads to the inefficient function of the system, since large resources are used for a very small time percentage, and so the cost of the system is

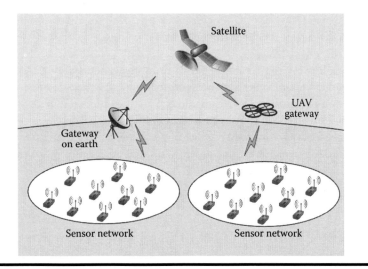

Figure 8.6   Gateway-based access architecture.

increasing. Therefore, fade mitigation techniques have been developed for satellite communication design at the Ka- and Q-frequency bands. Among these techniques, the adaptive coding and modulation, diversity techniques, and power control are included. In Figure 8.7, the exceedance probability (i.e., the complementary cumulative distribution function [CCDF]) of total attenuation for three different cities in Greece is shown for an operating frequency of 30 GHz.

Of the atmospheric phenomena that cause the degradation of a link's performance, rain is the dominant fading mechanism. However, rain presents spatial and temporal variations due to the spatial inhomogeneity of the rain medium. One way to compensate rain attenuation is by employing diversity techniques that take advantage of the decrease of correlation of rain attenuation induced on the links, as they are separated on the spatial domain. The spatial diversity techniques

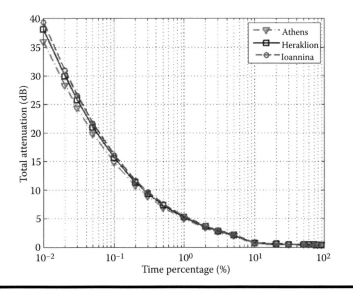

Figure 8.7   CCDF of rain attenuation at 30 GHz for three different cities in Greece.

can be divided into two categories (Kourogiorgas et al., 2011): site diversity and orbital diversity. Moreover, since in this access architecture the uplink is studied, the diversity techniques occur at the transmit side, that is, gateways.

According to the site diversity technique, two or more ground stations are employed for data transmission. Three techniques have been identified, and they are described in Paulraj et al. (2003), Lo (1999) and Ismail et al. (2011): maximal ratio transmission (MRT), switching transmit diversity (STD), and equal-power transmit diversity (EPTD). In particular, given that the SNR in clear-sky conditions is known, in the MRT technique the signals are processed at the transmitter and combined at the receiver. The signals before transmission are multiplied by a complex factor whose amplitude is given by

$$|w_i| = \frac{|h_i|}{\sqrt{\sum_{i=1}^{n}|h_i|^2}} \tag{8.11}$$

where h_i is the i channel and N is the total number of the links. The phase of w_i is equal to the negative of the complex channel. However, as observed in Equation 8.11, full knowledge of the channel is needed from all the transmitters, and so the complexity of the system is increasing. In EPTD, the amplitude of the factor w_i is equal to $1/\sqrt{N}$ and the phase of the transmitted link must be known to be multiplied at preprocessing at the transmitter side. Finally, in STD, the signal is transmitted by the gateway, which will give the highest SNR at the receiver side.

8.3.2.1 Communication to Satellite via Fixed Gateway

A fixed ground station can be used close to the cluster of sensors. In particular, the fixed gateway receives data from the sensors, and then these are transmitted to the satellite. Therefore, the link is considered a two-hop link. The sensors can transmit to the fixed gateway through a WSN protocol such as ZigBee (or through a mobile communication standard such as Long Term Evolution [LTE] for advanced sensor nodes). It is noted that wired communications can also be employed (e.g., optical fibers and cables).

As explained previously, rain attenuation is the dominant fading mechanism at higher frequency bands, such as the Ka-band. For the evaluation of a system's capabilities, first- and second-order statistics of rain attenuation are needed. For the first-order statistics of rain attenuation, different distributions are used, depending on the climatic region. On heavy rain areas, the gamma or inverse Gaussian distributions (Kanellopoulos et al., 2013; Kourogiorgas and Panagopoulos, 2013) are preferred, while for temperate climatic regions, the Weibull or lognormal distributions (Maseng and Bakken, 1981; Kanellopoulos et al., 2014) are more appropriate.

For the second-order statistics, one of the most widely used time series synthesizers for rain attenuation at temperate regions is the Maseng–Bakken (MB) model (Maseng and Bakken, 1981). According to the MB model, rain attenuation is described through the following stochastic differential equation (SDE):

$$da_t = a_t\beta_A\left[\sigma_a - \ln\left(\frac{a_t}{a_m}\right)\right]dt + a_t\sqrt{2\beta_A}\,\sigma_a d\hat{W}_t \tag{8.12}$$

where a_t is rain attenuation at time instance t, σ_a and a_m are the statistical parameters of the lognormal distribution of rain attenuation, and β_A is a parameter on which the dynamics of rain attenuation depends. Moreover, W_t is the standard Brownian motion (Karatzas and Shreve, 1991). In case that multiple links are considered, rain attenuation induced on different links is spatially correlated. For the generation of time series for multiple spatially separated links, the multidimensional SDEs have been proposed in Karagiannis et al. (2012).

Using the synthesizer of Karagiannis (2012), time series of received SNR can be derived as shown in Figure 8.8, considering the SNR in clear-sky conditions to be 28.3 dB. In the same figure, the time series of received SNRs for the various transmission techniques are shown.

As previously mentioned, an important metric is the outage probability as a function of the SNR threshold. Using the multidimensional synthesizer for generating rain attenuation on multiple links, the outage probability for a dual-site diversity system is shown in Figure 8.9. The operating frequency is equal to 20 GHz, and the gateways are located in Athens with a distance of 5 and 20 km between them for the two presented cases. The elevation angle is considered to be 30°, and the SNR in clear-sky conditions is 28.3 dB. It can be observed that the MRT techniques give the best performance, while EPTD and STD have similar behavior. Moreover, the increase of the distance between the gateways reduces the outage probability.

Moreover, in Figures 8.10 and 8.11, the outage probability and capacity are shown for a site diversity system with four gateways in a rectangular shape, with 10 km between them. The operating frequency is 30 GHz and the elevation angle 35°. Compared to the single link, the use of multiple gateways reduces the outage probability by a significant factor.

8.3.2.2 Communication to Satellite via Mobile Gateway

More and more applications are developed with sensors on mobile vehicles (e.g., cargo ships and trucks) for monitoring of the environment, measuring the humidity or temperature of the cargo, and so forth. In these cases, an earth station can be placed on the mobile vehicle to forward the data to the satellite. Furthermore, moving gateways, such as UAVs, can be employed for gathering

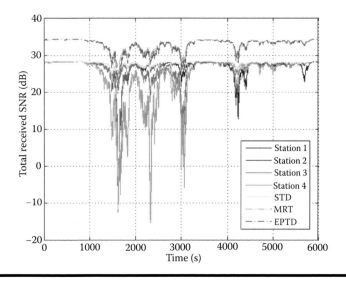

Figure 8.8   Time series of received SNRs.

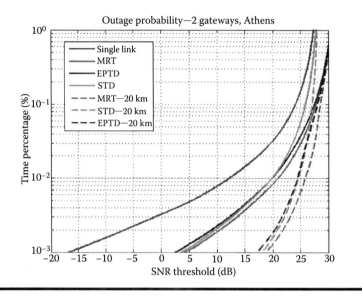

Figure 8.9 Outage probability for a dual-site diversity system in Athens.

and retransmitting data to the satellite from on-ground sensors. In both scenarios, mobile gateways communicate with the satellite, but in any use of mobile gateways, high operating frequencies (e.g., Ka-band) are obligatory for high-data-rate transmission to satellites. In Kourogiorgas et al. (2015), an analysis is presented for the coexistence of the services of ESOMPs and FSS at the same frequency bands for increasing the bandwidth of ESOMPs. The ESOMPs can be used as mobile gateways given that they can support the desired data rates. As explained previously, rain attenuation is the dominant fading mechanism at these frequencies; however, rain attenuation modeling, as explained below, is different than this for fixed terminals (i.e., gateways in our

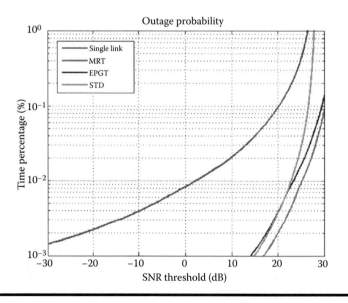

Figure 8.10 Outage probability for a site diversity system in Athens with four gateways.

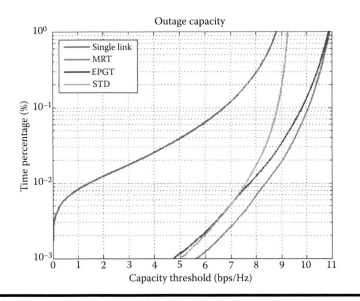

Figure 8.11 **Outage capacity for a site diversity system in Athens with four gateways.**

case). For the single link with mobile gateways, it is assumed that rain attenuation (*A*) follows the lognormal distribution, as also indicated in Arapoglou et al. (2012). Therefore, the exceedance probability of rain attenuation for mobile satellite links is given by

$$P_M(A) = \frac{1}{2} erfc\left(\frac{\ln A - \ln A_{m,M}}{\sqrt{2}S_{aM}} \right)$$

(8.13)

where $A_{m,M}$ is the median value of the lognormal distribution and S_{aM} is the standard deviation of the natural logarithm of rain attenuation for a mobile gateway.

In Matricciani (1995), the transformation of the rain attenuation statistics for fixed gateways to the transformation statistics for mobile gateways is given through the following expression:

$$P_M(A) = \xi P_F(A)$$

(8.14)

where P_F is the exceedance probability of rain attenuation for a fixed gateway and ξ is a parameter that depends on the gateway's speed (υ_M) and the velocity of the storm, given by

$$\xi = \frac{\upsilon_R}{\left| \upsilon_M - \upsilon_R \cos\varphi \right|}$$

(8.15)

where υ_R is the wind speed and φ is the angle between the mobile gateway's direction and the motion of the storm. Thus, the CCDF of rain attenuation at mobile gateways can be calculated, given the CCDF of rain attenuation for fixed services, which can be calculated from the ITU-R P.618-10 recommendation. Once the exceedance probability for mobile gateways is known, then through fitting procedures, the lognormal parameters $A_{m,M}$ and S_{aM} can be computed. The fitting procedures that can be implemented in this case are those proposed by either ITU-R P.1057-3 or Kourogiorgas and Panagopoulos (2013).

Moreover, in Arapoglou et al. (2012), a time series synthesizer for rain attenuation induced in a mobile satellite service (MSS) is proposed. Rain attenuation on MSS systems $\left(a_t^M\right)$ can be modeled as a diffusion process that is described by the following first-order SDE:

$$da_t^M = a_t^M \beta_M \left[S_{aM}^2 - \ln\left(\frac{a_t^M}{A_{m,M}} \right) \right] dt + a_t^M \sqrt{2\beta_M} \, S_{aM} \, dW_t \tag{8.16}$$

where dW_t are the increments of standard Brownian motion (Karatzas and Shreve, 1991) and β_M is the dynamic parameter of rain attenuation for mobile gateways. The latter is related to the dynamic parameter for fixed gateways (β_F) through (Arapoglou et al., 2012)

$$\beta_M = \frac{1}{\xi} \beta_F \tag{8.17}$$

An example is presented in Figure 8.12, where the CCDFs of rain attenuation for a fixed and a mobile gateway are shown, as well as the joint exceedance probability of attenuation for the two convergent links. The speed of the mobile gateway is 70 km/h, and the storm speed is 30 km/h. The mobile gateway is considered at the Mediterranean Sea, close to the island of Creta, Greece, with a latitude of 35.53°N and longitude of 21.66°E. Both links have the same elevation angle of 60°, and the separation angle is 60°, while the considered frequency is 28 GHz.

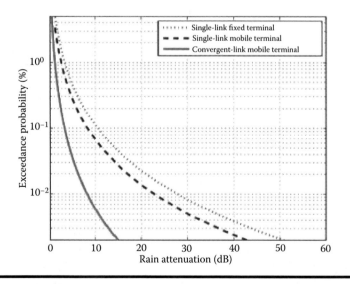

Figure 8.12 CCDF of rain attenuation for a fixed terminal (i.e., gateway), mobile terminal, and joint CCDF for a mobile terminal.

8.4 Major Issues in Satellite-Based WSNs

Beyond the aforementioned, several major issues should be taken into consideration in order for satellite-based WSNs to operate properly for both proposed architecture scenarios. First, synchronization among the wireless sensors is an issue of crucial importance for avoiding data and alert losses and inefficient energy consumption for retransmissions. Furthermore, the proposed network architecture should be compatible and satisfy certain QoS requirements (e.g., reliability, timeliness, robustness, availability, and security), which are always relevant to the type of target application. Last but not least, one of the major issues is to keep the satellite-based WSN connected. Therefore, a high level of connectivity should be sustained among the wireless sensors and, of most importance, between the gateway node and the satellite. In this section, these three major issues are analyzed and solutions for fulfilling their requirements are presented.

8.4.1 Time Synchronization

Time synchronization is a critical issue for every wireless communication system, and especially for distributed, decentralized systems such as WSNs, where the lack of central administration demands good collaboration and self-configuration by every single node. Especially in WSNs, synchronization is of utmost importance, mainly due to the fact that almost any action of sensor nodes (sensing, data fusion, cooperative communication, collaborative beamforming, etc.) requires synchronization among the sensors and between the WSN and the physical world. On the contrary, the limited resources of sensor nodes in conjunction with clock accuracy and strict precision requirements make synchronization more complex.

When designing time synchronization algorithms, wireless network limitations enforce certain requirements that need to be met. Some of these basic requirements are listed below (Sivrikaya and Yener, 2004; Lasassmeh and Conrad, 2010):

- **Scalability:** Synchronization scales well with an increasing number of nodes or high density in the network.
- **Accuracy:** The need for precision, or accuracy, may vary significantly, depending on the specific application and purpose of synchronization.
- **Energy efficiency:** Synchronization schemes should not consume a significant proportion of the limited energy resources contained in the sensor nodes. Energy resources should be kept for other important procedures, such as data processing and radio frequency (RF) transmissions.
- **Robustness:** In case of the failure of a few sensor nodes, the synchronization scheme should remain valid and functional for the rest of the network.
- **Lifetime/longevity:** Synchronization should be feasible for very long time periods.
- **Cost/size:** The necessary equipment for synchronization should be cheap and small.
- **Delay:** Synchronization schemes should take into consideration applications that demand an immediate response and high level of reliability when an important event suddenly occurs.

The intended protocols used for synchronization are different from each other in some aspects and similar to one another in other aspects. They can be classified according to the following categories (Sundararaman et al., 2005; Rhee et al., 2009):

- **Master–slave versus peer-to-peer synchronization:** In a master–slave protocol, one node is assigned as the master and the others as slaves. The slave nodes are synchronized and used as a reference for the local clock of the master node (Mock et al., 2000; Ping, 2003). In peer-to-peer synchronization, every node communicates directly with the other nodes in the network (Elson et al., 2002; Su and Akyildiz, 2005). These protocols are more robust and flexible, but less accurate.

■ **Internal synchronization versus external synchronization:** In internal synchronization, there is no global time reference and sensors tend to minimize the maximum differences between their local clocks. On the other hand, in the case of the Network Time Protocol (NTP) (Mills, 1991), the sensors' local clocks adjust their reference time in order to be synchronized with a standard time, such as Universal Time Coordinated (UTC).

■ **Probabilistic versus deterministic synchronization:** Probabilistic synchronization is a method that gives a probabilistic guarantee on the maximum clock offset with a failure probability, but this method can be very expensive. In deterministic synchronization, deterministic algorithms guarantee an upper bound on the clock offset with certainty (Arvind, 1994). Most algorithms in the literature are deterministic.

■ **Sender-to-receiver versus receiver-to-receiver versus receiver-only synchronization:** In the case of sender-to-receiver synchronization (SRS), the receiver is periodically synchronized with the sender's local clock via messages considering the sender's time stamp on the message plus the message delay. In the case of receiver-to-receiver synchronization (RRS), two receivers receive the same message in a single-hop transmission, and they exchange the time at which they received the same message and compute their offset based on the difference in reception times. Finally, in the case of receiver-only synchronization (ROS), a group of nodes can be simultaneously synchronized by only listening to the message exchanges of a pair of nodes.

■ **Clock correction versus untethered clocks:** In the first case, the local clocks of nodes participating in the network are corrected either instantaneously or continually to keep the entire network synchronized. In the second case, every node maintains its own clock as it is and keeps a time translation table relating its clock to the clock of the other nodes. Local time stamps are compared using the table.

8.4.2 QoS Requirements

WSNs are the most typical deputies of wireless multihop networks, mostly because WSNs have vast potential for future applications, which include security, energy, and environmental monitoring; efficient automation of industrial production; patient monitoring for better healthcare; and assisted living and home monitoring. WSNs have been used in many "smart" implementations, and they have also been proposed as a low-budget solution for monitoring crucial industrial and environmental parameters. Moreover, the efficient deployment of their technology is a major research subject in "green" communications. For all these reasons, it is obvious that the QoS requirements may highly differentiate from each implementation due to the wide range of potential applications. Therefore, a major concern regarding WSNs is the satisfaction of certain QoS requirements, given the challenging nature of the sensor nodes (Chen and Varshney, 2004; Reddy et al., 2006; Xia, 2008).

QoS regards the capability of the communication system to provide assurance that the service requirements of applications can be satisfied. Generally, in a satellite-based WSN, the service requirements depend on the type of target application. Thus, the QoS in WSNs can be characterized by reliability, robustness, availability, timeliness, and security, among others. Some QoS parameters may be used to measure the degree of satisfaction of these services, such as throughput, delay, jitter, and packet loss rate. However, a very significant issue arises for how the QoS is assessed due to the nature of WSNs and the possible service requirements. For example, in applications involving event detection and target tracking, the failure to detect information or extracting the wrong or incorrect information regarding a physical event may arise for many reasons. The problem may stem from wrong design and deployment of the WSN or from communication failure, or even from limited functionality of sensors. Therefore, the QoS for a single requirement is susceptible to a variety of different factors that should be taken into account. Two basic perspectives of QoS in satellite-based WSNs are the following (Chen and Varshney, 2004):

1. **Application-specific QoS:** From this perspective, the target application determines the QoS parameter values and specifies the requirements for the deployment and implementation of the WSN. QoS may depend on parameters such as network coverage or sensor topology and resource allocation strategies that can be defined during network design.

2. **Network QoS:** From this perspective, how the underlying communication network can deliver the QoS-constrained sensor data while efficiently utilizing network resources is considered. Generally, the communication network can deliver the QoS-constrained sensor data in three basic ways: event-driven, query-driven, and continuous-delivery models (Tilak et al., 2002).

   - Event driven: Most event-driven applications in WSNs are interactive, delay intolerant (real time), mission critical, and non–end to end. In these applications, when a sensor detects an event immediately, the application takes an appropriate action as quickly as possible and as reliably as possible.

   - Query-driven: Most query-driven applications in WSNs are interactive, query specific, delay tolerant, mission critical, and non–end to end. In order to save energy, the sink node pulls on demand from the queries of data.

   - Continuous: In the continuous model, sensors send their data continuously to the sink at a prespecified rate. This method is used by real-time applications (voice, image, video, and data) where packet losses can be tolerated to a certain extent.

Even though WSNs inherit most of the QoS challenges from other wireless networks, there are particular characteristics that should be also taken into consideration on QoS characterization, such as

- **Severe resource constraints:** Sensors are in most cases cheap battery devices with limited processing and transmission capabilities and small memory buffers. Therefore, the constraints on resources involve energy, bandwidth, memory, buffer size, processing capability, and limited transmission power.

- **Low and balanced energy consumption:** In order to achieve a long-lived network, usually installed in a remote or inhospitable environment, energy consumption must be kept very low and be allocated in a well-balanced way among all sensors, avoiding a small set of sensor nodes from being drained out too soon.

- **Scalability:** A satellite-based WSN usually consists of hundreds or thousands of sensor nodes densely distributed over a terrain. Therefore, QoS support designed for WSNs should be able to scale up to a large number of sensor nodes.

- **Traffic pattern:** In most WSN applications, traffic mainly flows from a large number of sensor nodes to a small subset of sink nodes. Moreover, in heterogeneous WSNs, the aggregated traffic stems from different types of sensors with different traffic patterns.

- **Data redundancy:** Due to the high redundancy of raw data, data processing is necessary to maintain robustness and reliability, but it induces latency and complicates QoS design in WSNs.
- **Network dynamics:** Network dynamics may arise from node failures, wireless link failures, node mobility, and node state transitions due to the use of power management or energy-efficient schemes.

8.4.3 Connectivity

A satellite-based WSN is a complex system comprising different network devices that should cooperate efficiently in order to successfully achieve the required QoS. One of the major concerns for achieving high performance, reliability, and robustness is keeping all the network devices connected. Therefore, connectivity should be achieved on three different levels: among the low-cost sensor nodes, between the sensor nodes and satellite gateways, and finally, between the satellite gateways or sensor nodes and the satellite communication system.

The first level refers to the local WSN, and it is crucial because network connectivity is very susceptible to sensor failures, transmission capabilities, and the propagation environment. Low-cost sensors are very vulnerable devices, especially in harsh operation environments. As a consequence, one of the basic tasks for WSNs and their communication protocols is to keep a high level of connectivity by intervening and reconfiguring the sensor nodes when a loss of a connection or a sensor node failure occurs. The second level refers to the part of the satellite-based WSN system that consists of the sensor nodes of different WSNs and the satellite gateways. It is possible for a group of remote WSNs located in different places with their own satellite gateways to consist of systems with high scalability (Vassaki et al., 2014). In the case of direct access architecture, this level does not exist. The third level is the connectivity between the satellite gateways and the satellite system for the gateway-based access architecture, or between the sensor nodes and the satellite system for the direct access architecture. In that case, the notion of connectivity resorts to the link availability of the satellite communication channel that was rigorously presented previously.

Subsequently, some connectivity evaluation issues are presented for the local WSN network, assuming the other levels are more robust and reliable than the remote WSN. Furthermore, taking into consideration that WSNs are usually installed in harsh habitats, such as forests, industrial zones, and places with high security levels, the propagation environment plays a crucial role for connectivity evaluation. The impact of shadowing and large-scale fading phenomena along the communication path has been rigorously studied in Pitsiladis et al. (2012) for both correlated and uncorrelated communication channels. Results have shown a dominant impact on network connectivity. Furthermore, the significant impact of small-scale fading on connectivity is presented in Miorandi et al. (2008) for various channel models, and a method for evaluating the impact of interferences is presented in Dousse et al. (2005).

Another very important factor that should be taken into account when studying random WSNs is the networks' spatial distribution. In most cases, when the positioning of sensor nodes is unknown, spatial field processes such as the Poisson point field process are employed (Pitsiladis et al., 2010). The spatial homogeneous Poisson point process is very common in the literature because it always results in the worst-case scenario, and it is a reference for other methodologies. Finally, it is worth mentioning that the most acceptable and commonly used connectivity evaluation metric is the isolation probability of a random network node (Bettstetter and Hartmann, 2005). This metric is derived from the Poisson point process, and it can be calculated for a variety of channel models.

8.5 Future Applications and Challenges

The ever-rising demand for future applications that use remote monitoring and surveillance systems (especially in areas with no terrestrial communication infrastructures), which include energy and environmental monitoring, security, efficient automation of industrial production, patient monitoring for better healthcare, and assisted living and home monitoring, will greatly enhance WSNs and other familiar network technologies. Therefore, it is likely that large-scale and complex networks such as the proposed satellite-based WSNs will be an inevitable solution for future applications and systems (e.g., M2M systems).

The challenges for future satellite-based WSNs include the achievement of higher levels of QoS by increasing the bit rate transmissions and offering a better user experience by reducing latency and introducing friendly application interfaces. All of the aforementioned demand higher throughputs, robustness, and reliability. A logical solution is to resort to higher frequencies where larger bandwidths exist, especially for the satellite connections and to enhance compatibility, cooperation, and interoperability between the different network technologies. Furthermore, it is very important to increase the sensor capabilities in terms of processing and buffering and reliability without significantly increasing their cost. New resource allocation strategies should be implemented to achieve better bandwidth allocation and reduce energy consumption.

8.6 Conclusion

This chapter studied the use of satellite communication technologies in WSNs. Satellite-based WSNs can be used in various scenarios, such as environmental monitoring, critical and emergency infrastructure monitoring, and surveillance and monitoring, as they constitute an appropriate designing solution for M2M systems. Toward this direction, the state-of-the-art technologies for WSNs and satellite systems are briefly presented. Afterwards, two possible architectures are proposed for the practical implementation of satellite-based WSNs. The first architecture deals with the direct communication of the sensors with the satellite. Particularly, the high-capability sensors individually communicate directly with the satellite, while the low-capability sensors employ the collaborative beamforming technique to increase their transmission range and reach the satellite. According to the second architecture, the sensors communicate with the satellite through a fixed or mobile gateway, which collects and forwards the sensors' data to the satellite. Finally, some major issues appearing in those systems are discussed, including synchronization, QoS, and connectivity concepts, as well as other challenges and future applications. The management of WSNs using satellite communication technologies is very promising, especially due to the rapid development of M2M systems; however, an efficient system and protocol design is necessary to bring this idea closer to practice.

Acknowledgment

This work was supported in part by Georgia Society of Radiologic Technologists (GSRT)–funded projects Cooperation 2011—JASON and Innovative Enterprise Clusters—ACRITAS.

References

Abdi A., W. C. Lau, M.-S. Alouini, M. Kaveh, A new simple model for land mobile satellite channels: First- and second-order statistics, *IEEE Transactions on Wireless Communications*, vol. 2, no. 3, pp. 519–528, 2003.

Abramson N., Spread Aloha CDMA data communications, U.S. Patent 5,537,397, July 16, 1996.

Arapoglou P.-D., K. P. Liolis, A. D. Panagopoulos, Railway satellite channel at Ku band and above: Composite dynamic modeling for the design of fade mitigation techniques, *International Journal of Satellite Communications and Networking*, vol. 30, pp. 1–17, 2012.

Arvind K., Probabilistic clock synchronization in distributed systems, *IEEE Transactions on Parallel and Distributed Systems*, vol. 5, no. 5, pp. 474–487, 1994.

Baronti P., P. Pillai, V. W. Chook, S. Chessa, A. Gotta, Y. F. Hu, Wireless sensor networks: A survey on the state of the art and the 802.15.4 and ZigBee standards, *Computer Communications*, vol. 30, no. 7, pp. 1655–1695, 2007.

Bettstetter C., C. Hartmann, Connectivity of wireless multihop networks in a shadow fading environment, *Wireless Networks*, vol. 11, pp. 571–579, 2005.

Boyinbode O., H. Le, M. Takizawa, A survey on clustering algorithms for wireless sensor networks, *International Journal of Space-Based and Situated Computing*, vol. 1, no. 2–3, pp. 130–136, 2011.

Bui H.-C., J. Lacan, M.-L. Boucheret, An enhanced multiple random access scheme for satellite communications, in *Wireless Telecommunications Symposium (WTS)*, London, 2012, pp. 1–6.

Casini E., R. De Gaudenzi, O. del Río Herrero, Contention Resolution Diversity Slotted ALOHA (CRDSA): An enhanced random access scheme for satellite access packet networks, *IEEE Transactions on Wireless Communications*, vol. 6, no. 4, pp. 1408–1419, 2007.

Celandroni N., E. Ferro, A. Gotta, G. Oligeri, C. Roseti, M. Luglio, I. Bisio et al., A survey of architectures and scenarios in satellite-based wireless sensor networks: System design aspects, *International Journal of Satellite Communications and Networking*, vol. 31, no. 1, pp. 1–38, 2013.

Chen D., M. Nixon, A. Mok, *WirelessHART Network*, Springer, Berlin, 2010, pp. 45–61.

Chen D., P. K. Varshney, QoS support in wireless sensor networks: A survey, in *Proceedings of International Conference on Wireless Networks (ICWN)*, Las Vegas, NV, 2004, pp. 227–233.

Chen K.-C., S.-Y. Lien, Machine-to-machine communications: Technologies and challenges, *Ad Hoc Networks*, vol. 18, pp. 3–23, 2013.

Chini P., G. Giambene, S. Kota, A survey on mobile satellite systems, *International Journal of Satellite Communications and Networking*, vol. 28, no. 1, pp. 29–57, 2009.

Choudhury G. L., S. S. Rappaport, Diversity ALOHA—A random access scheme for satellite communications, *IEEE Transactions on Communications*, vol. 31, no. 3, pp. 450–457, 1983.

Clazzer F., C. Kissling, Enhanced Contention Resolution Aloha—ECRA, in *Proceedings of 2013 9th International ITG Conference on Systems, Communication and Coding (SCC)*, January 21–24, 2013, pp. 1–6.

Cochetti R., *Mobile Satellite Communications Handbook*, John Wiley & Sons, Hoboken, NJ, 2014.

DeGaudenzi R., O. del Rio Herrero, G. Acar, E. Garrido Barrabes, Asynchronous Contention Resolution Diversity ALOHA: Making CRDSA truly asynchronous, *IEEE Transactions on Wireless Communications*, vol. 13, no. 11, pp. 6193–6206, 2014.

del Río Herrero O., R. DeGaudenzi, A high-performance MAC protocol for consumer broadband satellite systems, in *Proceedings of 27th AIAA International Communications Satellite Systems Conference*, Edinburgh, Scotland, 2009, p. 512.

del Río Herrero O., R. DeGaudenzi, High efficiency satellite multiple access scheme for machine-to-machine communications, *IEEE Transactions on Aerospace and Electronic Systems*, vol. 48, no. 4, pp. 2961–2989, 2012.

Digital Video Broadcasting (DVB), Interaction channel for satellite distribution systems, EN 301 790 V1.4.1, European Telecommunication Standardization Institute, Sophia Antipolis, France, 2005–2009.

Digital Video Broadcasting (DVB), Second generation DVB interactive satellite system (RCS2), Part 2: Lower layers for satellite specification standard, DVB Bluebook A155-2, European Telecommunication Standardization Institute, Sophia Antipolis, France, 2010.

Dousse O., F. Baccelli, P. Thiran, Impact of interferences on connectivity in ad hoc networks, *IEEE/ACM Transactions on Networking*, vol. 13, no. 2, pp. 425–436, 2005.

Elson J., L. Girod, D. Estrin, Fine-grained network time synchronization using reference broadcasts, in *Proceedings of Fifth Symposium on Operating Systems Design and Implementation (OSDI 2002)*, New York, 2002, vol. 36, pp. 147–163.

Feng J., C.-W. Chang, S. Sayilir, Y.-H. Lu, B. Jung, D. Peroulis, Y. C. Hu, Energy-efficient transmission for beamforming in wireless sensor networks, in *7th Annual IEEE Communications Society Conference on Sensor Mesh and Ad Hoc Communications and Networks (SECON)*, Boston, 2010, pp. 1–9.

Franchi A., A. Howell, J. Sengupta, Broadband mobile via satellite: Inmarsat BGAN, in *IEEE Seminar on the Critical Success Factors—Technology, Services and Markets*, London, 2000, pp. 23–30.

Ganeriwal S., R. Kumar, M. B. Srivastava, Timing-sync protocol for sensor networks, in *Proceedings of 1st International Conference on Embedded Networked Sensor Systems*, Los Angeles, CA, 2003, pp. 138–149.

Howard S. L., C. Schlegel, K. Iniewski, Error control coding in low-power wireless sensor networks: When is ECC energy-efficient? *EURASIP Journal of Wireless Communications and Networking*, vol. 2006, no. 2, pp. 1–14, 2006.

Hui J., P. Thubert, Compression format for IPv6 datagrams in 6LoWPAN networks, draft-ietf-6lowpan-hc-13, Internet Engineering Task Force, Fremont, CA, 2010.

IEEE (Institute of Electrical and Electronics Engineers), IEEE 802.15.4-2006 standard for information technology part 15.4: Wireless medium access control (MAC) and physical layer (PHY) specifications for low rate wireless personal area networks (LRWPANs), IEEE 802.15.4-2006, IEEE, Piscataway, NJ, 2006.

Ismail A., S. Sezginer, J. Fiorina, H. Sari, A simple and robust equal-power transmit diversity scheme, *IEEE Communications Letters*, vol. 15, no. 1, 2011.

ITU-R (ITU Radiocommunication Sector), Propagation data and prediction methods required for the design of earth-space telecommunication systems, Recommendation P.618-10, Geneva, Switzerland, 2009.

ITU-R (ITU Radiocommunication Sector), Attenuation in vegetation, Recommendation P.833-7, Geneva, Switzerland, 2012.

ITU-R (ITU Radiocommunication Sector), Probability distributions relevant to radiowave propagation modelling, Recommendation P.1057-3, Geneva, Switzerland, 2013.

Kanellopoulos S. A., C. Kourogiorgas, A. D. Panagopoulos, S. N. Livieratos, G. E. Chatzarakis, Channel model for satellite communication links above 10 GHz based on Weibull distribution, *IEEE Communications Letters*, vol. 18, no. 4, pp. 568–571, 2014.

Kanellopoulos S. A., A. D. Panagopoulos, C. Kourogiorgas, J. D. Kanellopoulos, Slant path and terrestrial links rain attenuation time series generator for heavy rain climatic regions, *IEEE Transactions on Antennas and Propagation*, vol. 61, no. 6, pp. 3396–3399, 2013.

Karagiannis G. A., A. D. Panagopoulos, J. D. Kanellopoulos, Multi-dimensional rain attenuation stochastic dynamic modeling: Application to earth-space diversity systems, *IEEE Transactions on Antennas and Propagation*, vol. 60, no. 11, pp. 5400–5411, 2012.

Karatzas I., S. E. Shreve, *Brownian Motion and Stochastic Calculus*, Springer-Verlag, Berlin, 1991.

Karl H., A. Willig, *Protocols and Architectures for Wireless Sensor Networks*, John Wiley & Sons, Hoboken, NJ, 2007.

Kissling C., Performance enhancements for asynchronous random access protocols over satellite, in *IEEE International Conference on Communications (ICC)*, Kyoto, Japan, June 2011, pp. 1–6.

Kourogiorgas C., P.-D. Arapoglou, A. D. Panagopoulos, Statistical characterization of adjacent satellite interference for earth stations on mobile platforms operating at Ku and Ka band, *Wireless Communications Letters, IEEE*, vol. 4, no. 1, pp. 82–85, 2015.

Kourogiorgas C., A. D. Panagopoulos, A physical-mathematical model for predicting slant path rain attenuation statistics, *IET Microwave, Antennas and Propagation*, vol. 7, no. 12, pp. 970–975, 2013.

Kourogiorgas C. I., A. D. Panagopoulos, J. D. Kanellopoulos, On the earth-space site diversity modeling: A novel physical-mathematical outage prediction model, *IEEE Transactions on Antennas and Propagation*, vol. 60, no. 9, pp. 4391–4397, 2011.

Lasassmeh S. M., J. M. Conrad, Time synchronization in wireless sensor networks: A survey, *Proceedings of the IEEE SoutheastCon 2010 (SoutheastCon)*, Concord, NC, 2010, pp. 242–245.

Liva G., Graph-based analysis and optimization of contention resolution diversity slotted ALOHA, *IEEE Transactions on Communications*, vol. 59, no. 2, pp. 477– 487, 2011.

Lo T. K. Y., Maximum ratio transmissions, *IEEE Transactions on Communications*, vol. 47, no. 10, 1999.

Loo C., A statistical model for a land mobile satellite link, IEEE *Transactions on Vehicular Technology*, vol. VT-34, pp. 122–127, 1985.

Maini A. K. and V. Agrawal, *Satellite Technology: Principles and Applications*, John Wiley & Sons, Hoboken, NJ, 2011.

Maral G. and M. Bousquet, *Satellite Communications Systems: Systems, Techniques and Technology*, 5th ed., Wiley, Hoboken, NJ, 2009.

Maseng T., P. M. Bakken, A stochastic dynamic model of rain attenuation, IEEE *Transactions on Communications*, vol. 29, no. 5, pp. 660–669, 1981.

Matricciani E., Transformation of rain attenuation statistics from fixed to mobile satellite communication systems, *IEEE Transactions on Vehicular Technology*, vol. 44, no. 2, pp. 565–569, 1995.

Mills D. L., Internet time synchronization: The network time protocol, *IEEE Transactions on Communications*, pp. 1482–1493, 1991.

Miorandi, D. The impact of channel randomness on coverage and connectivity of ad hoc and sensor networks, *IEEE Transactions on Wireless Communications*, vol. 7, no. 3, pp. 1062–1072, 2008.

Mock M., R. Frings, E. Nett, S. Trikaliotis, Continuous clock synchronization in wireless realtime applications, in *Proceedings of 19th IEEE Symposium on Reliable Distributed Systems (SRDS '00)*, Nurnberg, Germany, 2000, pp. 125–133.

Montenegro G., N. Kushalnagar, J. Hui, D. Culler, Transmission of IPv6 packets over IEEE 802.15.4 networks, RFC 4944, Microsoft Corporation, Intel Corp., Arch Rock Corp., 2007.

Morello A., V. Mignone, DVB-S2: The second generation standard for satellite broad-band services, *Proceedings of the IEEE*, vol. 94, no. 1, pp. 210–227, 2006.

Ochiai H., H. Imai, Collaborative beamforming, in *New Directions in Wireless Communications Research*, V. Tarokh (ed.), Springer, Berlin, 2009, pp. 175–197.

Ochiai H., P. Mitran, H. V. Poor, V. Tarokh, Collaborative beamforming for distributed wireless ad hoc sensor networks, *IEEE Transaction son Signal Processing*, vol. 53, no. 11, pp. 4110– 4124, 2005.

Panagopoulos A. D., P.-D. M. Arapoglou, P. G. Cottis, Satellite communications at Ku, Ka and V bands: Propagation impairments and mitigation techniques, in *IEEE Communications Surveys and Tutorials*, vol. 6, no. 3, pp. 2–14, 2004.

Paolini E., G. Liva, M. Chiani, High throughput random access via codes on graphs: Coded slotted Aloha, in *IEEE International Conference on Communications (ICC)*, 2011, pp. 1–6.

Paris J. F., Closed-form expressions for Rician shadowed cumulative distribution function, *Electronics Letters*, vol. 46, no. 13, pp. 952–953, 2010.

Paulraj A., R. Nabar, D. Gore, *Introduction to Space-Time Wireless Communications*, Cambridge University Press, Cambridge, UK, 2003.

Ping S., Delay measurement time synchronization for wireless sensor networks, IRB-TR-03-013, Intel, Santa Clara, CA, 2003.

Pitsiladis G. T., A. D. Panagopoulos, P. Constantinou, Connectivity evaluation and error performance of millimeter-wave wireless backhaul networks, *Annals of Telecommunications*, vol. 65, no. 11–12, pp. 795–802, 2010.

Pitsiladis G. T., A. D. Panagopoulos, Ph. Constantinou, A spanning-tree-based connectivity model in finite wireless networks and performance under correlated shadowing, *IEEE Communications Letters*, vol. 16, no. 6, pp. 842–845, 2012.

Poulakis M. I., S. Vassaki, A. D. Panagopoulos, Satellite-based wireless sensor networks: Radio communication link design, in *7th European Conference on Antennas and Propagation (EuCAP)*, Gothenburg, Sweden, 2013.

Reddy T. B., I. Karthigeyan, B. S. Manoj, C. S. R. Murthy, Quality of service provisioning in ad hoc wireless networks: A survey of issues and solutions, *Ad Hoc Networks Journal*, vol. 4, pp. 83–124, 2006.

Rhee I. K., J. Lee, J. Kim, E. Serpedin, Y. C. Wu, Clock synchronization in wireless sensor networks: An overview, *Sensors*, vol. 9, no. 1, pp. 56–85, 2009.

Richharia M., L. D. Westbrook, *Satellite Systems for Personal Applications Concepts and Technology*, John Wiley & Sons, Hoboken, NJ, 2010.

Roberts L. G., Aloha packet system with and without slots and capture, *SIGCOMM Computer Communication Review*, vol. 5, no. 2, pp. 28–42, 1975.

Shelby Z., C. Bormann, *6LoWPAN: The Wireless Embedded Internet*, vol. 43, John Wiley & Sons, Hoboken, NJ, 2011.

Shelby Z., S. Chakrabarti, E. Nordmark, Neighbor discovery optimization for low power and lossy networks (6LoWPAN), IETF draft-ietf-6lowpan-nd-18, Internet Engineering Task Force, Fremont, CA, 2011.

Sivrikaya F., B. Yener, Time synchronization in sensor networks: A survey, *IEEE Network*, vol. 18, pp. 45–50, 2004.

Song J., S. Han, A. K. Mok, D. Chen, M. Lucas, M. Nixon, WirelessHART: Applying wireless technology in real-time industrial process control, in *IEEE Real-Time and Embedded Technology and Applications Symposium (RTAS '08)*, St. Louis, MO, 2008, pp. 377–386.

Su W., I. F. Akyildiz, Time-diffusion synchronization protocol for wireless sensor networks, *IEEE/ACM Transactions on Networking*, vol. 13, no. 2, pp. 384–397, 2005.

Sundararaman B., U. Buy, A. Kshemkalyani, Clock synchronization for wireless sensor networks: A survey, *Ad Hoc Networks*, vol. 3, no. 3, pp. 281–323, 2005.

Tilak S., N. B. Abu-Ghazaleh, W. Heinzelman, A taxonomy of wireless micro-sensor network communication models, *ACM SIGMOBILE Mobile Computing and Communication Review*, vol. 6, no. 2, pp. 28–36, 2002.

Vassaki S., G. T. Pitsiladis, C. Kourogiorgas, M. Poulakis, A. D. Panagopoulos, G. Gardikis, S. Costicoglou, Satellite-based sensor networks: M2M sensor communications and connectivity analysis, in *International Conference on Telecommunications and Multimedia (TEMU)*, Heraklion, Greece, 2014, pp. 132–137.

Xia F., QoS challenges and opportunities in wireless sensor/actuator networks, *Sensors*, vol. 8, no. 2, pp. 1099–1110, 2008.

Zheng J., M. J. Lee, A comprehensive performance study of IEEE 802.15.4, Sensor Network Operations (2006), pp. 218–237, 2004.

ZigBee Alliance, ZigBee specifications, version 1.0, 2005. Available at http://www.zigbee.org/.

ZigBee Alliance, ZigBee specifications, ZigBee Document 053474r17, January 2008. Available at www. zigbee .org.

Chapter 9

Use of Wireless Sensor Networks in Smart Homes

Oktay Cetinkaya and Ozgur Baris Akan

Contents

Abstract

This chapter includes a brief discussion of the utilization of wireless sensor networks (WSNs) in smart homes (SHs). Existing scenarios, related automation solutions, and available applications of SHs are summarized. Wired and wireless communication protocols performed are discussed in a comparative way, and a detailed classification of network types, topologies and requirements, node architectures, and routing protocols is also provided. The chapter ends with the analysis of security and privacy considerations, emerging technologies, research challenges, and future directions of WSNs-enabled SH applications.

9.1 Introduction

Wireless sensor networks (WSNs) have come into prominence and recently started to attract certain environments' interest, such as academia, industry, and standard developing organizations [1]. With the improvements of advanced sensory structures, flexible and reliable communication technologies, and the offering of extended application diversity, WSNs are becoming more feasible and preferable day by day. In general, a WSN can be defined as a bunch of spatially distributed autonomous sensors to sense, measure, and monitor the physical and environmental parameters of a medium and to carry the gathered information through the network to a coordinator or an access point over wireless links [2,3]. WSNs provide numerous advantages in terms of cost, flexibility, installation, monitoring, management, and reliable data collection. Typical applications focus on environment, traffic, and healthcare monitoring; process management; military surveillance; transportation; logistics; disaster assessment; earth sensing, and building automation [2,3]. WSNs are formed by sensor nodes getting together in groups on tens to hundreds or even thousands. Each node consists of four collaboratively working units to sense, evaluate, and transfer the needed physical parameter of the environment. In a WSN, node deployment varies depending on the specific application. As an example, for a military operation, nodes are organized as uniformly distributed by dropping randomly from an aircraft. However, when a particular measurement is needed from a specific area, similar to home automation applications, nodes are placed in fixed, well-planned points, which is called regular deployment. For a tracking system, such as logistics transportations, mobile sensor nodes are utilized to compensate the negative effects of deployment-oriented problems.

Since existing power plants and systems fail to satisfy the energy demand and scarcely keep up with the constantly improving technology nowadays, planning and management of energy have undoubtedly become rather important. To maintain the energy need of existing systems, research is aimed at improving the usage and savings of available resources and power plants. The smart grid, an integrated infrastructure of the electrical grid, is one the best candidates for using energy in an efficient way to overcome the existing utilization-induced problems by providing enhancements in the production, transmission, and distribution of electricity. The smart grid can also be described as a modernized communication system between the supplier and consumer that is built by integrated circuits, smart meters, controlling units, and monitoring systems embedded in current power grids in order to track and update the grid data by providing more sustainable, reliable, and qualified transmission, and advanced user privacy. At the lower levels, smart cities and smart homes sustain the operations of smart grids just as extensively.

Information and communication technologies are utilized by smart cities in order to analyze and bring together the vital information of core systems in running cities. A smart city can also respond logically to a variety of needs, such as environment protection, public safety and city services, daily livelihood, and industrial and commercial activities [4].

A home or any quasi-residential area can be referred to as the simplest element of a smart grid due to being last in the transmission line. A smart home is a dwelling of ubiquitous or pervasive computing with automated and controlled components. There are several synonyms used to refer to a smart home: intelligent or adaptive home, home automation, and smart or aware house [5]. To illustrate, smart systems can be assumed to be triangular, where smart homes lie at the lowest level, to construct the base; smart cities are in the middle, between smart homes and grids; and smart grids fill the top section.

Smart systems require an exchange and transmission of control and sensor information among all components, systems, and their related peripheral units. To enable that, a network of collaboratively working devices should be utilized. Even if it is possible to constitute this as a wired network, wireless technologies are mostly preferred to satisfy today's needs. To sense, monitor, and control the environment, sensing and actuation capable circuitries, i.e., nodes, are deployed in the dwellings as fixed or mobile which communicate over a wireless protocol and compose a network in collaboration with other devices. From this point, it is obvious that smart homes and WSNs are interdependent systems that regulate energy utilization, measure physical environmental conditions, control the usage of electrical appliances and resources, and monitor a subject's movements, motions, and vital signs.

This chapter surveys existing studies and research opportunities of wireless sensor networks in smart home applications from the perspective of node and network structures, routing and communication technologies, and privacy considerations. We examine the state-of-the-art approaches related to energy and communication constraints, and the requirements of wireless home sensor network technologies. We clearly indicate the emerging and promising technologies for future homes and finalize the chapter with concluding remarks.

9.2 Smart Home Scenario

Smart homes and home automation systems are one of the first application fields of developing wireless sensor networking technologies. These networks provide ease and flexibility to users in terms of monitoring, management, and control of appliances, environmental parameters, and previously arranged scenarios. Small battery-powered embedded devices that have low-power radio frequency (RF) transceivers, low-computation-capable processors, and compact designed sensing circuitries are typical components of WSNs in a severely constrained form. RF communication enables adjustable device movement—removal from or addition to the network—and lower installation expenses due to not requiring any extra wiring, unlike its wired counterparts. As the sensing parts are the entities where the required physical parameters are gathered, the processor is in charge of evaluating this received information and managing both the system and communication duties. To provide proper system working, every component sustains its operation in an efficient manner. When the overall efficiency is concerned, each part of the system should work correspondingly, which necessitates the construction of homes as "smart." A home is required to be equipped with exactly three things in order to be considered smart: internal network, intelligent control, and automation system [5]. Since smart home networking is utilized over wireless links and corresponding devices, intelligent control is handled by advanced gateways such as smart meters. In addition, the automation part refers to internal and external system and service connections. Smart home applications are generally focused on comfort, healthcare, and security operations; some WSN-enabled use cases and corresponding examples are given below [1,5,6] (Figure 9.1):

Lighting control: Lighting management, which refers to a light switch or dimmer of any user outlet placed on a wall, decreases the necessity of new wired connections while providing both ease of use and flexibility. A remote control option is also available for activation of the

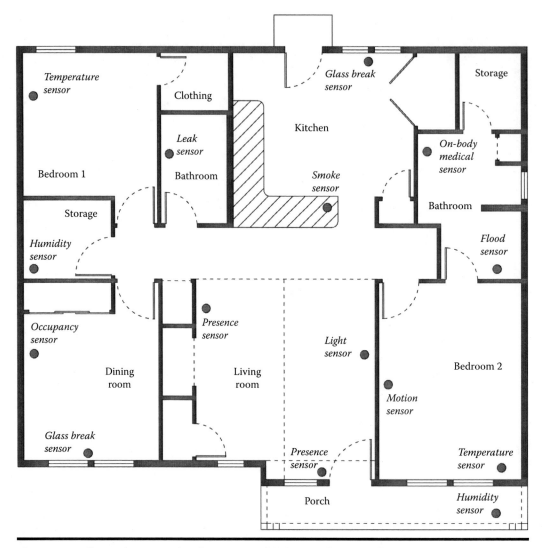

Figure 9.1   Illustrative example of a WSN-enabled smart home: node types and deployment.

lights via wired or wireless systems. Moreover, due to the luminance sensors' ability to detect the presence of people in an inadequately illuminated environment, lights can be turned on or off automatically without the influence of human beings.

Remote control: Infrared (IR) technology was used as a form of wireless communication in household appliance management. Nevertheless, short-distance coverage and line-of-sight necessity have resulted in the development of radio frequency (RF) operative communication protocols. There are several related technologies that enable the remote control of connected devices over wireless links.

Energy management: As the sensors are able to collect and process the parameters of temperature, humidity, light, and presence, and correspondingly control the utilization of window shades, doors, and heating, ventilating, and air conditioning (HVAC) systems, energy usage is directly linked to these gadgets. Power metering and managing units, that is, smart plugs, can be used in smart home applications for demand-responsive regulations, energy consumption monitoring, and prevention of standby-derived redundant energy usage by informing the users or any other related institution [7].

Remote care: In-home medical systems can provide aids to patients and disabled or elderly citizens. In this matter, various body and health-related parameters, such as blood pressure, hormone or sugar level, body temperature, or heart rate value, can be reported by wireless sensors for diagnosis and decision-making procedures. In any circumstance where the sensors detect an abnormal value, alarms can be instantly triggered to call for help from hospitals or health centers.

Security and safety: High-level security systems may include several sensors, such as glass break, motion, presence, and smoke, to identify the potential risks and control of trigger mechanisms for proper action management. There are such systems that directly connect with fire departments, police stations, or private security services.

9.3 Sensor Networks

9.3.1 Node Structure

As its name suggests, sensor nodes are the main constituents of WSNs. They come together to compose a network where the communication and management tasks are handled over a wireless medium. A typical wireless sensor node is composed of four main systems or components, as shown in Figure 9.2, and these can be classified as sensing, processing, communication, and energizing units [8]. In certain cases, such as when the processor does not contain any internal memory, the sensor needs to be equipped with an external storage that changes the total number of systems to five. A sensor node may include multiple sensors or actuators, depending on application, and can communicate over wired or wireless systems, or combination of these. A sensor node stores and executes the communication protocols as well as data processing algorithms. As the received data could be directly transferred to the upper-level components, it is also possible to evaluate them within network processing mechanisms [3]. All these features vary according to the application needs and requirements, with certain trade-offs. As the total number of connected devices varies, some nodes need to be organized as coordinators to manage the data flow and conduct the assigned tasks in the network. For small indoor environments such as homes, buildings, offices, and subway stations, sensor nodes are purposed for well-defined tasks to probe

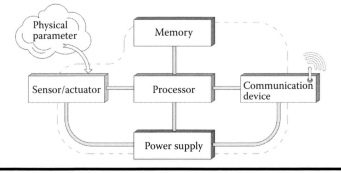

Figure 9.2 Typical schematic of a sensor node.

the area and inform the user or any other authenticated people. So, a wise deployment becomes crucial to increase reliable data gathering. Typically, the process starts with measurement of the environmental parameters. Then, these analog inputs are converted into digital bits by an analog-to-digital converter (ADC). The received useful and more comprehensible data are evaluated by the processor for further transactions. As the processor can temporarily store the gathered data in memory, the data can be saved permanently for future usage. All the operations conducted on a sensor node are managed and pursued by the processor, with the energy provided by the power supply. Considering certain constraints, all the units of a sensor node should work collaboratively to sustain the operation as long as possible. As a result, design efforts are focused on how to build a more robust, efficient, and flexible node to increase the overall system reliability [8,9].

9.3.1.1 Sensing

To observe and control the physical parameters of the environment, sensors and cognate devices such as actuators are utilized. Sensing circuitries can be categorized into three groups: passive omnidirectional, passive narrow-beam, and active sensors [2]. Omnidirectional sensors do not manipulate the environment during the measurement, and the information is gathered regardless of direction. They operate with tiny batteries to amplify the generated analog signals. Light, temperature, and humidity levels or any other related physical condition is sensed with this type of sensory circuit. The notion of direction is well defined in passive narrow-beam sensors. The sensor interacts with a previously stated particular area to obtain the needed information. A camera is the simplest example of a narrow-beam sensor. Active sensors continuously probe the environment where the measurement occurs. This type of sensor requires more energy than other sensor classes; therefore, it needs to be energized with external sources in some cases. Sonar sensors and radars can be counted as members of this group. As may be noticed from the definitions, passive sensors best fit home automation applications. For an indoor environment, commonly lux (for the level of light), humidity, temperature, motion (for occupancy or presence), glass break, dry contact (for flood or leak detection), open/close (for activating/deactivating lights when the doors and/or windows are opened/closed), smoke or gas, and passive infrared (PIR) sensors are deployed [9]. ADC is responsible for the conversion of the analog signals into their digital equivalents through two steps: quantization and sampling. After this process, any information representing a physical parameter related to the environment becomes more logical and perceptible by the rest of the node, which will be conveyed to the processing part.

9.3.1.2 Processing

When the processor is considered, there are a variety of options to implement. Microcontrollers, digital signal processors (DSPs), application-specific integrated circuits (ASICs), and field-programmable gate arrays (FPGAs) are the best-known types [3]. A processor is basically responsible for executing the node, performing tasks, collecting and processing data, and deciding where and when to send these gathered or generated information. It should be general purpose; flexible to interconnect different systems, devices, and peripherals; easily programmable; and low power consumptive. The processing unit of the sensor node includes a nonvolatile and an active temporary memory to store the program instructions and sensed data, respectively, a processing chip, and an internal clock for synchronization and timing operations. Microcontrollers or microcontroller units (MCUs) are frequently preferred for the processing, computing, and managing duties of sensor nodes when the design goal is to achieve flexibility. Besides being of compact construction, low cost, and low power consumptive, MCUs have debugging and programming easiness as a result of using high-level programming languages that make them more preferable. However, when an application requires more powerful or more cost- or energy-efficient capabilities, custom-made processors such as FPGAs and DSPs are more likely to be implemented. Texas Instruments' MSP430 and Atmel's Atmega series are well-known types of microcontrollers.

As the name suggests, DSPs are used for processing discrete signals with the help of digital filters to minimize the negative sides of hardware and circuitry-oriented noises and modify the spectral characteristics of signals. As mentioned before, the analog data gathered by sensory measurements from the environment are converted into digital signals with the help of ADCs and transferred to the DSP for evaluation. After processing, the meaningful information can be transmitted to another node, an access point (AP), or directly to the coordinator or supervisor over a communication module. The simple and easily implementable structure of control and flow components such as adders and multipliers makes DSPs more applicable for this type of constituent-required applications, and also, with the addition of specific algorithms, DSPs become the most suitable candidate for signal processing–required applications. Despite the many benefits DSPs possess, there are some drawbacks as well. Being relatively expensive and having an inflexible structure and ongoing deployment complications both restrict the capabilities and decrease the number of possible applications of DSPs.

ASICs are customized for specific tasks and applications in general. Rather than manage the whole circuit, they work as a complement of the mentioned processors to handle low-level duties. Being small, having higher bandwidths, performing much better, and consuming low power are the main good benefits of ASICs. The only drawback they hold is having relatively higher production costs as a corollary of being the final product of a complex design process.

FPGAs perform better in terms of programmability. This flexible and relatively complex-constructed hardware class has higher bandwidths than DSPs, is faster than both microcontrollers and DSPs in general, and is able to support parallel programming. Besides these specifications, their cost and complex structure restrict their application suitability.

In all, microcontrollers are the best candidates to meet the requirements of wireless indoor networking applications because they are low cost, low power consumptive, and flexible. Thus, they are mostly preferred in home energy management, control, and monitoring operations.

9.3.1.3 Communication

Sensed, processed, and evaluated data are sent or received by communication devices and modules. Time- and energy-effective transmission of the information between these devices and linked subsystems is becoming crucial, especially for the limited capable sensor node derived from tiny gadgets and the huge networks formed by themselves. There are various communication types and corresponding mediums to transfer the information, such as electromagnetic propagation, radio frequency (RF), optical, and ultrasound systems. An optical medium needs line of sight (LoS) for communication and tends to be affected by weather conditions. A small amount of energy is enough for both the generation and detection of the light and sending it distances of kilometers at high speeds. Ultrasound signals are preferred when a long-distance coverage area needs to be handled, when consuming low power where RF and optical systems are not applicable. As might be expected from this definition, it is best suited for underwater communication. Electromagnetic radiation at radio frequencies is the best service to employ a sensor network for indoor environments. RF provides high-data-rate and long-distance communication in non–line of sight (n-LoS).

There are four possible operational states for the RF transceivers chosen and managed by the protocol stack—transmit, receive, idle, and sleep—which are diversified with respect to energy consumption, active parts, and operation requirements. When a transceiver is put into the transmit state, the antenna continuously radiates energy, the transmitting circuitry becomes active, and the total power consumption fluctuates in the order of microwatts. In the receive state, only the receiving part is activated to collect information from other nodes or transmitters, and the level of power consumption is close to the transmit state. For the idle mode, the transceiver is ready to receive, but it is not willing to do something. Although some parts of the receiver circuitry are deactivated, the transceiver consumes a considerable amount of power. In the sleep mode, significant parts of the transceiver are switched off, and as an expected result of that, it is not able to receive something immediately. Power consumption is limited to milliwatts for this state. To leave the sleep mode, recovery time and start-up energy terms are sufficient [2]. In addition to being a time- and energy-costly operation, the major problem for state switching is the trade-off between energy consumption and packet transmission reliability. Interface, channel capacity, supported frequency band, communication range, and data rates are the main evaluative capabilities for the RF transceivers.

9.3.1.4 Powering

The main idea of node powering is to provide as much energy as possible by using tiny batteries that satisfy the minimal needs to operate the node with the least cost and recharge time, smallest volume and weight, and maximum longevity and robustness. However, node sizes restrict both the capabilities and the design goals of WSNs harshly, and energy becomes a scarce resource as a direct result of that. Wireless sensor nodes are so tiny to accommodate high-capacity power supplies, considering the complexity of the tasks they carry out. Being small is also a constraining factor for the deployment of renewable energy and recharging mechanisms. Therefore, the provided batteries have to durable and stand as long as possible to prevent possible failures. In parallel, sensory nodes have to sustain their operation in an energy-efficient manner, because replacing or recharging the batteries may or may not be possible [2]. When considering the requirements of an energy supply, attention is given to the load capacity, self-discharge time, voltage stability, shelf life, and recharging efficiency. A source should be adequate to meet these requirements, and there are two practical energy suppliers for a wireless sensor node, referred to as primary and secondary batteries. The primary ones are generally known as nonrechargeable, while the secondary equivalents are equipped with this capability. However, it only makes sense in combination with some form of energy harvesting operations. Benefitting from the freely available resources of the environment to charge the batteries becomes more logical, feasible, and applicable in parallel with ongoing developments in advanced systems and material science. Scavenging energy is a good way to enable a node's sustainability for long periods. Light with solar cells, vibration, noise, temperature gradient, and pressure with piezoelectric materials are among the best-known types of converting ambient dynamics into energy sources.

In addition to energy harvesting, some power management operations may be applied for obtaining the longevity by increasing the efficiency of energy usage, as discussed previously. To illustrate, it is not essential to operate the sensors as fully functional all the time if there is nothing to do. In these cases, the node can be switched to the power-safe mode in order to prevent redundant energy consumption. However, at this point, a problem comes up with the question of when and how to wake up and turn back to the active mode again. This results in a trade-off between the energy efficiency and quality of service (QoS). To manage the power usage, some operation modes, such as active, idle, and sleep for instruction execution, and turning on or off for both data transmission and sampling, are implemented in the energy-consumptive parts of the sensor nodes. But, as it is known, saved energy should be bigger than the overhead to handle the power management, when the operation performance is also acceptable. Instead of sending and carrying all sensed data through the network, evaluating the same type of information about a particular parameter and then conveying the gathered final one to the destination can help reduce the energy consumption. That operation is called in-network processing, which is based on the idea that transmitting is much more costly than evaluation.

9.3.2 Networking

9.3.2.1 Networking Types, Topologies, and Components

A wireless sensor network is formed by the combination of numerous nodes. These units are able to undertake three different tasks during operation, which can be categorized as gateways, end points, and routers [3]. In other words, there are three types of devices for a typical WSN: source, sink, and relay. Source nodes are the entities where the sensing and data collection operations are carried out. The needed information regarding a physical parameter of the environment is first sensed and then transferred by these types of devices. A source node is not sufficiently measured by just one kind of metric; it can be equipped with multiple sensors to sense several parameters from the related area, depending on the application. For conveying the data to the required place or device, relay nodes are utilized. These nodes are responsible for the transportation of useful information through the network until it reaches its destination. Sink nodes are the entities where the gathered data are actually needed and used. Actuators are the specialized sinks for the particular tasks. The sinks can be formed by three different modes. A sink may belong and connect to the network directly; it may be an external entity, but still be connected to the WSN; or it may be a member of another external network. However, it is somehow connected to the WSN over a gateway.

These mentioned device classes come together and compose the network in different topologies, namely, star, mesh, and cluster or tree [10], where the typical distribution of each structure is shown in Figure 9.3. In star topology, each node is directly connected to the coordinator. Even though this connection serves a simple communication, it restricts the overall achievable network range. To overcome this problem, the tree (cluster) topology can be utilized. For this architecture, each node maintains a single communication path to the coordinator, similar to star topology; however, it uses the other nodes as relays to form this path. The common drawback of this topology is possible connection losses when a relay node expires or shuts down. To handle this shortcoming, mesh structures are constructed. For the mesh topology, the nodes have different paths

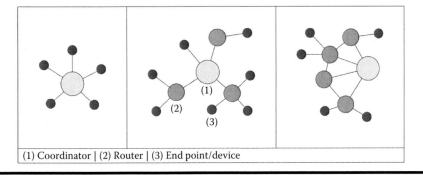

(1) Coordinator | (2) Router | (3) End point/device

Figure 9.3   Typical device connections of the star, tree, and mesh topologies, respectively.

to connect the coordinator, which enables rerouting of the data in case of communication failures. This attempt helps to extend the lifetime and increase the robustness of the network simultaneously and is quite reliable; however, it suffers from increasing levels of in-network latency as a direct result of multihopping caused by the data during travel to the destination.

For an indoor environment, all of the mentioned topologies are applicable, depending on the area that needs to be covered, application requirements, and node specifications. In addition to these wireless networking structures, a wired system can also be used to constitute a sensor network, especially in indoor environments such as homes, offices, and buildings. To explore these areas, sensing- and actuation-capable gadgets, that is, nodes, are deployed as fixed or semimobile and connected to each other with cables, wires, or derivatives of these forms to communicate over a protocol, and they compose a network with the integration of other subdevices. Network topologies are similar in some respects to their wireless counterparts. Mostly, peer-to-peer (P2P), line, and bus topologies are utilized for data flow in wired networks. The X10, Universal Powerline Bus (UPB), KNX, and INSTEON technologies are typical examples of these systems. A more illustrative comparison is detailed in the upcoming sections.

9.3.2.2 Communication Protocols for WSNs in Smart Homes

There are several standardization bodies in the field of WSNs. With regard to this, communication protocols operated in smart homes are developed in huge numbers as wired, wireless, or a combination of these systems. As the Institute of Electrical and Electronics Engineers (IEEE) focuses on the physical (PHY) and medium access control (MAC) layer definitions, interest groups, alliances, and, more specifically, the Internet Engineering Task Force (IETF) have tried to develop higher-level technologies. As a result, there exist several communication opportunities that are collaboratively driven by the different industrial consortiums. The PHY and MAC layer parameters, characteristic specifications, and both advantages and disadvantages of the related existing wired and wireless communication technologies are discussed below and comparatively scrutinized in Tables 9.1 and 9.2.

9.3.2.2.1 Wired Technologies

9.3.2.2.1.1 X10 X10 can be accepted as the initial general-purpose communication and networking standard for home management and control. Although it was first launched as power line (PL) based, it has evolved to a hybrid technology through time with the addition of an RF protocol. As X10-PL operates in 120 kHz, the carrier frequency of the supportive RF protocol varies between 310 and 433 MHz, depending on the region that actualizes the connectivity of wireless appliances or components of the control system. Besides the RF extension, some INSTEON-based wireless gateways are also applicable to the existing X10 systems, which increases the number of possible scenarios and range of applications [5].

The communication of X10 technology can be detailed as follows. Data are transmitted through the alternating current (AC) power line as a series of 1 ms bursts of 120 kHz when each half-wave reaches the zero point of the 50/60 Hz sine signal. A burst indicates logic 0 if it is absent, or inversely, in the case of presence, it signifies logic 1. A standard transmission means a 44-bit message spreading over 22 cycles of a 50/60 Hz waveform. Nine data bits, which are carried in complementarity, follow a 4-bit 1110 initial pattern in this design. The remaining 22-bit pattern is then sent again with the second half of the command [11].

Table 9.1 Summary of the Main Features of X10, UPB, KNX, and INSTEON Technologies

Feature/Protocol	X10	UPB	KNX	INSTEON
Operating frequency	120 kHz; 310–433.92 MHz	4–40 kHz	110/132 kHz; 868.3 MHz	131.65 kHz; 868–924 Mhz
Maximum data rate	20–60 bit/s; 9.6 kbit/s	480 bit/s	1200/2400 kbit/s; 16.384 kbit/s	13.165 kbit/s; 38.4 kbit/s
Nominal range	500–1000 m	80–500 m	~1000 m; 100 m	~500 m; 40 m
Modulation type	N/A	PPM	S-FSK; FSK	BPSK; FSK
Network topology	N/A; star	Peer to peer	Tree, line, star	P2P; mesh; dual mesh/band
Network size	256	~64,000	255; 65,536	65,536
Encryption	N/A	N/A	N/A; 128-bit AES	Rolling code encryption
Complexity	Complex	Complex	Simple; less complex	Less complex

Note: N/A, not available; PPM, pulse-position modulation; S-FSK, spread frequency shift keying.

Communicating over existing wires makes X10 easily implementable and cost-effective. However, there also exist some drawbacks that limit X10 employability, such as their relative slowness (data rates can reach up to 20 bit/s), limited function diversity, wiring complexity, incompatibility, absence of encryption, inevitable noises, interferences, and command losses [12].

9.3.2.2.1.2 UPB   Universal Powerline Bus (UPB) is a home automation protocol for communication among devices that use AC power line as the medium [13]. Although it takes its basis from X10, UPB uses higher-voltage and stronger signals to communicate, and it also has an improved data rate and considerable reliability in comparison with its ancestor. Not requiring extra wiring, hardware, or a power supply makes UPB appealing; it depends only on an embedded wired system that inevitably restricts both the operation area and application range as a result of its nonupgradable and inflexible structure.

The communication method of UPB consists of transmitting digitally encoded information over the electrical power line as a series of precisely timed electrical pulses called UPB pulses that are superimposed on top of the normal AC power waveform.

Regarding the long-distance communication, UPB is less responsive to AC line noises and signal reductions. The data rate is four times faster than that of its ancestor, X10, and by peer-to-peer connections, a UPB network can support nearly 64,000 devices in operation. Similar to X10 technology, UPB suffers from security issues as a result of not containing data encryption mechanisms [14].

9.3.2.2.1.3 KNX   KNX is an ISO/IEC 14543-3–based low-cost, flexible, secure, and compatible network communication protocol for smart buildings [15]. Instead of directly using the existing AC power line, KNX requires its own wired network, and there are three forms of it: star, line, and tree. A bus topology is able to support up to 256 KNX-oriented products, such as system

Table 9.2 Summary of the Main Features of Wi-Fi, ZigBee, Z-Wave, 6LoWPAN, Bluetooth, and BLE

Feature/ Protocol	Wi-Fi IEEE 801.11.n	ZigBee IEEE 802.15.4	Z-Wave	Bluetooth IEEE 802.15.1	BLE IEEE 802.15.1	6LoWPAN IETF RFC-6282
Operating frequency	2.4–5 GHz	868/915 MHz, 2.4 GHz	868/915 MHz	2.402–2.482 GHz	2.402–2.482 GHz	868/921 MHz, 2.4–5 GHz
Maximum data rate	11–54 Mbps	20/40 kbps; 250 kbps	9.6–40 kbps	0.7–2.1 Mbps	0.27 Mbps	10–40 kbps, 250 kbps
Nominal range	10–100 m	10–100 m	30–50 m	15–20 m	10–15 m	10–100 m
Modulation type	BPSK, QPSK, OFDM, M-QAM	D-BPSK, O-QPSK, QPSK	FSK, GFSK, narrowband	GFSK, CPFSK, 8-DPSK	GFSK	BPSK, O-QPSK, ASK
Network topology	Star, tree, P2P	Star, mesh, cluster tree	Mesh	Star	Star	Star, mesh, P2P
Network size	Thousands	65,536	232	8	N/A	~100
Encryption	RC4 stream and AES block cipher	128-bit AES	128-bit AES	AES block cipher	128-bit AES	128-bit AES
Coding	MC-DSSS, CCK, OFDM	DSSS (1 → 15), DSSS (4 → 32)	Manchester; NRZ	FHSS	Adaptive CCK	Header compression, DSSS
Channel bandwidth	20–25 MHz	0.3/0.6 MHz; 2–5 MHz	Fixed	1 MHz	8 MHz	2–5 MHz
Complexity	Complex	Simple	Complex	Complex	Simple	Less complex
Applications	Monitoring, Internet, data network	Wireless sensing, monitoring	Home automation, security	Wireless sensing, monitoring	Healthcare, beacon, fitness	Home, automation, IoT

Note: ASK, amplitude shift keying; CCK, complementary code keying; CPFSK, continuous phase frequency shift keying; D-BPSK, differential binary phase shift keying; DPSK, differential phase shift keying; M-QAM, M-ary quadrature amplitude modulation; MC-DSSS, multi-carrier direct sequence spread spectrum; NRZ, non-return-to-zero; O-BPSK, offset binary phase shift keying; OFDM, orthogonal frequency-division multiplexing; RC4, Rivest Cipher 4.

devices, control units, sensors, detectors, and actuators. KNX enables the remote control of lighting, heating, and cooling systems; gate entrance, security, and authorization transactions; and curtain- or shutter-derived home appliances, and it supports fault or malfunction tracking, such as central monitoring, managing, and controlling operations. With the 16.384 kbit/s data rate, KNX provides relatively faster, flexible, and energy-efficient solutions for home and building automation. The protocol defines several physical communication mediums, such as twisted-pair wiring (KNX-TP), power line narrowband networking (KNX-PL; PL110 and PL132), radio communication (KNX-RF; ZigBee, Z-Wave, and Wi-Fi), and Ethernet (KNX-IP), to expand the application range [16]. Besides these good specifications, KNX systems have some inevitable restrictions, such as installation costs that are nonnegligible, complexity that is increased by multiprotocol participation, and a relatively embedded structure beclouded by system upgrades.

9.3.2.2.1.4 INSTEON   INSTEON is a dual-band mesh home networking technology to bridge and connect the wireless and power line–based protocols and devices to each other [17]. It enables sensors, switches, lightening tools, remote controls, and any other electrical gadgets to communicate over power lines, radio frequencies, or both. INSTEON is compatible with X10 standard operative devices where the commands are conducted on AC lines, but the actions take place in needed areas, notwithstanding wires and cables, thanks to the RF support of INSTEON. Today, there are nearly 200 INSTEON-to-X10-enabled devices on the market. All the INSTEON-enabled devices are constructed the same, and they can transmit, receive, or repeat any message of the corresponding protocol, without requiring a network controller or a special routing algorithm. The only negative side of INSTEON technology is that it has considerably low data rates for communication [12].

9.3.2.2.1.5 Other Wired Systems   It is also possible to build an automation system over the traditional telephone line, digital subscriber line (DSL), integrated services digital network (ISDN) or Ethernet; however, these methods are not adequate to satisfy today's needs.

9.3.2.2.2 Wireless Technologies

9.3.2.2.2.1 Wi-Fi   Wi-Fi, or Wireless Fidelity, is an IEEE 802.11–based, low-cost, unlicensed wireless local area network (WLAN) technology [15]. It operates in the industrial, scientific, and medical (ISM) band with a carrier frequency of 2.4–5 GHz. Wi-Fi can be classified as a long-range solution for local area networking with 45 m indoor and 90 m outdoor coverage capabilities [14]. The majority of products that require wireless communication, such as laptops, cell phones, and Internet access devices (routers, modems, etc.), are equipped with Wi-Fi chipsets in parallel to an ongoing decrease of their unit price and ease of use. As the protocol was evolving to 802.11n from 802.11a/b/g, the number of hotspots, throughput, and data rate increased, so Wi-Fi became more feasible to serve high-bandwidth solutions for human-oriented applications. Wi-Fi supports point-to-multipoint (access point), point-to-point (ad hoc), and multipoint-to-multipoint (mesh) networking structures. The major limitations of Wi-Fi technology are high power requirements, interference from other devices as a result of using relatively crowded bandwidth, security, serious communication losses caused by obstacles, and lack of interoperability.

9.3.2.2.2.2 ZigBee   ZigBee is an IEEE 802.15.4–based, highly reliable, and low-cost wireless communication protocol that has low-power-consumption characteristics under operation [1]. It supports three different network structures—generic mesh, star, and cluster tree—which vary, depending on the specific application. Mesh networking enables routing protocols to reach the desired sensor nodes or devices by using intermediate network components as relays. That increases communication range while making the medium more reliable. ZigBee operates in 784, 868, 915, and 2400 MHz globally available ISM bands, with adaptive data rates that vary between 20 and 250 kbit/s [10,14]. This relatively low transfer rate implicitly provides multiyear battery life, requiring low energy. ZigBee reaches up to 70 m in LoS, depending on frequency, output power, and module characteristics. It is easier to utilize or implement a wireless personal area network, in contrast with Wi-Fi and Bluetooth.

ZigBee uses the physical and medium access control layer definitions of the IEEE 802.15.4 specification as a basis for higher-layer communications, as seen in Figure 9.4. IEEE 802.15.4 comprises various PHY opportunities for different regions in the world that operate in the 868 and 2400 MHz frequencies of the ISM band [15]. For the first one assigned to Europe, the data are spread by using direct-sequence spread spectrum (DSSS) at a rate of 20 kbit/s, with a carrier frequency of 868 MHz. One differential encoded data bit is spread to 15 chips. The chips are then physically transmitted with binary phase shift keying (BPSK) modulation. The maximum transmission power for this specification is limited to 25 mW, and the effective bandwidth is 600 kHz. The second PHY alternative operates in the 2.4 GHz ISM band with a rate of 20 kbit/s and can be used worldwide. The spreading method is again DSSS, similar to the first one, and 4 data bits are spread to 32 chips with the help of 16 quasi-orthogonal sequences. These chips are then transmitted with offset quadrature phase shift keying (O-QPSK) modulation with half-sine pulse shaping. The maximum transmission power for this specification is limited to 20 mW, and the effective bandwidth is 2 MHz. The MAC layer uses the physical channel to transmit the MAC frames and handles the channel access over a carrier sense multiple access with collision avoidance (CSMA/CA) procedure.

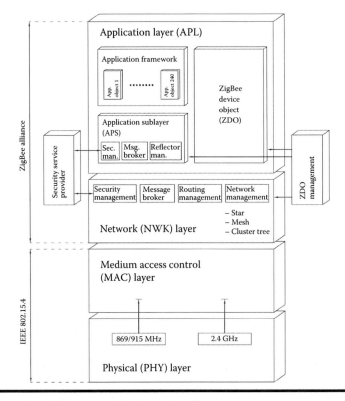

Figure 9.4 Protocol architecture of ZigBee technology.

Due to the small duty cycles, low data transfer rates, leading to less energy-consumptive small gadgets, power-save mode deployment, and energy-efficient modulation techniques, a ZigBee module has a longer process life, with tiny batteries preventing the unnecessary usage of external supplies [18]. Therefore, it becomes a cost- and energy-effective solution for monitoring, managing, and controlling duties. In addition to these advantages, fast and easy employability, flexibility, extra node capability, and manufacturer supplier independence make ZigBee more preferable. However, although ZigBee possesses many benefits, there are some drawbacks as well. Low data rate, license requirements, limited distance, and low data processing speed can be counted as well-known restrictions of ZigBee technology.

9.3.2.2.2.3 Z-Wave Z-Wave is a new wireless communication protocol that specializes in home automation [19]. In spite of coming out recently, it supports roughly 1000 different devices, and these can be controlled over a tablet, smartphone, or personal computer. This extremely low-power consumptive protocol makes any household product smart, serving secure, energy-efficient, convenient, and reliable scenarios. It is easy to implement a Z-Wave network in a home environment, the corollary of wireless communication, because it doesn't require any additional wiring or construction. The operating frequency of Z-Wave varies from 862.2 to 921.42 MHz, which helps to reduce the number of possible interferences from the mostly preferred frequencies and protocols, such as 2.4 GHz operative Wi-Fi, Bluetooth, and ZigBee. Unlike the collaboratively constructed technologies, for example, 6LoWPAN, the Z-Wave protocol is driven by Z-Wave alliance, as shown in Figure 9.5.

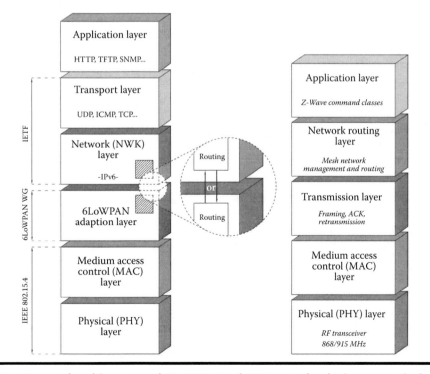

Figure 9.5 **Protocol architectures of 6LoWPAN and Z-Wave technologies, respectively.**

Considering that huge data transfer is not essential for controlling home appliances, Z-Wave has adequate and acceptable data rates up to 100 kbit/s. It enables routing protocol utilization with a supporting mesh network, which increases the communication range, quality, and reliability. Even if the mentioned specifications convince the user that it is a good option, there is still room for development. To illustrate, although the initial construction costs are low, Z-Wave-enabled devices are relatively expensive. It is not license-free, which means additional fees for commercial products have to be paid, and a network controller with an additional gateway must be provided to realize the connection between the end devices.

9.3.2.2.2.4 Bluetooth and BLE

Bluetooth can be defined as a IEEE 802.15.1–based wireless personal area networking technology. It uses short-wavelength radio waves, which limits the communication distance, and operates in the range of 2.4–2.485 GHz globally in the unlicensed ISM band [15]. There are three classes for Bluetooth, namely, 1, 2, and 3; their power requirements and typical ranges may vary, depending on the application. Bluetooth is frequently used in mobile phones, personal computers, and gaming consoles. There exists a typical master–slave interaction between devices, and a master can manage or support up to seven slaves. The relation or behavior of these classes can change during the communication, which means a master can turn into a slave, and a slave can manage the communication as a master. This helps to carry out healthy and reliable communication. A typical data rate of Bluetooth is 0.7–2.1 Mbit/s, and the range varies from 5 to 30 m. The chipset expenditure of Bluetooth goes down day by day, and that makes it a cost-effective solution for short-distance and low-data-rate applications [14]. In recent years, an improved version of Bluetooth, namely, Bluetooth Low Energy (BLE) or Bluetooth Smart, has been announced and used in smartphones, healthcare and fitness systems, beacons, and home entertainment. As its name suggests, BLE consumes considerably less power than Bluetooth without affecting the range; however, the data rate reaches up to only 0.27 Mbit/s. BLE has been used in home automation because it is cheap and low power consumptive. Data spreading is realized with frequency hopping spread-spectrum (FHSS) technology, and Gaussian frequency shift keying (GFSK) modulation is preferred for the data transmission. The notable drawbacks of both Bluetooth and BLE are their limited range, low data rate, unsecure structure, and having to use the device's battery.

9.3.2.2.2.5 6LoWPAN

6LoWPAN is a globally free and open standard for composing low-power wireless personal area networks (LoWPANs) over Internet Protocol version 6 (IPv6), which is defined in RFC 6282 by the Internet Engineering Task Force (IETF) [1]. Being based on IPv6 makes every device able to connect to the Internet, due to having a unique IP address, over open standards such transmission control protocol (TCP), user datagram protocol (UDP), hyper-text transfer protocol (HTTP), and constrained application protocol (CoAP). In other words, 6LoWPAN is a low-cost, low-power, battery-operated, and low-bandwidth wireless mesh networking technology that operates in the 868, 915, and 2400 MHz frequencies of the ISM band [20].

Data rates vary from 20 to 250 kbit/s, depending on the carrier frequency and mode specifications. 6LoWPAN uses IEEE 802.15.4's PHY and MAC layer (i.e., open systems interconnection [OSI] layers 1 and 2) definitions as a basis, similar to the ZigBee protocol. IETF is in charge of developing upper-layer technologies, as seen in Figure 9.5. As a natural consequence, data are spread and encrypted by using DSSS, at 2.4 GHz, and AES-128, respectively. Channel access is handled over a CSMA/CA procedure on the MAC layer, and two types of multihop/routing-capable devices, full and reduced function, similar to ZigBee technology, are supported for mesh networking. A full-function device (FFD) is able to communicate with all types of device classes in the network, unlike a reduced-function device (RFD), while supporting the full protocol. FFDs

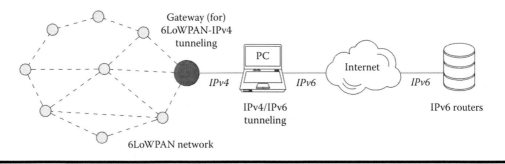

Figure 9.6   6LoWPAN protocol network model.

tend to consume more energy and require improved hardware, such as CPU/RAM, in contrast to RFDs, because they are responsible for the overall network management, address allocation assignments, evaluation and conveyance of the gathered information, and node–router joining or disjoining operations.

The main idea that lies behind this standard is the need to enable Internet connectivity for even the smallest device. IPv6 is preferred for this goal, due to its ability to have a heterogeneous network connection, worldwide free-to-use infrastructure, global scalability, and perhaps most importantly, great addressing space, numerically 2^{128} bits ~ 16 bytes, which is enough for the Internet of Things (IoT) [21]. Figure 9.6 shows both in-network data flow and a IPv4-to-IPv6 packet conversion structure. Considering the massive expansion of web-enabled devices, today's frequently used protocol for the Internet, IPv4, has only 2^{32} IP addresses and will eventually be inadequate. Therefore, IPv6 has been developed to last for decades, while enabling Internet access to wireless operated devices. 6LoWPAN is now mostly preferred in health and environment monitoring, security, logistics, and building automation.

9.3.2.3 Requirements, Characteristics, and Design Objectives of WSNs in Smart Homes

As mentioned before, WSNs enable numerous applications and scenarios for smart indoor areas. When considering these networks are formed by the combination of sensory nodes collaboratively getting together in huge numbers, there is no doubt that building a "healthy" WSN can only be done by constructing the nodes to be as good as possible [9]. In order to achieve that, some characteristic requirements, such as scalability, self-configurability, mobility, responsiveness, energy-efficient operation, and QoS, which compromises fault tolerance, programmability, reliability, maintainability, and longevity, should be fulfilled [22]. Some of these metrics are detailed below.

9.3.2.3.1 Scalability

Scalability refers to the ability to support network enlargement in terms of node quantity. In other words, a network should be operational regardless of the increasing number of deployed sensors. However, as the network size grows, usable bandwidth detractive overheads can occur, and data packet and message loss causative communication link failures may increase; therefore, more control packets could be required to follow the routing path and make the communication feasible [22]. Where a typical WSN can be formed by tens to thousands of sensor nodes, the average varies between two-digit numbers for an indoor environment. As the flexible structure of wireless home automation networks allows upgrading the system by adding extra sensors or any other related devices in certain quantities, they should be constructed to be durable as possible, even if the network size is constantly growing.

9.3.2.3.2 Reliability

Reliability is the ability to ensure reliable data transmission in continuously varying network dynamics. There is an inverse relation between scalability and reliability in WSNs; that means while the number of nodes is increasing, it becomes more difficult to ensure the reliability as expected.

Since the structure diversifies, the required number of control packets will increase, and the network eventually becomes unable to sustain the amount of overhead induced by the dynamics, which reduces the data transmission reliability. As expected, this breaking point is observed much earlier in a large-scale network. Therefore, scalability and reliability metrics are firmly coupled and typically prone to act against each other [22].

9.3.2.3.3 Longevity

Longevity can be defined as the ability to fulfill the previously assigned duties for as long as possible. Since the definition of lifetime varies among applications, there are some keywords for denoting it. For example, the time that elapses until the first node dies can define the lifetime of the network, coverage loss of a particular area, failure of the first event notification, and time until 50% of the nodes die—half-time expressions are also usable for the definition of this metric. For especially huge networks where mesh topologies and routing algorithms are employed, longevity becomes more important for the overall network sustainability.

9.3.2.3.4 Robustness

Robustness is the ability to withstand unpredictable node failures that occur as a result of environmental variations or design-based malfunctions. For a WSN, some nodes may expire in time due to the powering circumstances or any other restrictive characteristics, while the rest try to keep working under the coercive physical conditions of the operation area. The network should resist and sustain its operation by compensating the harmful effects of the system changing, and the level of this resistance is roughly defined by the robustness. In addition, *fault tolerance* or *resilience* can also be referred to as a measure of how strong a network is. With regard to that, there are two failure models, nodes and communication links, for the precise evaluation of robustness.

9.3.2.3.5 Responsiveness

Responsiveness means the ability of the network to rapidly adapt itself to unexpected changes in the topology. To achieve high responsiveness, more control packets are exchanged, which will inevitably decrease both the scalability and reliability. Typically, the latency of packet delivery in a dynamic environment reduces in a network with high responsiveness.

9.3.2.3.6 Mobility

Mobility can be defined as the ability to handle mobile nodes and changeable data paths in a network. Generally, a WSN that includes a bunch of mobile nodes should have high responsiveness to deal with mobility. That means it is not easy to design a large-scale and highly mobile wireless sensor network without considering the system constraints.

9.3.2.3.7 Maintainability

Maintainability is the ability to adapt to changes by utilizing self-monitoring-derived adaptive operations. For example, the system should provide lower operation quality when the energy source becomes scarce.

9.3.2.3.8 Programmability

Programmability refers to the ability to reprogram nodes in the field when it becomes compulsory. This metric improves the flexibility of the network as enabling proper system working.

9.3.2.3.9 Energy-Efficient Operation

As mentioned previously, energy can be the scarcest resource for WSNs in some cases. Therefore, it should be used as efficiently as possible, and the energy-efficient operation refers to this goal in terms of network requirements. To measure this metric, energy per correctly received bit or energy transporting one bit of information from the source to the destination can be considered.

A low-power wireless sensor network can be achieved by reducing the duty cycle of each node. However, as the wireless sensor node operates in the power-save mode—sometimes called sleep mode—much longer, the possibility of packet losses and communication failures increases. As mentioned before, although this mode saves a considerable amount of power, the recovery time and start-up energy are becoming crucial to maintaining efficient system work. This results in decreasing system responsiveness and reliability of the network from the absence of control packets and the increasing duration of packet delivery delays. At this point, more sophisticated synchronization techniques should be utilized to keep more nodes in the low duty cycle, without affecting the overall system responsiveness, scalability, and reliability.

9.3.2.4 Routing Protocols for WSNs in Smart Homes

Although the wireless systems deal with the drawbacks of their wired counterparts by providing satisfactory solutions in terms of simplification, QoS, and energy-efficient operation, there is still room for enhancement. To give an example, one common problem for wireless technologies can be denoted as the limited range of communication. This issue has essentially arisen due to the restricted transmitting power, path losses, and fixed or mobile obstacles. To overcome the lack of coverage-based transmission failures, multihop operations, that is, routing protocols, can be utilized [10].

These protocols are basically responsible for finding and maintaining the path from the sensors to the sinks, access points, or base stations. These architectures are based on the idea of using some intermediate nodes as relays to carry the sensor data from the source to the destination. Intermediate nodes forward the received packets to their destination, which is called store and forward multihopping or collaborative networking [2]. Even though the routing protocols expand the reachable communication distances, they result in trade-offs between the system characteristics and requirements. Therefore, multihopping should be carefully applied to the networks, considering the possible energy consumptions in the future, quality of the communication, total distance

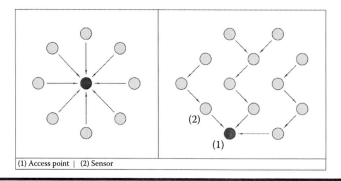

(1) Access point | (2) Sensor

Figure 9.7 **Single- and multihop communication schemes for sensor networks, respectively.**

to be traveled until reaching the destination, and robustness of the system. In general, the communication between the coordinator and the sensory nodes is handled through direct links (with a single hop) when the operation area is small enough, thanks to star or point-to-point networking topologies. However, this may be not feasible or applicable due to the constraint of energy, scale of network, and lack of unobstructed communication links. Although the area is limited and the nodes are deployed wisely, sometimes multihopping may become essential for the transmission. The schemes related to single- and multihop communication can be found in Figure 9.7.

There are several types of routing mechanisms that aim to minimize various cost functions, such as distance—minimum hop (shortest path), QoS (latency, throughput, packet loss, and error rate), robustness (link quality and stability), and energy consumption [23]. For indoor environments, mostly minimum distance-based algorithms are utilized; however, this approach might yield unsuccessful transmissions in scenarios where moving obstacles exist in the area. In the case of crowded indoor areas such as offices, subway stations, or supermarkets, human beings stand as the potential obstacles for the sensors deployed. Since nodes are fixed, they cannot reposition themselves to avoid the negative effects of human blockings. Therefore, this makes the sensors and thus the networks more prone to failure. Sensor networks provide the infrastructure for reliable transmission of data, but routing-based problems are likely to cause communication delay and an increasing number of retransmissions, which are not desired for WSNs with efficient transmission capabilities. So the multihopping techniques should be wisely selected and effectively utilized, depending on the specifications and requirements of the application. Routing protocols in WSNs can be categorized into four groups—sometimes five: the network structure or organization; the way routing paths are established, i.e., route discovery; the protocol operation; the initiator of communication; and (the fifth one) how a protocol selects the next hop on the route of the forwarded message. Nowadays, more advanced, specialized technologies for specific application requirements are preferred for transporting the data through the network [24]. Figure 9.8 and Table 9.3 summarize the classification and application distribution of the routing protocols conducted in WSNs.

9.3.3 Operating Systems for WSNs in Smart Homes

Regarding the unavailability of supportive fully operative structures, WSNs utilize less complex operating systems (OSs) rather than their general-purpose equivalents. They resemble embedded systems in that they are structured as application specific and are equipped with less capable components, in contrast with the commonly used operating systems. Therefore, it is possible to utilize embedded systems such as eCos or uC/OS in the WSNs. TinyOS can be referred to as the

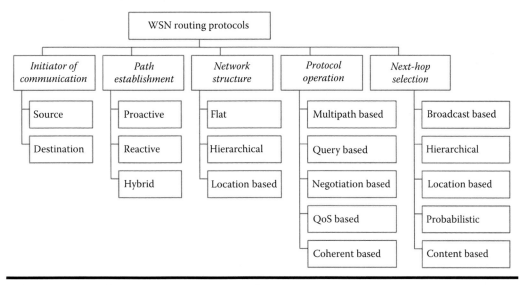

Figure 9.8   Classification of routing protocols in WSNs.

Table 9.3   Network Structure and Protocol Operation-Based Proposals

Classification	Category	Protocol
Network structure	Flat based	EAR, DD, SAR, MCFA, SPIN, ACQUIRE
	Hierarchical based	HPAR, TEEN, MECN, LEACH, PEGASIS, HEED
	Location based	SAR, APS, GAP, GOAFR, GEAR, GEDIR, MECN
Protocol operation	Multipath based	MMSPEED, SPIN, sensor, disjoint, braided
	Query based	SPIN, DD, COUGAR
	Negotiation based	SPAN, SAR, DD
	QoS based	SAR, SPEED, MMSPEED

Note: ACQUIRE, active query forwarding in sensor network; APS, Ad-hoc positioning system; DD, directed diffusion; EAR, energy aware routing; GAP, geographic adaptive fidelity; GEAR, geographic and energy aware routing; GEDIR, geographic distance routing; GOAFR, greedy other adaptive face routing; HEED, hybrid energy-efficient distributed clustering; HPAR, hierarchical power-active routing; LEACH, low energy adaptive clustering hierarchy; MCFA, minimal cost forwarding algorithm; MECN, minimum energy communication network; MMSPEED, multi path and multi speed; PEGASIS, power-efficient gathering in sensor information systems; SAR, sequential assignment routing; SPIN, sensor protocols for information via negotiation; TEEN, threshold sensitive energy efficient sensor network protocol.

first and most frequently preferred OS in WSNs based on an event-driven programming model instead of multithreading. It is designed for enabling capable, low-cost, and application-focused sensory nodes and combining highly efficient execution and component models and communication mechanisms. The components of TinyOS can be listed as command and event handlers, frames, and tasks. LiteOS is a relatively new OS that supports the C programming language and

UNIX-derived systems. Contiki also takes its basis from the C language, which uses a simpler programming style. RIOT provides multithreading and enables Internet connectivity for the nodes or related devices with supporting Internet of Things (IoT) protocols such as 6LoWPAN, IPv6, and TCP, while implementing a microkernel structure. Lastly, ERIKA Enterprise can be defined as an open-source and royalty-free OS derived in the C programming language. These systems undergo ongoing improvement to achieve energy-efficient operation, better-harmonized execution, and simplified administration [2,3].

Regarding the need for a graphical user interface (GUI) in smart home applications, smartphones have evolved to track, monitor, and manage the environment over a communication technology. Considering the influence of these gadgets to human lives, and also the usage intensity, smart home systems should be upgraded to be compatible with the operating systems running in smartphones. With this attempt, ease of use and flexible management of household appliances could be achieved.

9.3.4 Security and Privacy Considerations for WSNs in Smart Homes

As introduced previously, WSNs are able to monitor, collect, and analyze the essential data for people who live, work, or are situated in a smart environment, concerning their life-sustaining activities, movements, locations, and device and utility usage information, and the related physical parameters of the medium to control, manage, and secure this particular area by informing the users or any other authorized people or institutions for decision-making operations. Protection of privacy, including data, is usually handled over the existing wireless networking technologies. When considering that communication is realized over the open air, where numerous devices and networks are constantly transmitting through sharing and distributing information, smart environments will inevitably become vulnerable to security threats [5]. It could be easy to reach and probe the medium remotely by using powerful antennas to capture the sensors, and tamper with or modify the data by eavesdropping, jamming, or inserting malicious traffic [10]. To provide complete security, every sensor or any other subdevice of the network must be equipped with safety mechanisms to prevent attacks that might occur or be targeted over an insecure component [25]. Without any protection, the whole system could suffer from threats or malfunctions caused by external attacks that disrupt the services provided by the sensor networks. Security threats can be intended for the physical layer, routing, position information, and data aggregation, as well as be utilized for both eavesdropping and time synchronization affection. Security operations mostly focus on protecting environmental access, appliance usage, and users' privacy. There are two types of private information—data oriented and context oriented—that need to be protected or preserved [25]. For context-oriented privacy, mechanisms deal with the protection of the contextual information regarding the location and timing of both data generation and transmission processes. Data-oriented privacy can be treated by an external or internal adversary who is not an authorized member of the network. External attacks are generally intended for eavesdropping or listening to the data communication between sensory nodes since the internal ones are focused on node capturing and reprogramming for the collection of private information. Internal attacks could be more dangerous than those of external adversaries, due to the fact that it is relatively harder to detect these kinds of assaults with traditional methods. Therefore, the main effort of data-oriented privacy is focused on protecting information from internal adversaries. To rebuff privacy-oriented interventions, generally cryptographic encryption with authentication, and end-to-end encryption (between the sensors and access points) methods are utilized. Potential applications in smart environments, such as resource, appliance, and utility usage; presence, identification, and activity detection; and vital sign, location, and motion monitoring duties, require data-oriented protection rather than context-oriented privacy. Therefore, when dealing with security issues in smart networks, the requirements of integrity confidence, freshness, availability, and authenticity are observed to measure, compare, improve, and provide more secure and private environments [25].

The data should be encrypted to becloud the content leaks since the processor of the system must be capable of and intended for performing the required cryptographic operations itself or with the contribution of included cryptographic boosters. As an example, the IEEE 802.15.4 standard, which provides the physical and MAC layer definitions for the ZigBee- and 6LoWPAN-derived technologies for the construction of higher-level communication layers, serves three different security opportunities, classified as no security, noncryptographical access to control lists, and symmetric key security by employing AES-128 (Advanced Encryption Standard). These and the other quasi-enhancements contribute to building more secure systems in parallel with increasing the sizes of the networks and the amount of interaction diversity.

9.3.5 Emerging Technologies, Challenges, and Requirements for Future Smart Homes

When considering the total amount of spent time in residential areas by people, future homes will be designed with the all required services, such as entertainment, communication, energy planning, utility usage, medical monitoring, and security. The developers and investors will collaboratively play a key role for this progress [5].

Recently, smart grid systems have emerged for the intelligent control of electricity, which provides bidirectional communication between the suppliers and consumers. With this attempt, a supplier has evolved to control home appliances indirectly to guarantee the system works properly by providing uninterrupted electricity supply. To make intelligent energy control feasible, smart meters, that is, smart plugs [26], are utilized as an integral part of the smart grids. These systems allow the users to control and manage the connected devices remotely, and monitor the energy usage to regulate all kinds of consumptions with demand-responsive and redundant energy preventive algorithms. Also, the negative effects of grid or threat-induced malfunctions are reduced with employed mechanisms to obtain more reliable communication and operation. In the near future, the integration of smart grids, smart homes, and smart plugs will play a crucial role for the regulation of energy demand.

The other justification to deploy WSNs into residential areas is enabling healthcare systems for the elderly and disabled people. Requiring less manpower for controlling and monitoring the vital signs of patients, these systems have become more preferable. The growing number of the population will result in a lack of services and staff for the care and treatment of illnesses in hospitals; therefore, WSN-based local healthcare services will be expected to receive more emphasis in the future.

One of the compelling issues in existing smart homes is the lack of intercompatibility. Although there have been numerous communication and networking technologies employed so far, except for a small portion, these systems are not able to work collaboratively. That causes problems when a combination of different protocols is needed. To overcome this drawback, developers, institutes, interest groups, and alliances have started to work together for standardization. In the short run, more flexible and intercompatible home automation standards will influence the market.

The other restrictive issue for wireless home automation networks is possible threats to user privacy. Since reliable data transmission may not be provided at every step of the communication, private information may be leaked. Especially with the integration of the IP-based protocols in WSNs, the risk level of violation has increased rapidly. To maintain confidentiality, data encryption and cryptography mechanisms are currently being used in the security of communication systems.

To overcome the problem of energy scarcity in WSNs, energy harvesting operations may be maintained. Scavenging energy can provide longevity to the nodes while implicitly increasing the overall network sustainability. For an indoor environment, solar cells can be mounted in the places intensely exposed to sun, and for the crowded environments such as subway stations or hospitals, piezoelectric materials may be furnished at the ground to gather energy from the pressure variations. Vibrations, noises, and temperature gradients are the other existing proposed methods for converting ambient dynamics into energy sources [27]. Considering that the efficiency of energy harvesting operations is increasing continuously, there is no doubt that future smart homes will benefit heavily from improvements in these technologies.

9.3.6 Conclusion

In this chapter, we surveyed the existing literature on sensor networking–based indoor automation systems. Detailed information was given about the current situation of small-scale extensions of smart grids, namely, smart homes and their applications. Besides the older wired proposals, the most relevant emerging communication protocols tailored to wireless smart home sensor networks were investigated comparatively. Node architectures, network types, topologies and requirements, design objectives, routing protocols, and restrictive issues were also discussed. After these, the discussion part was finalized with the analysis of security and privacy considerations focusing on WSNs in smart home implementations, and emerging and promising technologies for future smart indoor areas. In all, this chapter aimed to survey the significant and recent papers on wireless sensor home networks and highlight the research potential and possible approaches to problems in smart home applications.

References

1. C. Gomez and J. Paradells. Wireless home automation networks: A survey of architectures and technologies. *IEEE Communications Magazine*, vol. 48, no. 6, pp. 92–101, 2010.
2. H. Karl and A. Willig. *Protocols and Architectures for Wireless Sensor Networks*. John Wiley & Sons, Hoboken, NJ, 2007.
3. W. Dargie and C. Poellabauer. *Fundamentals of Wireless Sensor Networks: Theory and Practice*. John Wiley & Sons, Hoboken, NJ, 2010.
4. K. Su, J. Li, and H. Fu. Smart city and the applications. In *International Conference on Electronics, Communications and Control 2011 (ICECC 2011)*. Institute of Electrical and Electronics Engineers, Ningbo, China, 2011.
5. M.R. Alam, M.B.I. Reaz, and M.A.M. Ali. A review of smart homes—Past, present, and future. *IEEE Transactions on Systems, Man, and Cybernetics, Part C: Applications and Reviews*, vol. 42, no. 6, pp. 1190–1203, 2012.
6. F. Viani, F. Robol, A. Polo, P. Rocca, G. Oliveri, and A. Massa. Wireless architectures for heterogeneous sensing in smart home applications: Concepts and real implementation. *Proceedings of the IEEE*, vol. 101, no. 11, pp. 2381–2396, 2013.
7. O. Cetinkaya and O.B. Akan. A DASH7-based power metering system. In *Proceedings of 13th Annual Consumer Communications and Networking Conference 2015 (CCNC 2015)*. Institute of Electrical and Electronics Engineers, Las Vegas, NV, 2015.
8. F. Karray, M.W. Jmal, M. Abid, M.S. BenSaleh, and A.M. Obeid. A review on wireless sensor node architectures. In *9th International Symposium on Reconfigurable and Communication-Centric Systems-on-Chip 2014 (ReCoSoC 2014)*, Montpellier, France, 2014, pp. 1–8.

9. D. Basu, G. Moretti, G.S. Gupta, and S. Marsland. Wireless sensor network based smart home: Sensor selection, deployment and monitoring. In *Sensors Applications Symposium 2013 (SAS 2013)*. Institute of Electrical and Electronics Engineers, Galveston, TX, 2013, pp. 49–54.

10. S. Singhal, A.K. Gankotiya, S. Agarwal, and T. Verma. An investigation of wireless sensor network: A distributed approach in smart environment. In *Second International Conference on Advanced Computing and Communication Technologies 2012 (ACCT 2012)*, Rohtak, India, 2012, pp. 522–529.

11. Simply Automated, Inc. X-10 to UPB migration document, version 1.1. Simply Automated, Carlsbad, CA, 2003. http://www.simply-automated.com/documents/X10ToUPB%20V1.1a.pdf.

12. INSTEON. Whitepaper: Compared, version 2.0. INSTEON, 2013. http://cache.insteon.com/pdf/INSTEONCompared.pdf.

13. Simply Automated, Inc. The UPB system description, version 1.2. Simply Automated, Carlsbad, CA, 2005. http://www.simply-automated.com/documents/UpbDescriptionV1.2a.pdf.

14. C. Saad, B. Mostafa, El A. Cheikh, and H. Abderrahmane. Comparative performance analysis of wireless communication protocols for intelligent sensors and their applications. *International Journal of Advanced Computer Science and Applications (IJACSA)*, vol. 5, no. 4, pp. 76–85, 2014.

15. N. Langhammer and R. Kays. Performance evaluation of wireless home automation networks in indoor scenarios. *IEEE Transactions on Smart Grid*, vol. 3, no. 4, pp. 2252–2261, 2012.

16. Y. Kyselytsya and T. Weinzierl. Implementation of the KNX standard. In *KNX Scientific Conference*, Vienna, Austria, 2006.

17. INSTEON. Whitepaper: The details, version 2.0. INSTEON, Irvine, CA, Irvine, CA, 2013. http://cache.insteon.com/pdf/insteondetails.pdf.

18. Z. Xiao-yan, H. Ting-lei, L. Pin, and L. Zhao-lai. Research on smart living technology based on WSN. In *International Conference on Intelligent Computing and Integrated Systems 2010 (ICISS 2010)*, Gandhinagar, Gujarat, India, 2010, pp. 938–941.

19. B. Fouladi and S. Ghanoun. Security evaluation of the Z-Wave wireless protocol. Blackhat USA, vol. 24, 2013.

20. V. Kumar and S. Tiwari. Routing in IPv6 over low-power wireless personal area networks (6LoWPAN): A survey. *Journal of Computer Networks and Communications*, 2012, 10 pp.

21. G. Mulligan. The 6LoWPAN architecture. In *Proceedings of the 4th Workshop on Embedded Networked Sensors (EmNets '07)*, Association for Computing Machinery, New York, 2007, pp. 78–82.

22. S. Muthukarpagam, V. Niveditta, and S. Neduncheliyan. Design issues, topology issues, quality of service support for wireless sensor networks: Survey and research challenges. *International Journal of Computer Applications*, vol. 1, no. 6, pp. 1–4, 2010.

23. Y. Xu, S. Wu, R. Tan, Z. Chen, M. Zha, and T. Tsou. Architecture and routing protocols for smart wireless home sensor networks. *International Journal of Distributed Sensor Networks*, 2013, 14 pp.

24. H. M. Salman. Survey of routing protocols in wireless sensor networks. *International Journal of Sensors and Sensor Networks*, vol. 2, no. 1, pp. 1–6, 2014.

25. K. Islam, W. Shen, and X. Wang. Security and privacy considerations for wireless sensor networks in smart home environments. In *16th International Conference on Computer Supported Cooperative Work in Design 2012 (CSCW 2012)*. Institute of Electrical and Electronics Engineers, Seattle, WA, 2012, pp. 626–633.

26. O. Cetinkaya and O.B. Akan. A ZigBee based reliable and efficient power metering system for energy management and controlling. In *Proceedings of International Conference on Computing, Networking and Communications (ICNC 2015)*. Institute of Electrical and Electronics Engineers, Anaheim, CA, 2015.

27. V.C. Lee. Energy harvesting for wireless sensor network. PhD dissertation, University of California, Berkeley, 2012.

Chapter 10

Realizing Cognitive Radio Technology for Wireless Sensor Networks

Tariq Jamil Saifullah Khanzada and Muhammad Farrukh Shahid

Contents

Abstract

Cognitive radio (CR) has revolutionized the next-generation communication technologies due to its dynamic characteristics and features. CR network provides cognitive behavior that enhances network performance in terms of data rates, bandwidth, latency, and throughput. This chapter provides insight into the realization of CR technology for wireless sensor networks and discusses related issues about the integration of both technologies.

10.1 Introduction

A wireless sensor network (WSN) is a newer technology that is being used to perform monitoring and surveillance tasks. The current development and evolution of electronic devices has risen to the point of amalgamation, providing an enriching capacity for the processing, sensing, and storing of data, compared to previous decades. The twenty-first century has observed rapid developments in sensor technology over the past 20 years. There are numerous applications of WSN, such as military battlefield surveillance and medical applications. In medical applications, WSN nodes can collect information about an individual's health, fitness, and energy expenditure. In the field of computer science and communication, WSNs have emerged and revolutionized many technologies. On the other side, the development of digital electronics has reached a point of integration that provides high capacity, storage, and data communication for sensor nodes, compared to previous decades [1].

The military applications have boosted the development of WSNs, which was the main objective. Since, WSNs have emerged in many fields and reached a high level of development. WSNs have also found applications in aerospace, agriculture, telemarketing, and disaster management. The biomedical field is also deploying WSNs to monitor patient health from a remote distance. WSNs consist of tiny nodes that are deployed in dynamic and hostile environments with no human existence, and therefore they must be tolerant of failure and the loss of connectivity of individual nodes. The nodes are small, low cost, and portable, finding applications in many areas related to health and weather monitoring, foresting, and more.

10.2 Cognitive Radio Technology

Cognitive radio (CR) has emerged as a widely expected "big bang" for next-generation wireless communications. Spectrum regulatory committees in many countries have been shifting to the use of dynamic spectrum access, using CR technology and also laying down rules for its implementation. Most of today's radio systems are not intelligent and are not aware of their radio spectrum environment. These radios operate in a specific band of frequency. Careful investigations of spectrum over different time periods indicate that not all of the spectrum in space is used by the licensed user. There remain some vacant spaces, called white spaces [2,3]. Therefore, these vacant spaces of the spectrum can be used to enhance spectrum efficiency and accommodate more users at a time. A CR radio can intelligently scan the spectrum and find vacant spaces and efficiently utilize vacant spaces without creating interference with other licensed users. In this way, efficient spectrum usage can be achieved [3].

10.2.1 Definition of Cognitive Radio

Several definitions for cognitive radio can be found in the research literature. The official definition for cognitive radio systems, developed by ITU Radiocommunication Sector (ITU-R) WP1B in 2009, is as follows [4]:

> Radio systems which incorporate technology that learn and get the knowledge of its operating environment and based on that knowledge make certain decision to achieve specific tasks.

This definition is purposely generic so that a single term can describe cognitive radio system (CRS) technology under any radio communication service system. Depending on the specific radio communication service in question, it is possible that the identification of unique and detailed characteristics may be required. Figure 10.1 shows a general cognitive cycle for characterizing the operations of CRS. According to the definition, CRS systems obtain knowledge about environments and adjust according to the intended requirements, which they have learned from the results of the environment. The definition is broad and the detailed techniques for creating CRS functionalities are on the way to developments.

For the sake of effective spectrum usage, it has been divided into two bands: licensed and unlicensed. Similarly, the users are divided according to the division of frequency bands. The primary users are licensed band users, and the secondary users are unlicensed spectrum band users. An infrastructure that has a specific band allocation is known as a primary network, for example, cellular and TV broadcast networks. The primary networks comprise licensed users that have a license to operate on that specific spectrum band, and the primary base station is another component that also has access to the frequency band allocated for the primary network; for example, a base station transceiver system (BTS) in a cellular system. A licensed spectrum allocated for wireless services has vacant portions known as white spaces or holes.

A cognitive radio is a radio transceiver that is designed to intelligently detect whether the radio spectrum is currently in use, and to switch to the unused spectrum, without interfering with the transmission of licensed users. The mechanism behind the working of cognitive radio is simple: it observes its environment using different techniques of spectrum sensing to detect spectrum holes. It plans and monitors all the possible options of unused spectrum from which to choose. From the definition of cognitive radios, the main characteristics are cognitive capability and reconfigurability. Capabilities of cognitive radio make it able to identify the unused spectrum. Consequently, it leads toward the best spectrum selection and effective operating parameters. The cognitive capability provides spectrum awareness, whereas reconfigurability characteristics enable the radio to be dynamically programmed according to the radio environment. If we want to be more precise, we can say that cognitive radio can be programmed as transmission is required. The idea behind cognitive radio was to make wireless services more flexible. It uses a small part of the physical world to get information from its surroundings [5]. To use the licensed spectrum, cognitive radio should sense spectrum holes to avoid interference. Let's say a primary user is transmitting on a licensed band; when the user leaves, cognitive radio should move to the band and transmit on it. If

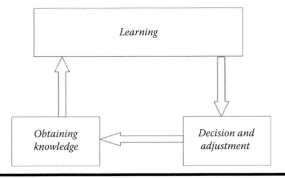

Figure 10.1 Generic cognitive cycle.

the primary user again wants to access that band, then the radio should be able to leave the band immediately to avoid interference. When we consider the whole network, cognitive radio senses all available channels, and all cognitive devices on a network sense independently. There are two techniques to detect primary users:

1. The signal of the primary user that is transmitted is either locally present or not.
2. Information from multiple users is incorporated to detect the primary user.

All the unused bands are at far distances over a wide frequency band. Spectrum management requires selection of the detected spectrum that best meets user requirements. Cognitive radio decides the best-available spectrum to achieve quality of service (QoS) for the required user. Management functionality is basically done by analyzing all available spectrum bands and then choosing the one that can give the best requirement.

In spectrum mobility, the best spectrum band analyzed in spectrum management is targeted. This is because we want to use the spectrum in a dynamic way by allowing radio terminals to operate in the best-available frequency band. We have come to know that intelligent channel selection for transmission of data and information according to spectrum sensing information should minimize interference with other users. The need to move from a certain band to another is because of worse channel conditions or a primary user is accessing that band. Spectrum sharing basically provides a good spectrum scheduling method. Spectrum sharing is a challenge when the spectrum is used by both licensed and unlicensed users. Interference is the major issue when a spectrum is approached by cognitive radio. When a specific licensed band is accessed by cognitive radio and it starts transmission over it, a licensed user that approaches this band has to vacate it. An electromagnetic spectrum consists of all the possible existing frequency bands. The radio band is from 300 kHz to 300 GHz. Different radio transmission technologies and applications are used on different radio bands. The radio band is regulated by the government, and in some cases, it is sold or licensed to private cellular operators. Now the need is to utilize this natural source in the best possible way, which requires spectrum management. The method of allocation of the frequency spectrum is fixed, done by either the government or telecommunication authorities. It would seem like the unlimited wideband of radio frequency is enough, but the growing technology is demanding more spectrum. Because of the demand for spectrum, new management techniques and systems are becoming popular. Considerable research work has been done on the issue of scarcity in the spectrum. In cognitive networks, secondary users have to monitor the presence of primary users on the spectrum. Cognitive radio has the ability to adapt dynamically according to its surroundings. It was a reasonable solution to cater to the problem.

On November 4, 2008, the Federal Communications Commission (FCC) unanimously agreed to open up unused broadcast TV spectrum for unlicensed users. In this way, cognitive radio technology became the practical solution for spectrum congestion. Specifically, the available TV band was considered due to the unused band being available most of the time. Cognitive technology is promising to provide improved spectrum usage, less delay, and good adaption to its surroundings.

10.2.2 Challenges Faced While Spectrum Is Sensed Cognitively

If a primary node is blocked out due to any obstacle, it becomes impossible for cognitive radio to locate that node. The noise effect present in an environment that is not known by cognitive radio is a hurdle in spectrum detections.

Multiple cognitive radios present in the same licensed band make it difficult to find accurate aggregated interference. As the spectral environment keeps changing, it is difficult for cognitive radio to move to another band. Basically, the spectral environment changes when a user moves from one place to another.

One of the most demanding features of cognitive radio is its detection capability. Detection must occur during a short period of time. For this purpose, orthogonal frequency division multiplexing (OFDM)–based cognitive networks are known that have excellent multicarrier sensing that reduces the overall sensing time significantly [6].

While cognitive radio is making decisions, the decision model should give the maximum quality of service. The cognitive radio's capability to interact with its environment to determine appropriate parameters should be in a manner that meets quality of service. In spectrum management, essential parameters need to be defined for quality of service. Interference is caused when a certain band is more crowded. Path loss increases with the increase in operating frequencies. If the transmission power of a cognitive radio remains the same, the transmission range decreases at high frequencies. Interference, path losses, link layer losses, and holding time are the parameters used to get information on the spectrum band on which cognitive radios are going to transmit. Cognitive radios have a number of applications, such as in moderate-bandwidth applications, broadband wireless networking, and multimedia wireless networking services; where high data rates are required, cognitive radio is very practical in these services. In advance network topologies based on enhanced system design description (SDD) techniques, cognitive can be deployed easily. In the future, cognitive radio technology would emerge from a laboratory application to a general-purpose programmable radio that would provide a smooth platform to wireless technology development. These cognitive radios can be used to make stand-alone network applications, such as military networks, emergency networks, multimedia networks in TV white space, and in-vehicle ad hoc networks for communication. These radios can even be used to change the shape and scheme of communication around us. For example, these may replace cellular network infrastructure for short-range communication and converge to it for long-range communication simultaneously. Cognitive radios are one of the big applications of military and public safety users nowadays. In addition to giving interoperability and guaranteed quality of service in extreme conditions, cognitive radios also offer the necessary autonomous operations for these users. The main feature of cognitive radio is that it has the ability to adapt or adjust itself with different standards and can communicate between different standards at the same time. Thus, by this concept of cognitive radio, it can give better quality of service to the users by sensing the environment and adjusting itself to that type of environment. Thus, both service providers and markets are giving a lot of attention to the use and development of cognitive radios, as well as software-defined radios. A software-defined radio is a technical platform to realize cognitive radios. Although there are many advantages of cognitive radio, there are still many technical hurdles that should be solved before the implementation of this technology in the real-world scenario. One of the many issues is that it cannot differentiate between primary users and spread-spectrum primary users, which may result in incorrect decision making by CR, and the spectrum may remain empty or create interference with primary users.

The transceiver or radio of any communication device is also called the interface of that device because it acts as a gateway between the device and network. On the basis of the number of transceivers, having a communication device, we can identify communication devices separately as single interface devices and multiple devices. In single-interface devices only a single transceiver provides communication to that device. In our daily routine, we face lot of single interface devices. Wi-Fi modems and televisions are the most popular examples of single-interface devices. Meanwhile, we use many multi-interface devices in our daily routine. The smart phone is a good example to understand multi-interface devices. In a smart phone we have isolated Global System for Mobile Communications (GSM) and Wi-Fi radio. Both of these interfaces communicate in single devices, and due to this reason, we may treat cell phones as multi-interface devices. Multi-interfaces may be heterogeneous and homogeneous. If all interfaces provide the same service and have the same features, then we say that the interfaces are homogeneous, but if the interfaces perform different functions and have different features, then the interfaces are heterogeneous. A cell phone is an example of a heterogeneous interface device. On the contrary, if we use multi-interfaces in a Wi-Fi modem, then it become a homogeneous multiple-interface device. The major discrimination between single-interface and multi-interface devices is that single-interface ones can operate on a single channel at a time. In a network of single interfaces, all nodes are necessary to communicate on a single channel to be a part of that network. If a node chooses any other channel, then it will not be reachable anymore. In contrast, a multi-interface device can communicate simultaneously on multiple channels. Within a network, we can assign multiple channels to communicate, using an effective channel assignment technique. With an effective channel assignment technique, the channels are assigned to nodes in such a way that no node gets out of the network. Using multiple channels increases the data rate dramatically because a channel has a limited capacity to carry the traffic, and in multiple channels, we increase the capacity of the network. Simply put, multichannel implementation is a demand that results from providing high data rates. One disadvantage of using multiple channels in a network is that it establishes a multihop connection in the network.

By studying cognitive radio surveys, we can see that in the cognitive environment, we have a large number of channels available in the network. Using single-interface devices, we cannot exploit more than one channel, so the availability of a large number of channels becomes useless. To carry out multiple channels in a cognitive environment, the deployment of multiple-interface devices is a good solution. By examining this scenario, the cognitive environment and multiple-interface devices seem to converge at a single point. This is a key point, which motivated us to work on this problem. Briefly, this is the integration of cognitive radio and multiple-interface devices. The CR has the following characteristics:

Agility and flexibility: The CR system possesses the ability to adopt new waveforms and other operational characteristics.
Sensing: The CR system senses the environment and make decisions accordingly.
Learning and adaptability: The CR system learns a specific pattern, recognizes values, and adapts new characteristics based on learning outcomes.

Figure 10.2 demonstrates the improved cognition cycle (CC). CC is a standout among the most imperative ideas utilized as a part of cognitive radio innovation. The cognition cycle delineates how the cognitive radio reacts to outer boosts inside its radio surroundings. The cognitive radio facilitates and watches its working surroundings in the watch state. It then arranges itself per the sensing result. Contingent upon whether the result of the sensing requires a quick need, desperation, or a typical move, the orient state can travel to the act, choose, and arrangement states separately. In the arrangement state, most jolts are managed deliberatively instead of responsively. An approaching system message would regularly be managed by producing an arrangement, which is the ordinary way. The arrangement stage ought to additionally incorporate thinking over the long haul. Ordinarily, think reactions are preplanned, while receptive reactions are found by being educated or prearranged. In the choose state, the radio settles on one of the different arrangements. The result of the choice prompts an activity, for example, asset distribution in the demonstration state. In the demonstration, a specific pick activity is executed, while the outcome of the picked activity is learned in the learn state. Learning is an element of alternate conditions of the cognizance cycle.

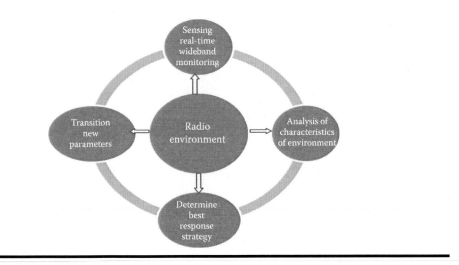

Figure 10.2   CC simplified diagram.

10.3 Emerging Wireless Sensor Networks

To monitor temperature, noise levels, object movement, and other conditions, WSNs are emerging technology nowadays. WSNs are a collection of low-cost devices that perform the functions of monitoring. WSNs have been deployed in various fields, including medical, oil field, surveillance, and rescue applications. WSNs can be configured in a centralized or distributed manner. The sensor nodes in a practical wireless sensor networks need not be uniformly distributed over the region, but they form a multihop network that communicates through mesh networking in order to complete a particular set objective. While the nodes could be few in number, there is no particular limit as to the number of sensor nodes that should constitute the sensor network. A given wireless sensor network could be made up of hundreds of thousands of sensor nodes deployed to monitor certain ambient conditions in a particular geographic region. Figure 10.3 shows the simplified architecture of WSNs. It consists of sinks, sensor nodes, and the gateway. The gateway performs the task of data aggregation, data distribution, and overall network monitoring.

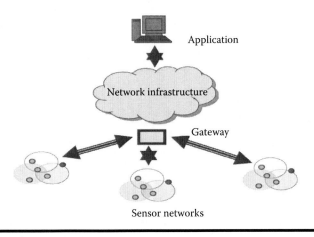

Figure 10.3   Simplified WSN architecture.

Table 10.1 WSN Characteristics

Characteristics	Description
Simple hardware structure	Sensors in WSNs have a simple, low-cost, and less complicated hardware structure.
Low energy consumption	Sensor nodes in WSNs consume less power.
Self-organization	WSN nodes have the capability to organize in a uniform manner in order to share and transmit information.

The architecture shown in Figure 10.3 is widely used because of its low energy consumption, simplified working, and improved QoS. Table 10.1 shows characteristics of WSNs.

10.3.1 Implementation Scenarios

CRSs are used to enhance the efficiency of the use of various resources in wireless communication systems. Spectrum has been identified as a scarce resource; hence, efficient utilization is of primary importance. In terms of spectrum, various deployment scenarios are envisaged for CRSs, which depend on the types of spectrum bands considered.

Specifically only four key deployment scenarios are added in the sequel [7,8]:

1. License spectrum
2. Unlicensed spectrum
3. Primary–secondary setting
4. Bands for CR on which they exist

10.4 Cognitive Radio–Based WSN Design Scenario

There exists a distinct similarity between WSN and CR from the operations point of view. In order to fully realize CR-based WSN, various issues must be addressed. The characteristics shown in Table 10.1 cannot be directly applied to CR-WSN. The CR-WSN should be capable of adopting the functionalities of WSN given in Table 10.1. In order to mitigate interference to the primary user (PU) while improving the performance of the secondary user (SU), WSN must use low-cost, battery-driven, and low-computation-capability sensor nodes. Each sensor node in WSN consists of simple hardware with a single transceiver. On the other side, in the CR network spectrum, sensing is a complex task. This requires sensor nodes in WSN to be equipped with sensing hardware. The coordinators are deployed throughout CR networks to avoid expensive hardware, excessive load, and energy consumption [9]. Table 10.2 shows CR-WSN issues.

10.5 Noncontiguous OFDM (NC-OFDM)

The radio spectrum is managed and restricted by the radio regulatory agencies in the current era. The frequency spectrum is allocated to those services that obtain a license from spectrum

Table 10.2 CR-WSN Issues

Issues	Anticipated Solutions
Sleep/wake-up strategy	In [3], the authors formulate two strategies, SR and AW. SR requires strict synchronization in order to communicate information among nodes in CR. AW requires challenging wake-up and sleep strategies, as it does not require synchronization.
Quiet period effects on CR-WSN	In CR-WSN, sensor nodes do not transmit packets during quiet periods. This increases end-to-end delay. In [4], the author shows that end-to-end delay can be minimized if the number of hops between the nodes and sink is reduced. For this purpose, a high-transmission-range channel may be employed.
Spectrum sensing	The real-time transmission sensing is less energy-efficient. In [5], the authors address issues related to spectrum sensing. They suggest that a good sensing algorithm is that which uses the least amount of energy for PU detection.
Backup nodes	In case of hardware damage or power loss of any node deployed in CR-WSN, the authors in [5] formulate backup coordinator nodes that may be used in the CR-WSN to perform CR-WSN functions consistently.

regularity bodies. Open access to the spectrum is available for technologies using very low powers for transmission. Ultra-wideband (UWB) technology is among those that use low power for transmission. The overlay sharing approach and transmissions of desired power levels are generally not authorized and are termed free access to the open spectrum. A few small portions of the frequency spectrum, known as unlicensed frequency bands, are available where the overlay sharing approach is commonly used. This led to a wide variety of new wireless technologies and services, for example, wireless local area network (LAN) IEEE 802.11 and Bluetooth for wireless personal area networks (PANs). However, inflexible restrictions on spectrum access are a hindrance to the development of new radio services that can be used to improve health, safety, the work environment, the education of people, and the quality of leisure time. Change in the current status of a licensed frequency spectrum is complex and time-consuming, as it demands a concerted effort by government regulatory agencies, technology developers, and service providers to achieve efficient and timely deployment. At a particular location for a specific time period, 90%–95% of licensed spectrum is unused due to the tedious process of change in the licensed frequency spectrum [10]. The current radio regulatory regime is too complex to manage emerging wireless applications, which results in wastage of precious spectrum. According to the FCC, frequency spectrum is underutilized most of the time [10]. In wireless communication, single-carrier modulation, multicarrier modulation, and spread-spectrum modulation are used in many applications for the transmission of user data. A single carrier is used in single-carrier transmission to modulate user data, whereas in multicarrier modulation, multiple carriers at different frequencies carry the information by sending multiple bits on each subchannel. The multicarrier modulation schemes offer many advantages over single-carrier modulation, such as simplicity of the designs of equalizers, bandwidth efficiency, and high data rates. In wireless channels, multipath propagation of signals causes fading effects. Fading effects generate intersymbol interference (ISI) that limits high-data-rate wireless communication. The problem of ISI is eliminated by using multicarrier modulation schemes (MCMs). In CR, techniques find suitable applications for high data rates in multipath fading channel conditions. To improve the overall spectrum utilization, temporarily unused spectrum bands can be used by opportunistic radios Hence, new spectrum allocation techniques are needed to maximize the benefits of the limited spectrum resource by detecting the unused spectrum bands in a given time and location. To overcome this problem, the noncontiguous OFDM (NC-OFDM) technique aims to solve the spectrum utilization problem. If 20 MHz is the requirement for a communication system, the bandwidth may be available as sections of 5 and 10 MHz each, that is, adopted from white spaces [11]. This situation is shown in Figure 10.4, detecting the unused spectrum bands in a given time and location.

The noncontiguous transmission technique is required for using white spaces in a spectrum. The noncontiguous block of subcarrier is used to transmit the data symbols. In OFDM, it is possible to use subcarriers in a noncontiguous manner, as subcarriers can be tuned to on or off in the vicinity of the primary user transmissions, and thus the spectral white spaces can be filled up efficiently. The modified form of OFDM is called noncontiguous OFDM (NC-OFDM). The NC-OFDM system uses noncontiguous blocks of subcarriers to transmit data symbols and deactivates the subcarriers used by the licensed users to avoid interference, as illustrated in Figure 10.4, consequently known to be noncontiguous OFDM.

10.5.1 NC-OFDM System Architecture

The NC-OFDM general transceiver is shown in Figure 10.5. In this architecture, a high-speed data stream, $x(n)$, is modulated using M-ary phase shift keying, and then using a serial-to-parallel (S/P) block, the modulated data stream is split into N slower data streams (Figure 10.6) [12].

The subcarriers in the NC-OFDM transceiver may be active or inactive, depending on the requirements. The subcarriers reside in unoccupied spectrum, which can be determined using sensing and estimation algorithms. The modulated data symbols are processed through inverse fast Fourier transform (FFT). A guard interval with a length greater than the channel delay spread is added prior to the transmission to each NC-OFDM symbol using the cyclic prefix (CP) block in order to mitigate the effects of intersymbol interference (ISI). The baseband NC-OFDM signal, $s(n)$, is passed through the transmitter radio frequency (RF) chain after the serial-to-parallel function. The RF section amplifies the signal and upconverts it to the desired center frequency. On the

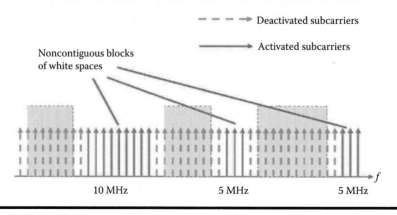

Figure 10.4   Noncontiguous block of white spaces.

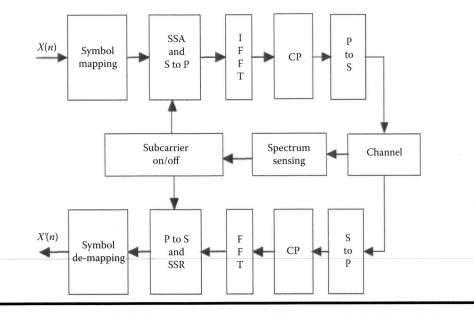

Figure 10.5 **NC-OFDM block diagram. (From J. D. Guffey, OFDM physical layer implementation for the Kansas University agile radio, Technical report, University of Kansas Information and Telecommunication Technology Center, Lawrence, February 2008.)**

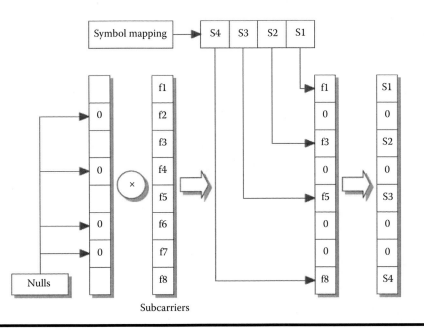

Figure 10.6 **NC-OFDM subcarrier selection. (From J. D. Guffey, OFDM physical layer implementation for the Kansas University agile radio, Technical report, University of Kansas Information and Telecommunication Technology Center, Lawrence, February 2008.)**

receiver side, downconversion of the signal is performed, followed by serial-to-parallel conversion, and then the cyclic prefix is removed. After removal of the cyclic prefix, FFT is performed and demodulation is done. From this system overview, we observe that the IFFT and FFT blocks are integral parts of the transceiver [13]. The conventional ith OFDM time-domain symbol, assuming a rectangular window $R(t)$, is [14]

$$y_i(t) = \sum_{m=0}^{k-1} X_{i,km} e^{j2\pi km\Delta ft} \quad -T_{cp} \leq t \leq T_s \tag{10.1}$$

in which T_{cp} is the cyclic prefix duration. The total OFDM signal, $s(t)$, can be written as

$$s(t) = \sum_{i=\infty}^{\infty} y_i(t - i(T_s + T_{cp})) \tag{10.2}$$

The frequency-domain representation of the k_mth subcarrier $C_{i,km}(f)$ is given by

$$C_{i,km}(f) = \frac{1}{T_s + T_{cp}} \int_{=\infty}^{\infty} R(t) X_{i,km} e^{j2\pi kmt/T} e^{-j2\pi ft} \, dt$$

$$= X_{i,km} Sinc\left(fm\left(1 + \frac{T_{cp}}{T_s}\right)\right) e^{j2\pi fm\left(1 - \frac{T_{cp}}{T_s}\right)} \tag{10.3}$$

where

$$Sinc(x) \equiv \sin\frac{(\pi x)}{\pi x} \quad \text{and} \quad f_m = k_m - fT_s$$

It also denotes the interference coefficient from the subcarrier k_m to the frequency f, which is outside the allocated frequency band. The power spectral density (PSD) of $s(t)$ is expressed as [9]

$$\psi(f) = (T_s + T_{cp}) E\left\{\left|\sum_{m=0}^{k-1} C_{ikm}(t)\right|^2\right\}$$

$$= (T_s + T_{cp}) E\left\{\left|\sum_{m=0}^{k=1} X_{i,km} e^{j2\pi fm\left(1 - \frac{T_{cp}}{T_s}\right)}\right|\right\} \sin c(f_m\left(1 + \frac{T_{cp}}{T_s}\right)|^2 \tag{10.4}$$

NC-OFDM introduces a frequency-domain precode to make the time-domain OFDM signal continuous, after smoothing two consecutive OFDM symbols by the precoder. The *i*th *N*-continuous time-domain OFDM symbol $y_i(t)$ and its first derivatives satisfy [10]

$$\overline{y_i^{(v)}}(t)\big|_{t=-T_{cp}} = \overline{y_{i-1}}(t)\big|_{t=T_s} \tag{10.5}$$

The precoder process can be summarized as

$$\overline{X_i} = (I_k - P)X_i + P\phi^H \overline{X_{i-1}} \quad i \geq 0$$

$$\overline{X_i} = X_0 \quad \text{for } i = 0$$

where

$$P = \phi^H A^H (AA^H)^{-1} A\phi$$

10.5.2 NC-OFDM-Based CR Network

A CR network identifies white spaces in the spectrum and grants access to secondary users in a licensed band. The CR network used unused part of the spectrum. The main objective of the NC-OFDM-based CR network is to maximize the system throughput while keeping the PU and SU interference levels to a minimum level. IEEE 802.22 is an example of a CR network in which TV channels are reused. In it, the primary users are TV channels, and the secondary users may be some other standard-based systems [15]. OFDM possesses sensing and spectrum-shaping capabilities, which makes it an ideal candidate for the CR network. Table 10.3 shows a summary of OFDM strength and CR network requirements [16].

The ability to sense, measure, and learn the environment from the operating conditions is the main function of CR. The parameters need to be sensed, and measuring may include channel parameters, interference level, system throughput, and white spaces in a spectrum. After scanning of the spectrum, it is desirable to shape the waveform parameters to avoid interference between PU and rented users.

Table 10.3 OFDM Strength for CR Network

CR Requirement	OFDM Strength
Spectrum sensing	Spectrum sensing can be achieved using FFT of OFDM.
Efficient spectrum utilization	By turning off some subcarriers, spectrum shaping is easily achieved.
Interoperability	Many standards use OFDM as their physical transmission scheme. Hence, interoperability becomes easy.
Advanced antenna techniques	Multi-input and multioutput (MIMO) are also used in OFDM to reduce receiver complexity.

References

1. K. A. Yau, P. Komisarczuk, and P. D. Teal, Cognitive radio-based wireless sensor networks: Conceptual design and open issues, in *Second IEEE Workshop on Wireless and Internet Services (WISe 2009)*, 2009.
2. S. Haykin, Cognitive radio: Brain-empowered wireless communications, *IEEE Journal on Selected Areas in Communications*, vol. 23, pp. 201–220, 2005.
3. C. Chong and S. Kumar, Sensor networks: Evolution, opportunities, and challenges, *Proceedings of IEEE*, vol. 91, pp. 1247–1256, 2003.
4. Y. Xu, Y. Sun, Y. Li, Y. Zhao, and H. Zou, Joint sensing period and transmission time optimization for energy-constrained cognitive radios, *EURASIP Journal on Wireless Communications and Networking*, vol. 2, p. 16, 2010.
5. J. Jia, Z. He, J. Kuang, and H. Wang, Analysis of key technologies for cognitive radio wireless sensor networks, in *6th International Conference on Wireless Communications Networking and Mobile Computing*, China, 2010.
6. P. Yi, L. Hua, X. Tao, and D. Qing Zhi, The research of CR-based WSNs architecture, in *International Conference on E-Business and E-Government (ICEE)*, 2010, pp. 2179–2182.
7. S. M. Kamruzzaman, M. Hamid, and M. Wadud, An energy-efficient MAC protocol for QoS provisioning in cognitive radio ad hoc networks, *Journal of Radio Engineering*, vol. 19, no. 4, 2010.
8. M. D. Gallagher et al., Facilitating opportunities for flexible, efficient, and reliable spectrum use employing cognitive, ET Docket No. 03-108, February 15, 2005.
9. H. Arslan and M. E. Sahin, UWB cognitive radio, in *Cognitive Radio, Software Defined Radio, and Adaptive Wireless Systems* (Signals and Communication Technology), Springer, Berlin, 2007.
10. J. D. Guffey, OFDM physical layer implementation for the Kansas University agile radio, Technical report, University of Kansas Information and Telecommunication Technology Center, Lawrence, February 2008.
11. Spectrum Efficiency Working Group, Report of the Spectrum Efficiency Working Group, Technical report, Federal Communications Commission Spectrum Policy Task Force, Washington, DC, November 2000.
12. H. Gao, Comparison of SC-FDMA and NC-OFDM schemes for cognitive radio networks, in *Second International Conference on Computational Intelligence and Natural Computing Proceedings (CINC)*, September 13–14, 2010, vol. 2, pp. 320–324.
13. J. Min, G. Xue-mai, and W. Qun, An improved channel estimation method for NC-OFDM systems in cognitive radio context, in *2011 6th International ICST Conference on Communications and Networking in China (CHINACOM)*, August 17–19, 2011, pp. 147–150.
14. R. Kapoor and P. Kumar, Spectral agility of NC-OFDM transmission for dynamic spectrum access, in *5th International Conference on Computers and Devices for Communication (CODEC) 2012*, December 17–19, 2012, pp. 1–4.
15. D. Li, X. Dai, and H. Zhang, Sidelobe suppression in NC-OFDM systems using constellation adjustment, *IEEE Communications Letters*, vol. 13, no. 5, pp. 327–329, 2009.
16. J. Shao and X. Liang, A new approach for PAPR reduction of NC-OFDM system based on cognitive radio, in *5th International Conference on Wireless Communications, Networking and Mobile Computing 2009 (WiCom '09)*, September 24–26, 2009, pp. 1–4.

WSNs PROTOCOLS AND ALGORITHMS

Chapter 11

Energy-Efficient Data Collection Techniques in Wireless Sensor Networks

Sukhwinder Sharma, Rakesh Kumar Bansal, and Savina Bansal

Contents

Abstract

To increase a wireless sensor network's lifetime, energy-efficient strategies are proposed by researchers from time to time. Most of these are developed under the assumption that the energy requirement during data collection is much less than that during communication. However, for many practical application scenarios, this assumption does not hold, especially when specific sensors are used for monitoring complex phenomena. The network lifetime of WSNs can be substantially improved with energy-efficient data collection techniques, which has the potential to make these networks more compliant for future wireless applications seeking huge energy consumption. This chapter deals with energy-efficient data collection techniques for wireless sensor networks.

11.1 Introduction

A wireless sensor network (WSN) consists of a large number of sensor nodes spread over a wide physical area. Each sensor node, having one or more sensors, is able to collect, compute, and communicate with other nodes. Sensor nodes are composed of small and cost-effective sensing devices with a wireless radio transceiver. The advantage of using small devices is the ability to monitor an environment that does not have infrastructure, such as power supply, a wired Internet connection, or human interaction. Sensor nodes are capable of sensing parameters such as temperature, pressure, and humidity from the physical environment, which is generally termed the sensing field. The sensed data is then processed at the node level or cluster level and communicated to the sink or base station, referred to as the collection point. Rapid deployment, self-organization, high sensing fidelity, flexibility, low cost, and fault-tolerant characteristics of WSNs [1] make them a very promising sensing technique for various applications. These sensor networks are very useful for collecting information from difficult and inaccessible terrains. There is an increasing number of real-life applications for sensor deployments that include environment monitoring, agriculture, military, public security, and warning, besides business competitiveness, improvement, and quality of life improvement. Wireless sense and control technology is being utilized to bridge the gap between the physical world of humans and the virtual world of electronics. The dream is to automatically monitor and respond to forest fires, avalanches, hurricanes, faults in countrywide utility equipment, traffic, hospitals, and many more wide areas, with billions of sensors.

Wireless sensor network design is, however, constrained by various research issues [1,2], which include fault tolerance, scalability, operating environment, sensor network topology, hardware constraints, transmission media, and power consumption. Various researchers are working toward addressing these issues. It has been agreed upon that energy consumption is the major obstacle toward realizing this technology. Sensors have limited battery capacity for their operation, thereby hampering the usage of complex and computational-intensive algorithms and applications, which require long battery life. Sensor nodes can be powered by energy harvesting technology [3], but energy obtained from the external environment is dependent on environmental conditions and is rarely continuous. In a few applications, batteries can be replaced or recharged, but in the majority of applications, where sensor beds are located in remote areas, replacing or recharging a battery is not possible. Efficient utilization of energy is the most vital design issue for these networks.

To increase network lifetime, energy-efficient management strategies are proposed by researchers from time to time [4]. Most of these work under the assumption that the energy requirement during data acquisition is much less than that while communicating data. However, many practical application scenarios have shown that this assumption does not hold, especially where specific sensors are used for monitoring complex phenomena. The protocols running on sensor networks must utilize the resources efficiently to achieve long network lifetime.

11.1.1 Data Collection Architecture for WSNs

Data collection plays an important role in WSNs and targets collecting the sensed data from all sensors deployed in the network and transferring it to the base station or user for further processing. Data transmission is very expensive because transmitting 1 bit of data is equal to performing thousands of processing operations in a sensor node. Therefore, it is felt that to reduce energy consumption, the number of samples or amount of data to be transmitted should be minimized, even at the cost of increasing data processing [5].

In WSN data collection (Figure 11.1), energy is consumed at the sensor, node, cluster, and network levels. At the sensor level, energy is consumed by sensors while sensing samples from the environment or sensing field. At the node level, energy is consumed during aggregation, compression, or manipulation of samples sensed by one or more sensors within a single node and between neighbor nodes. At the cluster level, energy is consumed during data aggregation between multiple

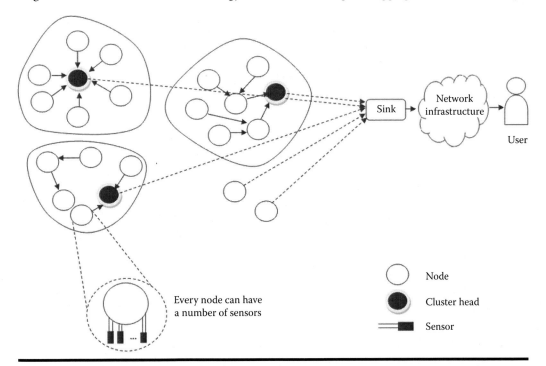

Figure 11.1 Data collection in wireless sensor network.

nodes of a cluster. At the network level, energy is consumed between multiple cluster heads, relays, and sinks. Along with this, energy is consumed due to the mobility of mobile relays, sinks, and base stations. Network lifetime in WSNs can be prolonged with energy-efficient data collection techniques to make them suitable for future wireless applications with huge energy requirements.

11.1.2 Energy-Efficient Data Collection Techniques

By considering the levels of energy consumption and mobility, various energy-efficient data collection techniques [6–11] can be grouped into three main categories (Figure 11.2):

1. **Based on sensing subsystem:** In this approach, energy management is carried out at the sensor level based on the sensing subsystem. Efforts are being made to reduce the amount or frequency of energy-expensive samples at the sensor level.
2. **Based on networking subsystem:** Here, the energy management approach is applied at the node, cluster, and network levels. The main stress is on reducing energy consumption during network activities at the node, cluster, and network levels by designing and implementing protocols and techniques for efficient energy management.
3. **Based on mobility:** In this category, energy management is done at the relay and sink levels. Techniques based on the mobility of network components such as mobile relays and sinks are being explored.

11.2 Data Collection Techniques Based on Sensing Subsystem

A sensor node (Figure 11.3) is a tiny device that includes three basic components: a sensing subsystem for data acquisition from the physical surrounding environment, a processing subsystem for local data processing and storage, and a wireless communication subsystem for data transmission. In addition, a power source supplies the energy required by the device to perform the programmed task. This power source often consists of a battery with a limited energy budget.

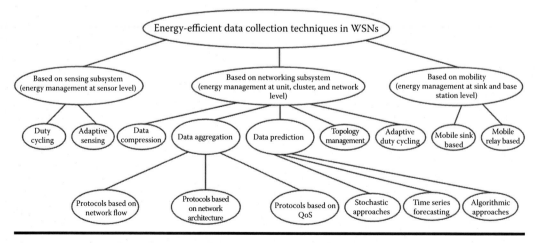

Figure 11.2 Overview of energy-efficient data collection in WSNs.

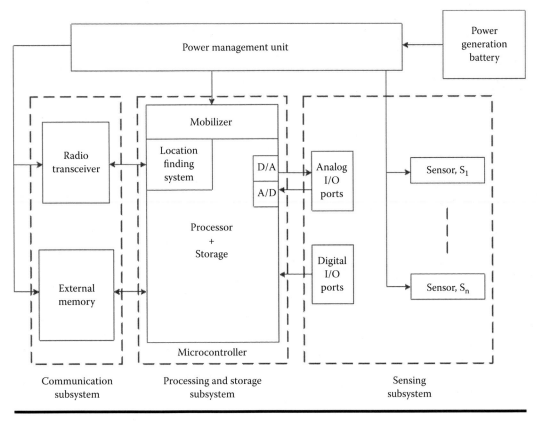

Figure 11.3   Typical sensor node.

Each sensor node is equipped with one or more sensors. A variety of mechanical, thermal, biological, chemical, optical, and magnetic sensors may be attached to the sensor node to measure the desired parameters of the environment. The sensing subsystem performs the sensing operations. The processing subsystem performs the node-level computations on sensed data. The communication subsystem performs the communication of data with adjacent nodes or sinks. Here, energy efficiency may be achieved by reducing the number of energy-expensive samples at the sensor level. This technique cuts down the energy consumption of sensors when data acquisition (i.e., sampling) cost is nonnegligible. There are two main approaches for energy management at the sensor level: (1) duty cycling and (2) adaptive sensing.

11.2.1 Duty Cycling–Based Approach

Duty cycle may be defined as the percentage of time that an entity spends in an active state as a fraction of the total time under consideration. Based on duty cycling, WSNs can be classified into four broad categories [12]: event driven, time driven, query based, and hybrid. In event-driven WSNs, data is only generated when an event of interest occurs. In time-driven WSNs, data is periodically sent to the sink at a constant interval of time. In query-based WSNs, data is collected according to the end user's demand. In hybrid WSNs, a combination of two or more of these techniques is employed.

An event can be continuous or happen gradually. Events can be complex where measurements are done for some unusual activity. Events can be further classified into two categories: system events and environmental events. System events are concerned with architectural or topological changes; for example, a mobile node entering a cluster area. Environmental events are concerned with the occurrences of unusual changes across the monitored environment. In the event-based data collection (e.g., REED [13]), the sensors detect and report events to the base station; for example, in battlefields, sensors can be used to detect enemies' vehicles and their movement. An event-driven model is valuable for detecting events as soon as they occur over a specified region. This prevents the network from unnecessarily monitoring and hunting for data collection, which in turn helps in energy conservation.

In time-driven approaches, a sensor node collects data from the environment, in which it is deployed, after some time interval set by the end user. It provides continuous monitoring of the sensing field and may involve handling of millions of nodes. The life of the nodes could be increased by sending the nodes to the sleep mode when not in use or not capturing data or between transmissions. They are limited to a specific set of applications where consistent changes occur across the network, for example agricultural applications. There is a probability of getting redundant data, which results in wastage of resources such as the battery. There are problems of synchronization with the global clock.

In the query-based approach, users can query the network and get a response. Queries can be sent on demand or at fixed intervals. Users can issue a Structured Query Language (SQL) query to get the desired response from the sensor nodes.

A hybrid approach is a combination of two or more of the above-mentioned approaches. A hybrid protocol that adaptively switches between time-driven and event-driven data collection was proposed in [14]. The event-driven approach works until the event of interest occurs, and after the event becomes invalid, the protocol starts behaving like a time-driven protocol. During this period, sensor nodes continuously report data to the sink. A hybrid framework also deploys event-driven and query-based approaches [15]. An energy-efficient hybrid data collection architecture has been proposed in [16].

11.2.2 Adaptive Sensing–Based Approach

Adaptive sensing [17] provides an energy-efficient approach that minimizes the number of active nodes according to the predictability of their measurements in order to extend the lifetime. Sensor nodes with limited bandwidth and buffers may cause dropped packets due to excessive data transfer while sensor nodes are collecting and communicating data all the time. So, it requires some means to collect and communicate only the useful data. In an adaptive sensing system, lifetime is extended by keeping redundant nodes in the passive mode so that they can be used later. Macronodes collect data periodically, and a low-order model is fitted to compute a prediction area for each sensor. This prediction area is used to measure the redundancy level of the sensors covered in it. The redundant nodes are put into the passive mode by macronodes, while keeping the distortion in the output low. Adaptive sensing can be implemented by exploiting three different approaches [18]: (1) hierarchical sensing, (2) adaptive sampling, and (3) model-based active sensing.

11.2.2.1 Hierarchical Sensing

In hierarchical sensing [19], there are sensor nodes equipped with different sensors, each having its own accuracy and power consumption. Each sensor is used to measure the same physical quantity. Under hierarchical sensing, there are two subapproaches: triggered and multiscale sensing. Triggered sensing activates a more accurate and power-consuming sensor after an event has occurred, whereas in multiscale sensing [20], high-resolution sensors are activated only when they are explicitly requested.

11.2.2.2 Adaptive Sampling

Adaptive sampling techniques are aimed at dynamically adapting the sampling rate by exploiting correlations among the sensed data or information related to the available energy. Adaptive sampling strategies may dynamically adapt the sensor sampling rate based on the spatial or temporal correlation among acquired data; this is called activity-driven adaptive sampling [21]. Whenever the sensor node is able to harvest energy from the environment, knowledge about the residual and forecasted energy coming from the harvester module is exploited to optimize power consumption at the node level and is termed harvesting-aware adaptive sampling [22].

11.2.2.3 Model-Based Active Sampling

The model-based active sampling [23] approach builds a model of the sensed phenomenon on top of an initial set of sampled data. Once the model is available, the next data can be predicted by the model instead of sampling the quantity of interest, hence saving the energy consumed for data sensing. Whenever the requested accuracy is no more satisfied, the model needs to be updated or reestimated, to adhere to the new dynamics of the physical phenomenon under observation.

11.3 Data Collection Techniques Based on Networking Subsystem

In WSN, nodes can be organized as clusters and networks. They can collect data without any intervention and send collected data to a sink or base station. Factors that determine the energy efficiency of a sensor network are the network architecture, data aggregation mechanism, and underlying routing protocol. There are five approaches for energy management at the network subsystem level (Figure 11.4): data compression, data aggregation, data prediction, topology management, and adaptive duty cycling.

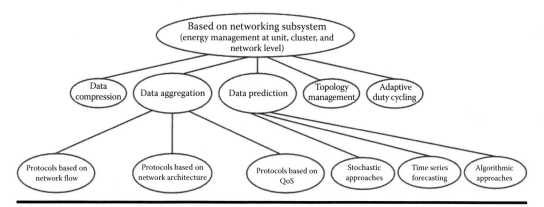

Figure 11.4 Energy-efficient data collection techniques based on networking subsystem.

11.3.1 Data Compression

Data transmission is a primary factor of energy consumption in WSNs. Various researchers have focused on reducing the number of bits to be transferred over the network. In order to reduce the amount of sensing data, we need to compress the data in the network. Compression-based data collection techniques given by various researchers [24–27] make it possible to directly acquire just the important information about the sensing field by not acquiring that part of the data that would eventually just be "thrown away" by lossy compression. So, compression techniques provide support for energy-efficient data collection. There are a large number of algorithms that have integrated data compression into data collection.

Data compression means reducing the amount of data to be transferred on the communication medium with the purpose of decreasing the network traffic load [28]. There are three types of compression techniques: lossy, lossless, and unrecoverable. Lossy compression means some features of the original data will be lost once it is compressed. This technique is usually used for image, video, and audio compression. The compression rate is very high in lossy compression. Lossless compression means original data will be recovered as it is after decompression; there will be no loss of data. Huffman coding, arithmetic coding, and some image compression techniques, such as png and gif, are examples of lossless techniques. An unrecoverable compression means that the compression operation is irreversible. In other words, there is no decompression operation. For example, one can compress a set of numbers by taking their average value, but each of the original numbers cannot be derived from this average value. Earlier data compression techniques were developed keeping bandwidth-limited communication networks in mind and were not necessarily suitable for WSNs. However, in recent times, researchers have investigated this area as well.

11.3.2 Data Aggregation

Data aggregation means collecting data from various sensor nodes in the network. In data aggregation, any redundant information is removed and unique information is sent to the base station. There are three subapproaches based on which data aggregation can be done: (1) protocols based on network architecture, (2) protocols based on network flow, and (3) protocols based on quality of service (QoS).

11.3.2.1 Protocols Based on Network Architecture

Data aggregation techniques can be designed for wireless sensor networks based on network architecture. There are two types of network: flat network and hierarchical network. In flat networks, all nodes have the same computation capabilities and responsibilities. All nodes act in a similar manner. Whereas in hierarchical networks, one head is chosen from the nodes forming a hierarchy. This cluster head can have more computation power and has added responsibility compared to the normal node.

11.3.2.1.1 Flat Networks

In this, all nodes have the same battery and computation power. Data aggregation in a flat network is performed by different nodes along the multihop path. The sink sends a query to all the sensor nodes in the network, and the sensor node having the answering data sends back a response. The choice of a particular communication protocol depends on the specific application.

One-phase and two-phase push diffusion and pull diffusion are the techniques for data collection in flat networks. In push diffusion, nodes, on the occurrence of an event, send data to the sink. Diffusion is initiated by the node. The sensor protocol for information via negotiation (SPIN) [29] is one of the push diffusion techniques that incurs less energy consumption and is able to distribute 60% more data per unit than flooding. Two-phase pull diffusion or directed diffusion [30] is based on the data acquired by sensors. In this, the sink initially broadcasts an interest message. There is one gradient that specifies the data rate and the direction in which to send the data. Intermediate nodes and all the other nodes keep track of all the data that is sent. One-phase pull diffusion [31] overcomes the disadvantage of two-phase pull diffusion, having excessive overhead due to more sinks and sources, by skipping the flooding process of two-phase pull diffusion. In one-phase pull diffusion, sinks send interest messages. The sources do not transmit exploratory data. This results in a decrease in control overhead, conserving the energy of the sensors.

11.3.2.1.2 Hierarchical Networks

Owing to the flat structure, there is excessive communication in flat networks, which puts an extra burden on the sink node, resulting in faster depletion of the battery. Several hierarchical data aggregation approaches have been proposed. Hierarchical data aggregation involves data fusion at special nodes, which reduces the number of messages transmitted to the sink. This improves the energy efficiency of the network. In hierarchical network-based data aggregation, there are four types of networks: cluster based, chain based, tree based, and hybrid.

11.3.2.1.2.1 Cluster-Based Networks   In large networks, it is inefficient to send data directly to the single sink node. So, in cluster-based networks (Figure 11.5), various sensor nodes can be grouped into different clusters. In each cluster, a cluster head can be nominated that can collect and aggregate data from the corresponding cluster nodes and send it to the sink. Many techniques and protocols have been proposed by researchers in this area. Such protocols are of the low-energy

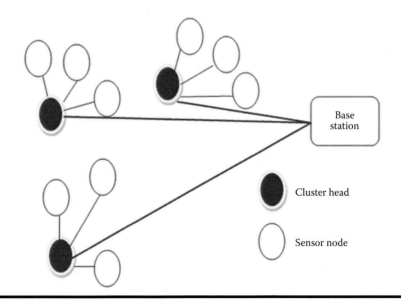

Figure 11.5   Cluster-based networks.

adaptive clustering hierarchy (LEACH) type [32]; variants of LEACH include E-LEACH [33], TL-LEACH [34], M-LEACH [35], V-LEACH [36], hybrid energy-efficient distributed clustering approach (HEED) [37], clustered diffusion with dynamic data aggregation (CLUDDA) [38,39], and energy-efficient multilevel clustering (EEMC) [40]. LEACH is the basis for many cluster-based techniques. LEACH uses a random approach for distributing energy consumption among the nodes. In this approach, the nodes are organized in the form of a cluster, and one node acts as a cluster head. Nodes, rather than sending data to the sink node, send it to their cluster head, which in turn sends it to the sink. Cluster head selection plays a major role in the network lifetime. In every round of data collection, a new cluster is formed and a new cluster head is chosen. A bad cluster head selection technique can cause the network to die soon. Nodes having more residual energy are probable contestants of the cluster head selection process.

11.3.2.1.2.2 Chain-Based Networks

In cluster-based networks (Figure 11.6), sensors transmit data to cluster heads. If a cluster head is away from sensors, then it consumes a lot of energy. Energy can be conserved if it communicates with its neighbors only. In chain-based networks, each sensor sends data to its nearest neighbor, and the neighbor sends data to its nearest neighbor, excluding the node that sent data to it. In this way, data is forwarded until the data reaches the leader, forming a chain. PEGASIS (a power-efficient data gathering protocol for sensor information systems) [41] is a chain-based data aggregation tool. It evenly distributes the network load among all nodes of the network. This approach distributes the energy load evenly among the sensor nodes in the network. A greedy algorithm is used by nodes to form a linear chain. In it, each node receives data from its neighboring node and adds its own data to it and sends it further. This process continues until it reaches its chain leader. The leader includes its own data in the packet and sends it to the sink. Energy expenditure becomes high when nodes are farther apart. In addition, transmission energies are not evenly distributed, but depend on the actual distances between the nodes and their neighbors; that is, nodes with distant neighbors dissipate more energy.

11.3.2.1.2.3 Tree-Based Networks

In a tree-based network (Figure 11.7), sensor nodes are organized into trees, where data aggregation is performed at intermediate nodes along the tree. One of the main aspects of tree-based networks is the construction of an energy-efficient data aggregation tree. Ding et al. [42] proposed heuristics to construct and maintain an energy-aware aggregation tree in a wireless sensor network. In this, the sink acts as the root of the tree. All data collected by leaf nodes is sent to the root node or sink via a preset path. Tan and Korpeoglu [43]

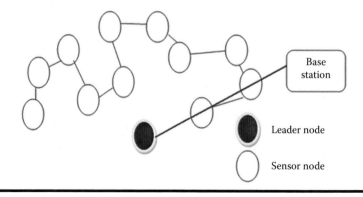

Figure 11.6　Chain-based networks.

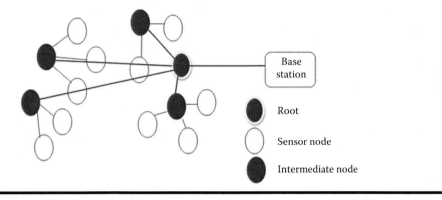

Figure 11.7   Tree-based networks.

proposed a power-efficient data gathering and aggregation protocol (PEDAP). The goal of PEDAP is to maximize the lifetime of the network in terms of number of rounds, where each round corresponds to aggregation of data transmitted from different sensor nodes to the sink. In [44], the authors proposed a tree-based energy-efficient protocol for sensor information (TREEPSI). Before the data transmission phase, WSNs will select a root node in all the sensor nodes. There are two ways to build the tree path. One is computing the path centrally by the sink and broadcasting the path information to the network. The other can be the same tree structure locally, by using a common algorithm in each node. The root uses a standard tree traversal algorithm for collecting data. Leaf nodes start sending data to their respective parent node. The parent node sends data to its parent node until data reaches the root node. The root node aggregates data and sends it to the sink. The path, once set, is used to send data until the root is alive. When the root node dies, a new root node is selected and the whole process is repeated to collect data. PEGASIS also uses this way to transmit data, but TREEPSI performs better than PEGASIS due to a shorter path. The secure data aggregation protocol (SDAP) [45] delivers data in a secure manner over aggregation trees. Tiny aggregation (TAG) [46] is designed for monitoring applications. TAG uses a tree-based routing scheme for aggregating data and sending it to the sink. TAG adopts the selection and aggregation facilities of the database query languages (SQL).

E-span [47] is an energy-aware spanning tree algorithm. The node with the highest residual energy is chosen as the root node. Parent nodes are also chosen based on decreasing residual energy from the root and distance from the root. In the E-span network, coverage is not high, as it uses distance as the major factor in selecting the parent node, because sometimes nodes that are closer to the root have low energy and are chosen as the parent node. Since they have low energy, they will die soon, and the child nodes are disconnected from the rest of the network.

11.3.2.1.2.4 Hybrid Networks   In order to benefit from the advantages of cluster-based, tree-based, and chain-based schemes, hybrid approaches [48] have been suggested, which adaptively tune their data aggregation structure for optimal performance. The tributaries and deltas protocol [49] try to overcome the problems of both the tree- and multi-path-based structures, by combining the best features of both schemes. The result is a hybrid algorithm where both data aggregation structures may simultaneously run in different regions of the network. The idea is that under low-packet-loss rates, a data aggregation tree is the most suitable structure due to the possibility of implementing efficient sleeping modes with increased efficiency in representing and compressing the data. On the other hand, in the case of high-loss rates, or when transmitting partial results that are accumulated from many sensor readings, a multipath approach may be the best option due to its increased robustness.

11.3.2.2 Protocols Based on Network Flow

The main goal of network flow–based protocols is optimization of network lifetime subject to energy constraints on sensor nodes and flow constraints on information routed in the network. Protocols based on network flow includes data aggregation protocols for lifetime maximization. Kalpakis et al. [50] studied the maximum lifetime data gathering with aggregation (MLDA) problem employing efficient data aggregation algorithms. The goal of the MLDA problem is to obtain a data gathering schedule with maximum lifetime where sensors aggregate incoming data packets.

A clustering-based approach, called greedy clustering-based maximum lifetime data aggregation schedule (CMLDA) [50], was proposed to obtain efficient data gathering schedules in large networks. The CMLDA algorithm works by first clustering the nodes into groups of a given size. Each cluster's energy is set to the sum of the energy of the contained nodes. The distance between clusters is set to the maximum distance between any pair of nodes of two clusters. After the cluster formation, MLDA is applied among the clusters to build cluster trees. CMLDA then utilizes an energy balancing strategy within a cluster tree to maximize network lifetime. CMLDA has a much faster running time than MLDA, but does not work well on networks that have nodes spaced far apart. CMLDA works better on a dense network when nodes are deployed in groups in close proximity.

Xue et al. [51] investigated the data aggregation problem in the context of energy-efficient routing for maximizing system lifetime. The problem was modeled as a multicommodity flow problem [52], where the data generated by a sensor node is analogous to a commodity. Hong and Prasanna [53] focused on data gathering problems in energy-constrained network sensor systems. It was shown that the problem can be reduced to a restricted flow problem with a quota constraint, flow conservation requirement, and edge capacity constraint. A strongly polynomial time algorithm, restricted flow problem with edge capacities (RFEC), was proposed for this problem to find an optimal integer-valued solution that specifies the number of data packets to be transferred between any two neighboring sensors for each round.

11.3.2.3 Protocols Based on QoS

The protocols described so far focus on conserving energy and increasing network lifetime. However, in some real-time applications, quality of service (QoS), such as reliability, delay, packet loss, and throughput, is obligatory. Protocols based on QoS parameters are discussed in subsection 11.3.2.3.1.

11.3.2.3.1 Data Aggregation Protocols for Optimal Information Extraction

Sadagopan and Krishnamachari [54] examined the problem of maximizing data collection from an energy-limited store-and-extract wireless sensor network, which is analogous to the maximum lifetime problem of interest in continuous data gathering sensor networks. One significant difference is that this problem requires attention to "data awareness," in addition to "energy awareness." The maximum data extraction problem was formulated as a linear program, and a $1 + \omega$ iterative approximation algorithm was presented for it. As a practical distributed implementation, a faster greedy heuristic was developed that used an exponential metric based on the approximation algorithm. Simulation results showed that the greedy heuristic incorporating this exponential metric performs near optimally (within 1%–20% of optimal, with low overhead) and significantly better than other shortest-path routing approaches, particularly when nodes are heterogeneous in their energy and data availability.

Ordonez and Krishnamachari [55] presented models both for maximizing the total information gathered subject to energy constraints (on sensing, transmission, and reception) and for minimizing the energy usage subject to information constraints. Other constraints in their models correspond to fairness and channel capacity (assuming noise but no interference). Extensions of these models were also deliberated that can handle data aggregation, interference, and even node mobility.

11.3.2.3.2 Data Aggregation Protocols for End-to-End Reliability and Congestion Control

A QoS protocol that supports both periodic and event-based data collection is presented by Fonoage et al. [56]. A geographic routing mechanism combined with QoS support is used to forward packets in the network. Congestion control is achieved by using a ring or barrier mechanism that captures and aggregates messages that report the same event to the same sink. Othman and Yahya [57] proposed the energy-efficient and QoS-aware multipath routing protocol (EQSR), which maximizes network lifetime through balancing energy consumption across multiple nodes. It allows delay-sensitive traffic to reach the sink within an acceptable delay. It also reduces the end-to-end delay through spreading out the traffic across multiple paths, and increases the throughput by introducing data redundancy. He et al. [58] also proposed an aggregation scheme to maximize utilization of the communication channel.

11.3.3 Data Prediction

Data prediction is a fascinating area for energy management of WSNs. Data prediction means the model has the capability to forecast the values sensed by the nodes. A sink can directly respond to user queries by predicting the value without sending a request to the node and conserve energy. This is possible only if the model is accurate; else, data has to be retrieved by sending a query to the node. The model at the source node is used by the sink to ensure the model's effectiveness. For verification, source nodes just sample the data and compare it with the predicted value. There are three approaches for data prediction: (1) stochastic, (2) time series forecasting, and (3) algorithmic.

11.3.3.1 Stochastic Approaches

This approach is based on predicting data that may be analyzed statistically, but may not be predicted precisely. In this, a probabilistic model is used to predict the sensed value. A Kalman filter can be used on each node to estimate the next data [59]. A sink also uses the same model to make similar predictions. The predicted value is compared with the sensed value. If the difference is below the preestimated threshold, the predicted value is accepted; else, the sensed value is sent to the sink. Both the sensor and the sink use this value and simultaneously estimate the prediction model and update the filter weights. Kanagal and Deshpande [60] gave an approach to build an extensible database system. In this, the dynamic probabilistic model can be applied to data in the database. The Ken technique, proposed by Chu et al. [61], also uses a replicated dynamic probabilistic model to minimize communication from the sensor node to the base station.

11.3.3.2 Time Series Forecasting

In this approach, prediction of future values is based on the historic values. Time series data has a natural temporal ordering. Time series is an effective way to reduce the communication cost. There are three major classes under it: moving average (MA), autoregressive (AR), or autoregressive moving average (ARMA) models. These three classes depend linearly on the previous data points. These models are simple, but they can be used in many practical cases with good accuracy. Tulone and Madden [62] presented the probabilistic adaptable query (PAQ) system, based on the autoregressive model, to predict readings. The model is transmitted on the sink and can use it to predict sensor values without communicating with the node. The similarity-based adaptive framework (SAF) [63] improves the AR model. It can predict the trend of values over a time period. Inconsistent data can also be detected. Borgne et al. [64] proposed adaptive model selection (AMS), a generic algorithm for online model selection of time series prediction models for reducing the communication effort in wireless sensor networks. Their adaptive selection scheme is aimed at making sensor nodes smart enough to be able to autonomously and adaptively determine the best-performing model to use for performing data prediction. All nodes keep a set of models, but at a given instant, only one of them is used for data prediction. Complex models can lead to better prediction at the expense of a higher update cost, as they need more parameters to be described properly.

11.3.3.3 Algorithmic Approach

The algorithmic approach is used in application-specific scenarios. It relies on a heuristic or experience-based scheme about the domain explored. Various algorithms are used to build a model based on the characteristics of the system. Prediction-based monitoring (PREMON) for energy-efficient monitoring was proposed in [65]. It is inspired by the MPEG file, in which each frame separates from its neighbor by only a few parameters. The next frame can be predicted from the current and the previous frame. The performance evaluation carried out on a test bed of Rene Motes [66] reflected significant reduction in energy consumption (by more than five times) and increased the sensor lifetime. The buddy protocol [67] is based on the PREMON paradigm. In this protocol, each node forms a buddy relationship with its nearest neighbor, forming a buddy group. One buddy group leader is assigned, which answers all queries on behalf of the whole group. Only the group leader is turned on, and the rest of the nodes are turned off. This helps reduce the energy consumption in the network.

11.3.4 Topology Management

Wireless sensor network nodes are often unattended. Topology management is required in those areas where nodes are prone to damage and failure. This may result in topology change. Sometimes the node failure will result in partitioning of the area, preventing data exchange and communication. Topology management also helps save energy in the communication subsystem. A node's radio can be switched off completely to save power. Its goal is to coordinate the sleep transitions of all the nodes, while ensuring that data can be forwarded efficiently to the data sink. Topology affects many network characteristics, such as latency, robustness, and capacity. The complexity of data routing and processing also depends on the topology. When there are large numbers of nodes to be placed in some remote area, a network should be able to organize itself, and manual configuration is not possible. The network should be up all the time, even if some node fails at any time. Topology management consists of knowing the physical connections and logical relationships among the sensors and, at the same time, creating a subset of nodes actively participating in the network, thus creating less communication and conserving energy in nodes. Networks require constant monitoring in order to ensure consistent and efficient operations. The primary goal of topology management is to maintain network connectivity in an energy-efficient manner. Topology management is one of the key aspects of configuration management, which entails initial setup of the network devices and continuous monitoring and controlling of these devices. Sparse topology and energy management (STEM) [68], geographical adaptive fidelity (GAF) [69], and Naps [70] are a few topology management protocols.

Topology control plays a vital role in energy management, as the right number of nodes needs to be turned on and off at the right time, so that there is no packet loss due to fewer nodes being turned on and no energy loss due to extra nodes being turned on. STEM takes advantage of network density to increase the lifetime of the network. GAF is a location-driven protocol. In this, nodes need to coordinate to decide which node will go to sleep and for how long, but in STEM, nodes do not need to communicate. With these techniques, increased energy savings can be obtained at the cost of either deploying more nodes or allowing more setup latency per hop. These choices are essentially part of a multidimensional design trade-off, which is impacted by the specific application, layout of the network, cost of the nodes, the network lifetime, and many other factors. The Naps scheme is not based on geographic location and is very flexible, also providing connectivity at low densities. Topology management can increase the network lifetime, compared to the network always being on. So research in this direction can be very helpful, especially while exploring locations that are hard to reach.

11.3.5 Adaptive Duty Cycling

A longer network lifetime of wireless sensor networks is highly desired. Nodes are not transmitting data all the time. If they are active all the time, then their radio will consume energy. It has been observed that radio consumes a major portion of energy. Measurements have shown that a typical radio consumes a similar level of energy in the idle mode as in the receiving mode. It is important that nodes are able to operate in low duty cycles. Piconet [71] is a low-rate, low-range, ad hoc radio network. A piconet node can be used to provide a connection to the embedded network. Piconet provides a broad range of mobile and embedded computing objects with the ability to exploit an awareness of, and connectivity to, their environment. Sensors can use piconet to relay information about the state of the local environment or of a particular device. In piconet, one node randomly goes to sleep. When the node wakes up, it sends a beacon with its own ID. If other nodes want to talk to this node, they need to wake up and listen until receiving the beacon.

Sparse topology and energy management (STEM) reduces the energy consumption in the monitoring state to a bare minimum while ensuring satisfactory latency for transitioning to the transfer state. STEM uses two radios operating in different channels, one for data transmission and the other for node wake-up. The radio is turned off when there is no data to send. The sender wakes up the receiver by sending a wake-up tone STEM-T and beacon STEM-B.

In some medium access controls (MACs), which are time division multiple access (TDMA) based, they work on a low duty cycle. Their radio is turned on when they need to transmit or receive data. For example, Bluetooth [72] is turned on when we need to send or receive data; else, it is turned off, as it will unnecessarily consume energy in searching nearby devices. S-MAC [73,74] is a contention MAC with integrated low-duty-cycle operation that supports multihop operation.

11.4 Data Collection Techniques Based on Mobility

In mobility-based data collection techniques, data can be aggregated with the introduction of a mobile element in the network. It is helpful in isolated regions for data collection. These techniques are used where the sensing area is mobile or some nodes are mobile. Mobile elements can visit nodes in the network and collect data directly through single-hop transmissions. They help to reduce congestion in the network. Mobile elements can prevent the sink from being overloaded and spread energy consumption uniformly over the network. There are two ways in which data collected by mobile nodes can be managed: mobile sink based and mobile relay based.

11.4.1 Mobile Sink Based

In mobile sink–based collectors, sensor nodes are static in nature and the sink is mobile. There can be more than one mobile sink in the network. The mobile sink moves around the network to collect the data from the sensor nodes. Wang et al. [75] and Rao et al. [76] have proposed techniques for such data collectors.

11.4.2 Mobile Relay Based

In mobile relay–based collectors, the sink node is static, compared to the previous technique, where the sink was mobile. In this, there are some special nodes, called relays, that move around the network to collect data from the sensor nodes and send collected data to the static sink. These are not end points in the network; rather, they act as mobile forwarders. Various researchers [77,78] have proposed the data MULE system for mobile relay–based data collection.

11.5 Discussion

The duty cycling technique, where the sensor needs to be switched on and off based on the requirement and technique applied, is one of the initial approaches for energy-efficient data acquisition. However, for complex events, event-driven duty cycling may not work and time-driven duty cycling is limited to specific applications only. Query-based duty cycling does not provide suitable constructs to easily articulate spatiotemporal sense data characteristics, and it is difficult to formulate queries using current languages. Hybrid approaches offer a good substitute for energy-efficient data acquisition, though they are not guaranteed to work well for all applications. The choice of duty cycling approach to be used is more or less application specific.

Hierarchical sampling is used to measure the same physical quantity where nodes are equipped with different sensors, each characterized by its own accuracy and power consumption. Triggered sensing is suitable for object detection systems, and multiscale sensing is more appropriate for environmental monitoring applications. Activity-driven adaptive sampling is efficient, but it typically exploits either temporal or spatial correlation. Some approaches need to be adopted in the area of spatiotemporal adaptive sampling.

In hierarchical networks, data aggregation is performed by cluster heads, whereas in flat networks, it is performed by different nodes along the multihop path. So, even if one cluster head fails, the hierarchical network may still be operational, as other cluster heads are still alive and can carry on communication. But in flat networks, the failure of the sink node may result in the breakdown of the entire network because the sink node has to aggregate all the data. The routing structure is simple in hierarchical networks, but not necessarily optimal. On the contrary, flat networks guarantee optimal routing with additional overhead. Hybrid approaches that merge benefits of more than one network are providing a way for future researchers.

The shortest-path heuristic may not obtain the optimal solution because it searches over possible paths in the original graph instead of the residual graph. The MLDA algorithm achieves only 50% of the optimal system lifetime. CMLDA uses a linear programming approach, but it has high computational complexity for networks of large sizes. In the max concurrent flow algorithm, performance decreases with an increase in the number of sinks. RFEC cannot be used for realistic models.

Investigations made for QoS support in WSNs are somewhat less explored. QoS support in WSNs should also include QoS control, besides QoS assurance mechanisms. The major constraints in obtaining good QoS are fairness and channel capacity. Data aggregation is a solution to maintain robustness while decreasing redundancy in the data, but this mechanism also introduces latency and complicates QoS design in WSNs.

A stochastic approach is used for high-level operations, but it suffers from high computational cost. It can be deployed if powerful sensors are used. More research efforts are required in this area for efficiently utilizing this approach. The time series forecasting can satisfactorily work for simple models. They do not require the exchange of all sensed data until a model is available. Moreover, they provide the ability to detect outliers and model inconsistencies. But forecasting is not always possible. If a specific type of model is used, it should be suitable to represent the phenomenon of interest. These techniques need more attention, thereby leaving vast room for research. Algorithmic approaches are more application specific, and as a result, the research direction needs to focus on a certain class of application, for possible improvements in this direction.

The adaptive duty cycling technique is not fit for situations where a significant amount of data needs to be transferred. STEM is not suitable for multihop routing where data needs to be transferred between cluster heads at different levels. This problem can be solved using S-MAC, which is specially meant for multihop routing and supports long messages to send summaries efficiently. In the case of interactive user sessions, STEM can be used if multiple radios are available at each node. Such a protocol would ensure that queries are answered with low latency, albeit at higher cost.

Mobile-based techniques are a new area for research. These techniques help to reduce the traffic in the network. Two techniques—sink based and relay based—can help keep the sink from being overloaded. Moreover, energy consumption in the network becomes uniform. Still, there is a need to explore this area, so that all aspects of mobility can be explored and implemented in real-world scenarios.

11.6 Conclusion

Energy is a very critical resource in WSNs. Energy-saving techniques have attracted the attention of a lot of researchers, and wide-ranging research is being conducted to efficiently utilize the resources of WSNs. Energy needs to be conserved at each and every step in the network. It may be at the level of the sensor, node, network, or cluster. A collective effort on every part of the system is required for better energy conservation. Further, some of these techniques are also application specific. So, the correct combination of techniques, along with the application, must be devised for efficient energy utilization. Various types of energy-efficient data collection techniques were overviewed and presented in this chapter. The techniques discussed look promising, but there are still many challenges and issues that need to be addressed and resolved. Many new dimensions and directions are open for required research and improvement in WSN technologies and applications. The sensor network is an area of research where various technological advances of different domains of engineering can be exploited. Apart from technological challenges, sociological, ecological, demographical, and security-related issues are also open-ended questions that need to be answered by future researchers.

References

1. Akyildiz, I.F., Su, W., Sankarasubramaniam, Y., and Cayirci, E. (2002). Wireless Sensor Networks: A Survey. *Computer Networks*, vol. 38, no. 4, pp. 393–422.
2. Anastasi, G., Conti, M., Francesco, M.D., and Passarella, A. (2009). Energy Conservation in Wireless Sensor Networks: A Survey. *Ad Hoc Networks*, vol. 7, no. 3, pp. 537–568.
3. Kansal, A., Hsu, J., Zahedi, S., and Srivastava, M. (2007). Power Management in Energy Harvesting Sensor Networks. *ACM Transactions on Embedded Computing Systems*, vol. 6, no. 4, pp. 1–38.
4. Anisi, M.H., Abdullah, A.H., and Razak, S.A. (2011). Energy-Efficient Data Collection in Wireless Sensor Networks. *Wireless Sensor Networks*, vol. 3, pp. 329–333.
5. Pottie, G., and Kaiser, W. (2000). Wireless Integrated Network Sensors. *Communications of the ACM*, vol. 43, no. 5, pp. 51–58.
6. Hwang, S., Jin, G.J., Shin, C., and Kim, B. (2009). Energy-Aware Data Gathering in Wireless Sensor Networks. In *Proceedings of 6th IEEE Conference on Consumer Communications and Networking*, Las Vegas, NV, pp. 1–4.
7. Jin, G., Shin, C., and Kim, B. (2009). Energy-Aware Data Gathering in Wireless Sensor Networks. In *Proceedings of 6th IEEE Conference on Consumer Communications and Networking*, Las Vegas, NV, pp. 1–4.
8. Kalpakis, K., Dasgupta, K., and Namjoshi, P. (2003). Efficient Algorithms for Maximum Lifetime Data Gathering and Aggregation in Wireless Sensor Networks. *Computer Networks*, vol. 42, no. 6, pp. 697–716.
9. Tan, H.O., and Korpeoglu, I. (2003). Power Efficient Data Gathering and Aggregation in Wireless Sensor Networks. *SIGMOD Record Newsletter*, vol. 32, no. 4, pp. 66–71.
10. Yang, J., Li, Z., Lin, Y., and Zhao, W. (2010). A Novel Energy-Efficient Data Gathering Algorithm for Wireless Sensor Networks. In *Proceedings of 8th World Congress on Intelligent Control and Automation*, Jinan, China, pp. 7016–7020.
11. Zhu, Y., Wu, W., Pan, J., and Tang, Y. (2010). An Energy-Efficient Data Gathering Algorithm to Prolong Lifetime of Wireless Sensor Networks. *Computer Communications*, vol. 33, no. 5, pp. 639–647.
12. Alsboui, T., Hammoudeh, M., Bandar, Z., and Nisbet, A. (2011). An Overview and Classification of Approaches to Information Extraction in Wireless Sensor Networks. In *Proceedings of 5th International Conference on Sensor Technologies and Applications*, Saint Laurent du Var, France, pp. 255–260.
13. Abadi, D.J., Madden, S., and Lindner, W. (2005). REED: Robust, Efficient Filtering and Event Detection in Sensor Networks. In *Proceedings of 31st International Conference on Very Large Data Bases*, Trondheim, Norway, pp. 769–780.
14. Akyildiz, I.F., Melodia, T., and Chowdury, K.R. (2007). Wireless Multimedia Sensor Networks: A Survey. *IEEE Wireless Communications*, vol. 14, no. 6, pp. 32–39.
15. Lee, C.H., Chung, C.W., and Chun, S.J. (2010). Effective Processing of Continuous Group-By Aggregate Queries in Sensor Networks. *Journal of System Software*, vol. 83, pp. 2627–2641.
16. Cheng, L., Chen, Y., Chen, C., and Ma, J. (2009). Query-Based Data Collection in Wireless Sensor Networks with Mobile Sinks. In *Proceedings of 2009 International Conference on Wireless Communications and Mobile Computing: Connecting the World Wirelessly*, Leipzig, Germany, pp. 1157–1162.
17. Arici, T., and Altunbasak, Y. (2004). Adaptive Sensing for Environment Monitoring Using Wireless Sensor Networks. *IEEE Wireless Communications and Networking Conference*, vol. 4, pp. 2347–2352.

18. Jelicic, V. (2011). Power Management in Wireless Sensor Networks with High-Consuming Sensors. *Qualifying Doctoral Examination*, pp. 1–9.

19. Kijewski-Correa, T., Haenggi, M., and Antsaklis, P. (2006). Wireless Sensor Networks for Structural Health Monitoring: A Multi-Scale Approach. In *Proceedings of 17th Analysis and Computational Specialty Conference*, St. Louis, MO, pp. 1–16.

20. Singh, A., Budzik, D., Chen, W., Batalin, M.A., Stealey, M., Borgstrom, H., and Kaiser, W.J. (2006). Multiscale Sensing: A New Paradigm for Actuated Sensing of High Frequency Dynamic Phenomena. In *Proceedings of IEEE/RSJ International Conference on Intelligent Robots and Systems*, Beijing, China, pp. 328–335.

21. Alippi, C., Anastasi, G., Galperti, C., Mancini, F., and Roveri, M. (2007). Adaptive Sampling for Energy Conservation in Wireless Sensor Networks for Snow Monitoring Applications. In *Proceedings of IEEE International Conference on Mobile Adhoc and Sensor Systems*, Pisa, Italy, pp. 1–6.

22. Kansal, A., Hsu, J., Zahedi, S., and Srivastava, M. (2007). Power Management in Energy Harvesting Sensor Networks. *ACM Transactions on Embedded Computing Systems*, vol. 6, no. 4, pp. 1–34.

23. Deshpande, A., Guestrin, C., Madden, S., Hellerstein, J.M., and Hong, W. (2004). Model-Driven Data Acquisition in Sensor Networks. In *Proceedings of International Conference on Very Large Data Bases*, Toronto, Canada, vol. 30, pp. 588–599.

24. Kolo, J.G., Shanmugam, S.A., Lim, D.G., Ang, L.M., and Seng, K.P. (2012). An Adaptive Lossless Data Compression Scheme for Wireless Sensor Networks. *Journal of Sensors*, vol. 2012, pp. 1–20.

25. Puthenpurayil, S., Gu, R., and Bhattacharyya, S.S. (2007). Energy-Aware Data Compression for Wireless Sensor Networks. In *Proceedings of International Conference on Acoustics, Speech, and Signal Processing*, Honolulu, HI, vol. 2, pp. 45–48.

26. Liu, Z., Zhang, M., and Cui, J. (2014). An Adaptive Data Collection Algorithm Based on a Bayesian Compressed Sensing Framework. *Sensors*, vol. 14, pp. 8330–8349.

27. Donoho, D.L. (2006). Compressed Sensing. *IEEE Transactions on Information Theory*, vol. 52, pp. 1289–1306.

28. Anisi, M.H., Abdullah, A.H., and Razak, S.A. (2011). Energy-Efficient Data Collection in Wireless Sensor Networks. *Wireless Sensor Network*, vol. 3, pp. 329–333.

29. Kulik, J., Heinzelman, W.R., and Balakrishnan, H. (2002). Negotiation-Based Protocols for Disseminating Information in Wireless Sensor Networks. *Wireless Networks*, vol. 8, pp. 169–185.

30. Alippi, C., Anastasi, G., Francesco, M.D., and Roveri, M. (2009). Energy Management in Wireless Sensor Networks with Energy-Hungry Sensors. *IEEE Instrumentation and Measurement Magazine*, vol. 12, no. 2, pp. 16–23.

31. Krishnamachari, B., and Heidemann, J. (2004). Application Specific Modeling of Information Routing in Wireless Sensor Networks. In *Proceedings of IEEE International Performance, Computing and Communications Conference*, vol. 23, pp. 717–722.

32. Heinzelman, W.R., Chandrakasan, A.P., and Balakrishnan, H. (2002). An Application-Specific Protocol Architecture for Wireless Microsensor Networks. *IEEE Transactions on Wireless Communications*, pp. 660–670.

33. Xiangning, F., and Yulin, S. (2007). Improvement on LEACH Protocol of Wireless Sensor Network. In *Proceedings of International Conference on Sensor Technologies and Applications*, pp. 260–264.

34. Loscri V., Morabito G., and Marano S. (2005). A Two-Levels Hierarchy for Low-Energy Adaptive Clustering Hierarchy (TL-LEACH). In *Proceedings of IEEE Vehicular Technology Conference*, Dallas, pp. 1809–1813.

35. Liu, Y., Zhao, Y., and Gao, J. (2009). A New Clustering Mechanism Based on LEACH Protocol. In *Proceedings of International Joint Conference on Artificial Intelligence*, Pasadena, CA, pp. 715–718.

36. Yassein, M.B., Al-zou'bi, A., Khamayseh, Y., and Mardini, W. (2009). Improvement on LEACH Protocol of Wireless Sensor Network (VLEACH). *International Journal of Digital Content Technology and Its Applications*, vol. 3, no. 2, pp. 132–136.

37. Younis, O., and Fahmy, S. (2004). HEED: A Hybrid, Energy-Efficient, Distributed Clustering Approach for Ad Hoc Sensor Networks. *IEEE Transactions on Mobile Computing*, vol. 3, no. 4, pp. 366–379.

38. Chatterjea, S., and Havinga, P. (2003). A Dynamic Data Aggregation Scheme for Wireless Sensor Networks. In *Proceedings of 14th Workshop on Circuits, Systems and Signal Processing*, Veldhoven, The Netherlands, pp. 1–7.

39. Chatterjea, S., and Havinga, P. (2003). CLUDDA—Clustered Diffusion with Dynamic Data Aggregation. In *Proceedings of 8th Cabernet Radicals Workshop*, Ajaccio, France, pp. 1201–1210.

40. Jina, Y., Wanga, L., Kimb, Y., and Yanga, X. (2008). EEMC: An Energy-Efficient Multi-Level Clustering Algorithm for Large-Scale Wireless Sensor Networks. *Computer Networks*, vol. 52, no. 3, pp. 542–562.

41. Lindsey, S., Raghavendra, C., and Sivalingam, K.M. (2002). Data Gathering Algorithms in Sensor Networks Using Energy Metrics. *IEEE Transactions on Parallel and Distributed Systems*, vol. 13, no. 9, pp. 924–935.

42. Ding, M., Cheng, X., and Xue, G. (2003). Aggregation Tree Construction in Sensor Networks. In *Proceedings of IEEE 58th Vehicular Technology Conference*, Orlando, FL, vol. 4, no. 4, pp. 2168–2172.

43. Tan, H.O., and Korpeoglu, I. (2003). Power Efficient Data Gathering and Aggregation in Wireless Sensor Networks. *SIGMOD Record*, vol. 32, no. 4, pp. 66–71.

44. Satapathy, S.S., and Sarma, N. (2006). TREEPSI: TRee Based Energy Efficient Protocol for Sensor Information. In *Proceedings of IEEE International Conference on Wireless and Optical Communications Networks*, Bangalore, India, pp. 1–4.

45. Yang, Y., Wang, X., Zhu, S., and Cao, G. (2006). SDAP: A Secure Hop-by-Hop Data Aggregation Protocol for Sensor Networks. In *Proceedings of 7th ACM International Symposium on Mobile Ad Hoc Networking and Computing*, Florence, Italy, pp. 356–367.

46. Madden, S., Franklin, M.J., Hellerstein, J.M., and Hong, W. (2002). TAG: A Tiny Aggregation Service for Ad-Hoc Sensor Networks. In *Proceedings of 5th Symposium on Operating Systems Design and Implementation*, vol. 36, pp. 131–146.

47. Lee, W.M., and Wong, V.W.S. (2006). E-Span and LPT for Data Aggregation in Wireless Sensor Networks. *Computer Communications*, vol. 29, no. 13–14, pp. 2506–2520.

48. Vaidhyanathan, K., Sur, S., Narravula, S., and Sinha, P. (2004). Data Aggregation Techniques for Sensor Networks, Technical Report, OSU-CISRC-11/04-TR60. Ohio State University, Columbus, pp. 1–9.

49. Manjhi, A., Nath, S., and Gibbons, P.B. (2005). Tributaries and Deltas: Efficient and Robust Aggregation in Sensor Network Stream. In *Proceedings of 2005 ACM SIGMOD International Conference on Management of Data*, Baltimore, pp. 287–298.

50. Kalpakis, K., Dasgupta, K., and Namjoshi, P. (2003). Efficient Algorithms for Maximum Lifetime Data Gathering and Aggregation in Wireless Sensor Networks. *Computer Networks*, vol. 42, no. 6, pp. 697–716.

51. Xue, Y., Cui, Y., and Nahrstedt, K. (2005). Maximizing Lifetime for Data Aggregation in Wireless Sensor Networks. *ACM/Kluwer Mobile Networks and Applications (MONET): Special Issue on Energy Constraints and Lifetime Performance in Wireless Sensor Networks*, pp. 853–864.

52. Cormen, T.H., Leiserson, C.E., Rivest, R.L., and Stein, C. (2001). *Introduction to Algorithms*, 2nd ed. MIT Press and McGraw-Hill, Cambridge, MA, pp. 788–789.

53. Hong, B., and Prasanna, V.K. (2004). Optimizing System Lifetime for Data Gathering in Networked Sensor Systems. In *Proceedings of Workshop on Algorithms for Wireless and Ad-Hoc Networks*, Boston, pp. 1–10.

54. Sadagopan, N., and Krishnamachari, B. (2004). Maximizing Data Extraction in Energy-Limited Sensor Networks. In *Proceedings of IEEE INFOCOM 2004*, vol. 3, pp. 1717–1727.

55. Ordonez, F., and Krishnamachari, B. (2004). Optimal Information Extraction in Energy-Limited Wireless Sensor Networks. *IEEE Journal on Selected Areas in Communications*, vol. 22, no. 6, pp. 1121–1129.

56. Fonoage, M., Cardei, M., and Ambrose, A. (2010). A QoS Based Routing Protocol for Wireless Sensor Networks. In *Proceedings of 29th IEEE International Performance Computing and Communications Conference*, Albuquerque, NM, December 9–11, 2010, pp. 122–129.

57. Othman, J.B., and Yahya, B. (2010). Energy Efficient and QoS Based Routing Protocol for Wireless Sensor Network. *IEEE Journals of Parallel and Distributed Computing*, vol. 70, pp. 849–857.

58. He, T., Blum, B.M., Stankovic, J.A., and Abdelzaher, T. (2004). AIDA: Adaptive Application Independent Data Aggregation in Wireless Sensor Networks. *ACM Transactions on Embedded Computing Systems*, vol. 3, no. 2, pp. 426–457.

59. Jain, A., Chang, E.Y., and Wang, Y.F. (2004). Adaptive Stream Resource Management Using Kalman Filters. In *Proceedings of ACM International Conference on Management of Data*, Paris, France, pp. 11–22.

60. Kanagal, B., and Deshpande, A. (2008). Online Filtering, Smoothing and Probabilistic Modeling of Streaming Data. In *Proceedings of 24th International Conference on Data Engineering*, Cancún, México, pp. 1160–1169.

61. Chu, D., Deshpande, A., Hellerstein, J.M., and Hong, W. (2006). Approximate Data Collection in Sensor Networks Using Probabilistic Models. In *Proceedings of 22nd International Conference on Data Engineering*, Atlanta, GA, p. 48.

62. Tulone, D., and Madden, S. (2006). PAQ: Time Series Forecasting for Approximate Query Answering in Sensor Networks. In *Proceedings of 3rd European Conference on Wireless Sensor Networks*, Zurich, Switzerland, pp. 21–37.

63. Tulone, D., and Madden, S. (2006). An Energy-Efficient Querying Framework in Sensor Networks for Detecting Node Similarities. In *Proceedings of 9th International ACM Symposium on Modeling, Analysis and Simulation of Wireless and Mobile Systems*, Torremolinos, Malaga, Spain, pp. 291–300.

64. Borgne, Y.L., Santini, S., and Bontempi, G. (2007). Adaptive Model Selection for Time Series Prediction in Wireless Sensor Networks. *Signal Processing*, vol. 87, no. 12, pp. 3010–3020.

65. Goel, S., and Imielinski, T. (2001). Prediction-Based Monitoring in Sensor Networks: Taking Lessons from MPEG. *ACM SIGCOMM Computer Communication Review: Special Issue on Wireless Extensions to the Internet*, vol. 31, no. 5, pp. 82–98.

66. TinyOs: An Operating System for Networked Sensors. http://tinyos.millenium.berkeley.edu.

67. Goel, S., Passarella, A., and Imielinski, T. (2006). Using Buddies to Live Longer in a Boring World. In *Proceedings of 4th Annual IEEE International Conference on Pervasive Computing and Communications Workshops*, Pisa, Italy, pp. 342–346.

68. Schurgers, C., Tsiatsis, V., Ganeriwal, S., and Srivastava, M. (2002). Optimizing Sensor Networks in the Energy-Latency-Density Space. *IEEE Transactions on Mobile Computing*, vol. 1, no. 1, pp. 70–80.

69. Xu, Y., Heidemann, J., and Estrin, D. (2001). Geography-Informed Energy Conservation for Ad Hoc. *Proceedings of 7th Annual International Conference on Mobile Computing and Networking*, Rome, Italy, pp. 70–84.

70. Godfrey, P.B., and Ratajczak, D. (2004). Naps: Scalable, Robust Topology Management in Wireless Ad Hoc Networks. In *Proceedings of 3rd International Symposium on Information Processing in Sensor Networks*, Berkeley, CA, pp. 443–451.

71. Bennett, F., Clarke, D., Evans, J.B., Hopper, A., Jones, A., and Leask, D. (1997). Piconet: Embedded Mobile Networking. *IEEE Personal Communications Magazine*, vol. 4, no. 5, pp. 8–15.

72. Haartsen, J.C. (2000). The Bluetooth Radio System. *IEEE Personal Communications Magazine*, pp. 28–36.

73. Ye, W., Heidemann, J., and Estrin, D. (2002). An Energy-Efficient MAC Protocol for Wireless Sensor Networks. In *Proceedings of 21st Annual Joint Conference of the IEEE Computer and Communications Societies (INFOCOM 2002)*, New York, vol. 3, pp. 1567–1576.

74. Ye, W., Heidemann, J., and Estrin, D. (2004). Medium Access Control with Coordinated, Adaptive Sleeping for Wireless Sensor Networks. *IEEE/ACM Transactions on Networking*, vol. 12, no. 3, pp. 493–506.

75. Wang, Z.M., Basagni, S., Melachrinoudis, E., and Petrioli, C. (2005). Exploiting Sink Mobility for Maximizing Sensor Networks Lifetime. In *Proceedings of 38th Hawaii International Conference on System Sciences*, Big Island, HI, pp. 287.1–287.9.

76. Rao, J., Wu, T., and Biswas, S. (2008). Network-Assisted Sink Navigation Protocols for Data Harvesting in Sensor Networks. In *Proceedings of 2008 IEEE Conference on Wireless Communications and Networking*, Las Vegas, NV, pp. 2887–2892.

77. Jain, S., Shah, R., Brunette, W., Borriello, G., and Roy, S. (2006). Exploiting Mobility for Energy Efficient Data Collection in Wireless Sensor Networks. *ACM/Springer Mobile Networks and Applications*, vol. 11, no. 3, pp. 327–339.
78. Shah, R.C., Roy, S., Jain, S., and Brunette, W. (2003). Data Mules: Modeling a Three-Tier Architecture for Sparse Sensor Networks. In *Proceedings of 2nd ACM International Workshop on Wireless Sensor Networks and Applications*, San Diego, CA, pp. 30–41.

Chapter 12

A Pairwise Key Distribution Mechanism and Distributed Trust Evaluation Model for Secure Data Aggregation in Mobile Sensor Networks

Natarajan Meghanathan

Contents

Abstract

We propose a secure data aggregation (SDA) framework for mobile sensor networks whose topology changes dynamically with time. The SDA framework (designed to be resilient to both insider and outsider attacks) comprises a pairwise key establishment mechanism that runs along the edges of a data gathering tree and a distributed trust evaluation model that is tightly integrated with the data aggregation process itself. If an aggregator node already shares a secret key with its child node, the two nodes locally coordinate to refresh and establish a new pairwise secret key; otherwise, the aggregator node requests the sink to send a seed-secret-key message that is used as the basis to establish a new pairwise secret key. The trust evaluation model uses the two-sided Grubbs' test to identify outlier data in the periodic beacons collected from the child nodes and neighbor nodes. Once the estimated trust score for a neighbor node falls below a threshold, the sensor node "locally" classifies its neighbor node as a compromised or faulty (CF) node, and discards the data or aggregated data received from the CF node. This way, the erroneous data generated by the CF nodes can be filtered at various levels of the data gathering tree and is prevented from reaching the root node (sink node). Finally, we assess the effectiveness of our trust evaluation model through a comprehensive simulation study.

12.1 Introduction

A wireless sensor network (WSN) comprises hundreds to thousands of sensor nodes that collect data from the environment for a given task, and one or more sinks (aka base stations [BSs]) responsible for administering and collecting data from the sensor nodes [1]. WSNs are often deployed to detect a parameter or event of common interest (e.g., temperature, fire, and intrusion), and the sink needs only one representative data of the entire area being monitored. Hence, instead of requiring the sensor nodes to individually send their data (either directly or through multihop paths), it would be more efficient (resource-wise) to gather data from all of these sensor nodes and send only one aggregated version of the data (say, the minimum, maximum, or sum) to the sink. In this context, several data aggregation protocols (e.g., [2–4]) have been proposed for WSNs to eliminate the redundancy in data transmission and thereby reduce the communication and energy overhead at the sensor nodes.

Data aggregation in WSNs is typically conducted using a tree topology, as it is the most energy efficient in terms of the number of link transmissions; there is only one path from any node to the root leader node (no scope for duplicate packets). Each intermediate nonleaf node in a data gathering tree (DG-tree) acts as an aggregator, fusing the data collected from its immediate child nodes and forwarding the aggregated data to its own upstream parent node in the tree. This way, data is processed and fused at several hops on the way to the leader node, which eventually forwards the aggregated data to the sink.

However, hop-by-hop aggregation of data along the tree is prone to false data injection attacks. Once under his control, an adversary can reprogram a sensor node with malicious code to disrupt the normal functioning of the network. For example, a compromised sensor node could inject one or more spurious data packets that could corrupt the aggregated data that is on its way further up a DA-tree. Though several works in the literature (e.g., [5,6]) have addressed the problem of secure data aggregation, their focus has been on static sensor networks wherein the nodes do not move.

In this research, we focus on secure data aggregation in wireless mobile sensor networks (WMSNs), wherein the sensor nodes move randomly, independent of each other. With the proliferation of electronic devices such as smart phones and personal digital assistants that are embedded with sensors to measure the temperature, location, humidity, and other vital parameters of interest, there has been considerable interest in the sensor community to study data aggregation in the presence of node mobility. Due to the dynamically changing topology, the communication protocols and their secure variants developed for static sensor networks are not directly applicable for WMSNs. Nevertheless, several network disruption attacks (e.g., node capture and false packet injection), packet interception attacks, and denial-of-service attacks (e.g., energy depletion) are very much possible in WMSNs. In this context, we propose to develop a secure data aggregation framework for WMSNs that comprises a rigorous trust evaluation model and pairwise secret key establishment process for the mobile sensor nodes. Ours will be the first such comprehensive secure data aggregation framework for WMSNs. Existing work, if any exists [7], has dealt with the above two mechanisms only in an isolated fashion, and not together.

In the context of communications, *security* more often refers to preserving the privacy, confidentiality, integrity, and authenticity of the information being communicated [8], whereas *trust* is more like a measure of the credibility level of the node from which the information originates [9]. More specifically, our notion of trust here is like a reputation score, assigned by a node to each of its neighbor nodes based on the sequence of beacons (data) received from the neighbor nodes. We assume a compromised node (due to security attacks) or faulty node (due to some repair), referred to as a CF node, generates data in a range of values that is much broader than the range of values generated by normal nodes (non-CF nodes). Ours is a distributed trust evaluation framework, wherein each node i maintains an estimated average trust score for a neighbor node j based on the data received from node j and the rest of node i's neighbors over a period of time. If the estimated average trust score calculated by node i for a neighbor node j falls below a threshold, then node i does not consider data received from the neighbor node j for aggregation. Thus, while the pairwise key establishment framework vouches for secure aggregation with respect to confidentiality, integrity, and authentication, the distributed trust evaluation model facilitates a node to decide whether to accept data originating from a neighbor node for aggregation.

The following are the objectives of the proposed secure data aggregation (SDA) framework for WMSNs: (1) protect from outsider attacks through secure, energy-efficient in-network data aggregation that provides confidentiality, integrity, and authentication; (2) protect from insider attacks through an embedded trust evaluation model that can be used to detect and mitigate network disruption and energy depletion attacks, as well as identify malicious compromised nodes and faulty nodes; and (3) maximize the number of pairwise secret keys established between sensor nodes, facilitated through node mobility and communication as part of data aggregation.

The two key characteristic features that are embedded within the proposed SDA framework are as follows:

1. Pairwise key distribution: SDA facilitates establishment of pairwise secret keys between as many sensor nodes as possible. and to be robust from node capture and packet interception attacks, the pairwise secret keys need to be refreshed by the concerned nodes every time a new data aggregation tree (DA-tree) is set up. With node mobility [10], we foresee frequent reconfiguration of the DA-trees, and hence scope for establishment and renewal of pairwise secret keys between several node pairs.

2. Trust evaluation of sensor nodes: With the trust level of the sensor nodes evaluated as part of the data aggregation process itself, the SDA framework is robust to network disruption and denial-of-service attacks without any additional communication overhead other than that incurred for the secure data aggregation process designed to provide confidentiality, integrity, and authentication. Note that in this chapter, the terms *data aggregation* and *data gathering*, as well as the acronyms *DA-tree* and *DG-tree*, are used interchangeably. They mean the same thing.

The rest of the chapter is organized as follows: Section 12.2 reviews the literature on pairwise key establishment mechanisms for static and mobile sensor networks and outlines the unique characteristics of our proposed mechanism. Section 12.3 presents the design of our proposed DA-tree-based pairwise key establishment mechanism for mobile sensor networks. Section 12.4 presents the distributed trust evaluation model. Section 12.5 presents a detailed simulation study of the trust evaluation model with respect to stability and minimum-distance spanning tree–based data gathering algorithms and various parameters of the model. Section 12.6 outlines tasks for future work. Section 12.7 concludes the chapter.

12.2 Related Work on Pairwise Key Establishment and Our Contribution

Most of the literature available for pairwise key establishment in sensor networks corresponds to static sensor networks, where all nodes are static (do not move). We describe below a generic pairwise key establishment model for static sensor networks, followed by a discussion of selected strategies proposed to improve the generic model.

12.2.1 Generic Pairwise Key Establishment Model for Static Sensor Networks

The schemes proposed for static sensor networks go through the following two or three phases: (1) a key predistribution phase, (2) a direct key establishment phase (shared key discovery), followed by (3) a path-based key establishment phase. In the key distribution phase (prior to deployment), each node is assigned a fraction of the keys (called key ring) from a larger key pool with the premise that there is a reasonable chance that two neighbor nodes share at least one common key. After deployment, the neighbor nodes exchange their key rings to find a common key and share it, if any is found. If two neighbor nodes can not agree on a common key as part of the direct key establishment phase, they establish a path through one or more intermediate nodes such that there is a shared key between two successive nodes on the path. A secret key is then established between the neighbor nodes through such a multihop path. The integrity of the nodes constituting the path is critical to the confidentiality of the shared key established between the two neighbor nodes, and may further impact the secret key established in the individual neighborhood of these nodes. Several improvements to the above basic pairwise key establishment model for static sensor networks have been proposed in the literature, and we discuss here some of the representative schemes.

12.2.2 Eschenauer and Gligor Scheme

If the key rings are exchanged (during the key setup phase) using a simple local broadcast, a casual eavesdropper could identify the key sets of all the nodes in the network and could identify an optimal subset of nodes to compromise in order to discover a larger subset of the key pool. To counter such eavesdropper attacks, Eschenauer and Gligor (EG-scheme) [11] proposed a puzzle-based approach (e.g., the Merkle puzzle [12]), wherein a sensor node could issue a puzzle for each key in its key ring to the neighbor nodes. Any neighbor node that responds with the correct answer for a client puzzle is identified as correctly knowing the associated key.

12.2.3 q-Composite Key Scheme

To improve network resilience to node capture attacks, Chan et al. [13] proposed a q-composite key scheme, as a modification to the basic EG-scheme, wherein q common keys, with q > 1, are needed to overlap between the key rings of two nodes to set up a secure direct link. The motivation behind this scheme is that as the number of keys (in the key rings) required for overlap increases, it becomes exponentially harder for an attacker with a given key set to break a link. However, the trade-off is that the number of keys forming the key pool has to be reduced in order to preserve the probability that two nodes share sufficient keys to establish a secure link. This facilitates an attacker to gain a larger sample of the key pool by breaking fewer nodes.

12.2.4 Multipath Key Reinforcement

If two nodes A and B use a common key k in their key rings as the secret key for their communication link, the security of the link could be jeopardized if one or more nodes (other than A or B) that have the secret key k in their key rings are captured. To minimize the consequences of such node capture attacks, Anderson et al. [14] propose a multipath key reinforcement scheme wherein node A discovers multiple node-disjoint paths to its neighbor node B and sends a random sequence of bits (of the same length as the original secret key k), one sequence per node-disjoint path. If $v_1, v_2, \ldots, v_j$ are the random bit sequences sent along j node-disjoint paths from A to B, then the new secret key k' computed and used by both A and B for communication across their link would be a simple XOR of the original secret k with all the random bit sequences $v_1, v_2, \ldots, v_j$. That is, $k' = k \oplus v_1 \oplus v_2 \oplus \ldots \oplus v_j$. Otherwise, unless an attacker eavesdrops on all the j node-disjoint paths, it would be difficult to extract k' from k. However, the longer a node-disjoint path, the more vulnerable the path to reveal a random bit sequence (if the attacker manages to capture at least one intermediate node on a path, he can get access to the random bit sequence sent along that path). Hence, Chan et al. [13] suggest the use of a two-hop multipath scheme for deriving a new secret key from a shared secret key in the key rings. The route discovery overhead associated with setting up multihop node-disjoint paths would also be minimal in the case of a two-hop multipath scheme: the two end nodes can exchange their neighbor lists (in the underlying communication network topology) and choose to route the random bit sequences through those neighbor nodes with which they share a secret key in their key rings.

12.2.5 Issues with Pairwise Key Establishment in Mobile Sensor Networks

There are not many schemes proposed for key distribution in mobile sensor networks. The topology of the mobile sensor networks changes dynamically with time. Due to node mobility, there is a good chance that any two nodes could be neighbors at some point in time. As a result, we may have to run path-based key establishment at any time (and not just after the initial node deployment). For the path-based key establishment phase to be successful, we need connectivity of the underlying network of direct links. It is very difficult to select a proper size for the key rings and the global key pool so that there can be at least one common key between an appreciable number of neighbor nodes to establish the direct links between nodes (as part of the direct key establishment phase) leading to connectivity of the underlying network. In the literature on mobile sensor networks, we came across a location-based strategy [15] that requires sensor nodes to be aware of their location (postdeployment) and use it as part of the pairwise key establishment process. However, it is too energy draining for a sensor node to use GPS kinds of mechanisms to regularly find out about its location.

12.2.6 Unique Characteristics of Our Proposed Pairwise Key Establishment Scheme

There is no need to generate a key ring for every sensor node. Instead, we primarily assume every sensor node to share a secret key with the sink (see Section 12.3.1 for a detailed list of assumptions). Our proposed scheme does not require a sensor node to establish direct link or path-based keys with each of its neighbor nodes. Instead, we first set up a data aggregation tree (DA-tree) of the underlying sensor network and require an aggregator node (intermediate node) of the tree to take up the responsibility of establishing pairwise secret keys with its child nodes, if none exist so far, in coordination with the BS (sink). If an aggregator node already shares a secret key with a child node, it establishes a new secret key using the current shared secret key as the basis. By using the DA-tree as the underlying communication topology for pairwise key establishment, we avoid the communication overhead as part of path-based key establishment in the neighborhood of every sensor node (associated with the pairwise key schemes for static sensor networks). The communication between the aggregator node and the sink is facilitated through the paths that are part of the DA-tree itself, and there is no need to establish separate paths between the sink and each of the aggregator nodes for pairwise key establishment with their child nodes. Moreover, unlike the earlier mechanism proposed in the literature (discussed in Section 12.2.5), our proposed pairwise key establishment mechanism does not require the sensor nodes to be location aware.

12.3 Pairwise Key Establishment along a Data Aggregation Tree

12.3.1 Key Assumptions

The key assumptions are

1. The sink BS is secure and shares a 128-bit secret key with each sensor node.
2. The BS broadcasts any information to one or more sensor nodes in a secure fashion through the well-known μTESLA [16] protocol that also provides authentication.

3. Each node stores a key cache in its memory. The key cache of a node consists of the 128-bit pairwise secret keys the node shares with one or more of the other nodes in the network and the sink.
4. The DA-tree is securely formed each time when needed. There is no scope for attack or impersonation during its construction.

Table 12.1 lists the symbols that are used to describe the pairwise key establishment process.

12.3.2 Sequence of Steps in Pairwise Key Establishment

1. **Construction of the DA-tree:** The sink initiates the construction of the DA-tree by sending a trigger signal to the leader node (the node with the largest available energy) to broadcast a tree construction message to its neighborhood. When a sensor node receives the tree construction message for the first time, it sets the sender of the message as its predecessor upstream node in the tree and broadcasts the message further to its neighbors, and discards any further tree construction messages received. This way, the tree construction message is broadcast by every sensor node (but exactly once), so that it can propagate throughout the network to eventually establish a spanning topology tree rooted at the leader node. The child nodes send a parent-update message to inform their parent node that they have chosen the latter as their immediate upstream node in the DA-tree.
2. **Maintenance of the DA-tree:** Each intermediate node of the DA-tree monitors the links to its downstream nodes (expecting periodic beacons from the neighbor nodes). When a downstream link breaks, the upstream node sends a secure tree error message (encrypts and decrypts every hop on the path with the pairwise secret keys that would have been

Table 12.1 Explanation of Symbols Used to Describe the Pairwise Key Establishment Process

Symbol	Explanation
BS	Base station (aka control center or the sink)
$K_{Secret}^{BS,I}$	Secret key known only to the BS (sink) and a node with identity (ID) I
ID_I	Unique identify for a node I
Nonce(N_{I-J})	Random number (nonce value) generated by a node I to authenticate the response coming from node J; node J has to typically include the incremented nonce value in its response
$RN(I)$	Random number generated by the BS (sink) for a node with identity I
$K_{RN(I)*RN(J)}$	Temporary key that is basically the product of the random numbers generated by the sink for nodes I and J
$K_{Current}^{I-J}$	Pairwise secret key that is currently existing between two nodes I and J
K_{New}^{I-J}	New pairwise secret key between two nodes I and J set up as a result of the key establishment or refreshment and renewal process

established by then for the DA-tree using steps 3–5) to the leader node, which forwards the message further to the sink. The sink initiates another tree construction process (step 1).

3. **DA-notification message from the aggregator node to the sink:** If the aggregator node of the DA-tree does not share a secret key with one or more of the child nodes, then the aggregator node sends a DA-notification message (encrypted with the shared secret key) to the BS, containing the IDs of the child nodes with which it does not yet share a secret key. The DA-notification message also contains a nonce (randomly generated by the aggregator node) that needs to be incremented and included in the seed-secret-key message sent as a response by the sink BS.

4. **Seed-secret-key message from the sink to the aggregator node:** The sink BS responds to the DA-notification message with a seed-secret-key message that contains different components: one for the aggregator node and one for every requested child node (with which the aggregator node needs to establish a new pairwise secret key). The seed-secret-key message component meant for the aggregator node contains 64-bit random numbers generated by the BS (one for each of the aggregator nodes and its requested child nodes), as well as the incremented value of the nonce sent in the DA-notification message. The seed-secret-key message components meant for the individual child nodes include the 64-bit random number generated for the aggregator node as well as the aggregator node ID and a 64-bit random number generated for the individual child node. The seed-secret-key message component meant for a node is encrypted using the secret key shared by the sink BS with the individual node.

5. **Agreement on a new pairwise secret key:** Upon receipt of the seed-secret-key message from the sink BS, an aggregator node forwards to the appropriate child nodes their respective encrypted message components as is. A child node decrements the seed-secret-key component message (addressed for it) using the secret key shared with the BS and ensures that the message component includes the ID of the aggregator node that forwarded the seed-secret-key message. The child node then generates a 128-bit pairwise secret key (generated on the fly by the child node) as well as a nonce value (meant for the aggregator node to later send an incremented nonce value in its response to get authenticated for receiving the new pairwise secret key) and encrypts them using a temporary key, which is basically a product of the random numbers (generated by the sink BS and included in the seed-secret-key message component) for the individual child node and the aggregator node. The child node sends this encrypted pairwise secret key to the aggregator node as part of a new pairwise key establishment message. Upon receiving this message from a child node, the aggregator node computes on its own the temporary key (product of the random number sent to it by the sink BS and the random number meant for the appropriate child node) and decrements the message to extract the new pairwise secret key and the nonce value. The aggregator node responds with the new pairwise key acknowledgment message (encrypted with the new pairwise secret key) containing the incremented nonce value. After validating the incremented nonce value in the acknowledgment message, the child node is convinced of the establishment of the pairwise secret key with the aggregator node.

6. **Refreshment of a pairwise secret key:** If an aggregator node already shares a pairwise secret key with a child node established earlier due to its association in some data gathering tree (DG-tree) of the past, the aggregator node refreshes this association by establishing a new pairwise secret key using the currently shared pairwise secret key as the basis. In this pursuit, the aggregator node sends the child node a pairwise key refresh request message that contains a randomly generated nonce encrypted using the currently shared pairwise secret

key with the child node. The child node decrypts the message using the shared pairwise secret key, and sends the following as part of a pairwise key refresh response message: the incremented value of the nonce received from the aggregator node, a new 128-bit pairwise secret key (generated on the fly by the child node), and a new nonce value (to be used to validate the aggregator node's receipt of the new pairwise secret key)—all encrypted with the most recently shared pairwise secret key. The aggregator node decrypts the pairwise key refresh response message to extract the new pairwise secret key and confirms the receipt of the same by sending back (to the child node) an incremented value of the nonce (generated and included by the child node in the refresh response message) encrypted with the new pairwise secret key sent as part of a Pairwise key refresh acknowledgment message.

12.3.3 Example for Pairwise Key Establishment

We now illustrate the contents of the DA-notification message, seed-secret-key message, and the sequence of messages that will be exchanged to establish or refresh the pairwise secret key between an aggregator node and its child nodes. Note that the DA-notification message contains only the IDs of those child nodes with which the aggregator node does not already share a pairwise secret key. For the sake of illustration, we refer to the aggregator node as A and its child nodes as B, C, and D. We also assume that node A already shares a pairwise secret key with node B, and the key is to be refreshed (Figure 12.1). Node A requests the BS for a seed-secret-key only to establish pairwise secret keys with C and D (Figure 12.2 shows the message exchange between A and D to establish a pairwise secret key).

12.4 Distributed Trust Evaluation Model

We consider spanning tree–based data aggregation for mobile sensor networks. The data gathering tree (DG-tree) is formed using a spanning tree of the network that also includes the sink node as one of the constituent nodes. We run the breadth first search algorithm [17] on the spanning tree, starting from the sink node as the root node (leader node). As the edges of the spanning tree are traversed from the root node, the directions get assigned to the edges, leading to the nodes being classified as intermediate nodes and leaf nodes.

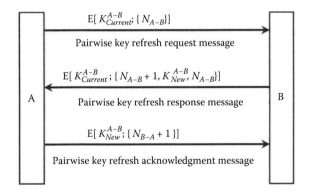

$$E[\,K^{A-B}_{Current}; \{N_{A-B}\}]$$

Pairwise key refresh request message

$$E[\,K^{A-B}_{Current}; \{N_{A-B}+1, K^{A-B}_{New}, N_{A-B}\}]$$

Pairwise key refresh response message

$$E[\,K^{A-B}_{New}; \{N_{B-A}+1\}]$$

Pairwise key refresh acknowledgment message

Figure 12.1 Message exchange: aggregator–child nodes to establish a new pair-wise secret key.

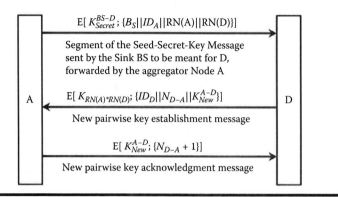

$$E[\,K_{Secret}^{BS-D}\,;\,\{B_S||ID_A||RN(A)||RN(D)\}]$$

Segment of the Seed-Secret-Key Message
sent by the Sink BS to be meant for D,
forwarded by the aggregator Node A

$$E[\,K_{RN(A)*RN(D)}\,;\,\{ID_D||N_{D-A}||K_{New}^{A-D}\}]$$

New pairwise key establishment message

$$E[\,K_{New}^{A-D}\,;\,\{N_{D-A}+1\}]$$

New pairwise key acknowledgment message

Figure 12.2 Message exchange: aggregator–child nodes to refresh/renew a pair-wise secret key.

Data aggregation starts from the leaf nodes. The nodes in the DG-tree are classified according to their position in the tree. The height of the DG-tree is the distance from the sink node to the leaf node that is located at the farthest distance (measured in terms of the number of edges). The root node is considered to be at level 0, the immediate downstream child nodes of the root node are said to be at level 1, and so on. A leaf node forwards its individual beacon data (also considered to be the aggregated data) to its immediate upstream parent node. For an intermediate node to aggregate data, it should have received the aggregated data from all of its immediate downstream child nodes. The aggregated data (in the case of a child node that is a leaf node, the aggregated data is the beacon data) at an intermediate node is the sum of the aggregated data received from its immediate downstream child nodes and its own beacon data; this aggregated data is forwarded by the intermediate node to its own upstream parent node in the DG-tree.

As the sensor nodes could become compromised (insider attacks) or faulty and generate erroneous data, it is imperative to weed out such corrupt data and facilitate the data aggregation process to bypass the compromised or faulty nodes (referred to as CF nodes). We propose a trust evaluation model that is based on assigning raw trust scores to sensor nodes based on the validity of the data generated by the nodes themselves (test for outliers). We assume a CF node to randomly generate data from a broader range, and that is likely to be different from the range for normal data that is expected of the node. A trust score is assigned for a node based on the outliers detected in its data sequence, and the trust score is dynamically updated for every round of data aggregation. A node whose sequence of data has more outliers, compared to the regular data, is likely to be untrustworthy and is flagged a CF node after its trust score falls below a threshold. As part of our research, we have also analyzed the impact of different operating parameters of the secure data aggregation (SDA) framework on the effectiveness of the trust evaluation model, measured with respect to the number of rounds (median value) incurred to detect the presence of CF nodes, as well as the average value of the aggregated data at the sink (wherein data aggregation occurs in the presence of the CF nodes).

An intermediate node considers the aggregated data received from a downstream child node in its data aggregation calculations only if the downstream child node has not been classified as a CF node. Aggregated data received from a downstream child node that is classified as a CF node is not considered for aggregation. To speed up the execution of the whole SDA framework, once a node is classified as a CF node, even the periodic beacon data received from that node is dropped, and the trust value for the CF child or neighbor node is not calculated.

To keep track of the number of individual nodes that have contributed to the aggregated data received from a downstream child node, we include a *numSDAUsedNodes* field in the header of the aggregated data packet. The value of the *numSDAUsedNodes* field in the header of the aggregated data packet is the sum of the values of the *numSDAUsedNodes* fields in the aggregated data packets received from the immediate downstream child nodes plus 1 (the 1 corresponds to the intermediate parent node that is aggregating the downstream child node data with its own beacon data). The *numSDAUsedNodes* field value for a leaf node is 1.

12.4.1 Example for Data Aggregation in the Presence of CF Nodes

We exemplify the data aggregation process and the treatment meted out to aggregated data received from the CF nodes in Figure 12.3. There are a total of 12 nodes in the network, including the root node of the DG-tree, which is the sink node (shown in green). The regular nodes (non-CF nodes) are shown in light blue, and the CF nodes are shown in orange. The numbers outside the nodes or circles indicate the raw sensed data at the individual sensor nodes and the *numSDAUsedNodes* values in the header of the aggregated data packet at the level corresponding to the node. All leaf nodes have a *numSDAUsedNodes* value of 1. Beacon data or aggregated data from nodes classified as CF nodes are ignored. This could be observed in the calculations done at the intermediate node to the left of the DG-tree in Figure 12.3.

The beacon data (of value 45) collected from a child CF node is ignored while calculating the aggregated data value (82 + 75 + 90 + 87 = 334) at the intermediate node, including the latter's own beacon data (of value 87). The *numSDAUsedNodes* value is 3 + 1 = 4 (where 3 corresponds to the sum of *numSDAUsedNodes* values at the child nodes, which are all leaf nodes, and 1 corresponds to the intermediate node itself). The right part of the DG-tree has no CF nodes; so, the data aggregated at the right child node of the sink node is 93 + 73 + 95 = 261, and the *numSDAUsedNodes* value is 2 + 1 = 3. Note that the sink node rejects the aggregated data value received from the middle child node (second child node, which had generated a data value of 140) that had already been classified as a CF node. The sink node aggregates data from the left and right child nodes (ignoring the aggregated data received from the middle child node) and uses the corresponding *numSDAUsedNodes* to determine the overall average value for the data sensed from the network field. In the above example, the aggregated data value at the sink node is 334 + 261 = 595

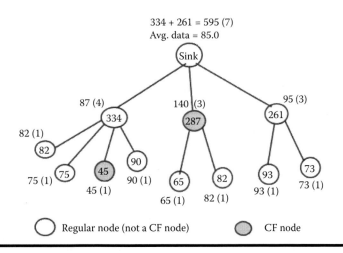

Figure 12.3 **Example to illustrate data aggregation in the presence of CF nodes.**

and the total value for *numSDAUsedNodes* is 4 + 3 = 7, leading to the overall average value for the data sensed from the network field to be computed as 595/7 = 85.0.

12.4.2 Computation of the Raw Trust Score

An intermediate parent node conducts the trust calculations on each of its immediate downstream child nodes for every beacon data received and updates an estimated average trust score (*est. avg. trust score*) based on the raw trust score evaluated on the result of the *TrustComputation* procedure run on the newly (latest) added beacon data (referred to as *insertedData*) to the *BeaconWindow*. The raw trust score for the *insertedData* is computed using the Grubbs' two-sided test [18].

The outlier test for *insertedData* is conducted as follows: We determine the minimum and maximum data (*MinData* and *MaxData*) in *BeaconWindow* as well as the mean and standard deviation (*SD*) of the values. We then determine the appropriate *t*-score for the sample size corresponding to the size of *BeaconWindow*. Table 12.2 lists the *t*-score reference values [18] used in our outlier detection calculations; the confidence interval is 95%. For a sample size (size of the beacon window) that is <5000 and not in Table 12.2, the appropriate *t*-score value is obtained through linear interpolation. Table 12.3 summarizes the notations used in the trust evaluation model.

The threshold value (G_{thresh}) for the Grubbs' test is calculated as follows:

$$G_{thresh} = \frac{|BeaconWindow|-1}{\sqrt{|BeaconWindow|}} * \sqrt{\frac{t\text{-}score * t\text{-}score}{|BeaconWindow|-2+(t\text{-}score * t\text{-}score)}} \qquad (12.1)$$

insertedData is an outlier if it is equal to either *MinData* or *MaxData* and satisfies one of the following two inequalities, as appropriate:

$$\frac{|Mean - MinData|}{SD} > G_{thresh} \quad \text{or} \quad \frac{|Mean - MaxData|}{SD} > G_{thresh} \qquad (12.2)$$

If *insertedData* is classified as an outlier, the *TrustComputation* procedure returns 0; otherwise, it returns 1. An intermediate parent node keeps track of the trust scores (stored in *TrustScoreBuffer*) computed on an intermediate downstream child node in the recent past and during the current association. The size of *TrustScoreBuffer* is limited by the *MaxTrustScoreBufferSize* variable, an input parameter in our simulations.

Table 12.2 *t*-Score Table Used for Two-Sided Grubbs' Test: 95% Confidence Interval

No. of samples	1	2	3	4	5	6	7	8
t-Score	12.706	4.303	3.182	2.776	2.571	2.447	2.365	2.306
No. of samples	9	10	15	20	25	30	40	60
t-Score	2.262	2.228	2.131	2.086	2.060	2.042	2.021	2.000
No. of samples	120	≥5000						
t-Score	1.980	1.960						

Source: G. Frank, *Technometrics*, vol. 11, no. 1, pp. 1–21, 1969. doi: 10.1080/00401706.1969.10490657.

Table 12.3 Explanation of the Notations Used in the Trust Evaluation Model

Notation	Explanation
|*Beacon Window*|	Maximum number of beacons that can be stored at any time for each of its neighbor nodes
t-Score	Standardized score for the data in the beacon window (sampled set); the mean and standard deviation of the actual data set is not known
G_{thresh}	Threshold value (calculated using Equation 12.1) for the standardized score of the data set in the beacon window; if the absolute standardized score of a raw individual data in the beacon window exceeds the threshold, the individual data is considered an outlier (Equation 12.2)
Mean	Mean of the data in the beacon window of a node
SD	Standard deviation of the data in the beacon window of a node
MinData	Minimum value of the data in the beacon window for a neighbor node
MaxData	Maximum value of the data in the beacon window for a neighbor node
insertedData	The latest data added to the beacon window for a neighbor node
historyWeight	Weighing parameter in the range [0 ... 1] used to calculate the weighted average of the estimated trust score for a neighbor node based on the current association with the node, as well as the previous associations
TrustScoreBuffer	Buffer of the raw trust scores maintained by a node for its neighbor node based on both their current association and previous associations
TrustThreshold	Minimum value of the estimated average trust score for a neighbor node, below which the node is classified as a CF (compromised or faulty) node

12.4.3 Computation of the Estimated Average Trust Score

The calculations to update *est. avg. trust score* are conducted every time a raw trust score is returned from the *TrustComputation* procedure and the number of raw trust scores accumulated in *TrustScoreBuffer* is at least half *MaxTrustScoreBufferSize*. *Est. avg. trust score* (maintained at an intermediate parent node for an immediate downstream child node) is the weighted average trust score of two averages of the raw trust scores (0s and 1s) collected in *TrustScoreBuffer*: one average is computed on all the raw trust score data accumulated due to previous associations between the two nodes on DG-trees that had existed before the current DG-tree; the second average is computed on all the raw trust score data accumulated due to the association between the two nodes on the current DG-tree. The weighted average is computed using a parameter *historyWeight*, whose values range from 0 to 1. The formulation is as shown below:

$$\text{Est. avg. trust score} = historyWeight*\text{Average (data from}$$
$$TrustScoreBuffer \text{ based on previous associations)} + (1 - historyWeight)*\text{Average}$$
$$\text{(data from } TrustScoreBuffer \text{ based on current association)} \quad (12.3)$$

If *est. avg. trust score* is below *TrustThreshold*, then the sensor node is classified as a CF node. We record the round number it took to identify a node as a CF node and keep track of the difference in the number of rounds since the node was set to be a CF node. A CF node is used primarily to determine or establish a data gathering tree, and not for data aggregation. The aggregated data or beacon data received from a CF node is not processed by its parent or immediate upstream node in the DG-tree.

12.5 Simulations

Simulations are conducted in a discrete-event simulator developed by us in Java. The same simulator was earlier used by us for research [19] in mobile sensor networks. We conduct simulations on a network topology of dimensions 100 × 100 m. There are a total of 100 nodes in the network, each operating at a fixed transmission range. The sink is located at one end of the network (at [100, 100]) and is also the root of the data gathering tree (DG-tree). The DG-tree used for data aggregation is based on the MST minimum-distance-based spanning tree (MST) or the LET link expiration time–based spanning tree (LET). We use the distributed algorithms proposed in our earlier work [19] to determine the MST- and LET-based DG-trees for mobile sensor networks.

Node mobility is according to the random waypoint model [20], according to which nodes move randomly, independent of each other: To start with, nodes are uniform-randomly distributed throughout the network topology; each node moves to a randomly chosen target location at a velocity uniform-randomly selected from the range $[0 \ldots v_{max}]$, and the node moves to the target location with the chosen velocity; after moving to the target location, the node chooses another value for the velocity from the range $[0 \ldots v_{max}]$ and moves to a new randomly chosen target location. Each node continues to move like this, independent of each other, until the end of the simulation time, which is 1000 s. The values for v_{max} used are 3 m/s (low mobility) and 10 m/s (high mobility). For each combination of v_{max}, number of nodes and sink location, we generate mobility profiles for nodes offline and input them to the data gathering algorithm under execution.

12.5.1 Simulation Parameters and Their Values

The following are the input parameters for the simulation:

- *TransRange*: The fixed transmission range for every sensor node.
- *MeanData*: The mean value of data generated by every sensor node.
- *STDData*: The standard deviation of data generated by every sensor node.
- *MaxBeaconWindowSize*: The maximum size of the beacon window (BW) maintained to store the beacons received from each node in the neighborhood.
- *MaxTrustScoreBufferSize*: The maximum size for the trust score buffer (TSB) maintained at an intermediate parent node for every immediate downstream child node. The TSB is used to store the raw trust score values (0 s and 1 s) calculated based on the beacon data received from the neighbor child nodes at the beginning of each round.
- *TrustThreshold*: The threshold value for the estimated average trust score below which the sensor node is classified as CF node.
- *CFProb*: The probability with which a non-CF node could become compromised or faulty at any round. Once a node becomes a CF node, it continues to remain so until the end of the simulation.

- *MaxCFNodes*: The maximum number of nodes that could become CF nodes.
- *historyWeight*: A weight parameter (ranging from 0 to 1) to capture the trade-off with respect to giving importance to the trust evaluation results collected on a child node at a parent node (in the trust score buffer) during their previous associations versus the trust evaluations during their current association.

The maximum BW size at each sensor node is set to be 10 and 50. The maximum size of the trust buffer (TSB) maintained by a sensor node for each of the other nodes (updated for the child nodes) is set to be 10, 30, and 50. The *check for CF status* is conducted for every round of data aggregation, once the TSB size reaches half of its maximum value (set to one of the above three values). The value for *TrustThreshold* (the minimum value for the estimated average trust score below which the node is classified as a CF node) is set to values of 0.5, 0.7, and 0.9. The *historyWeight* parameter is set to values of 0.3, 0.5, 0.7, 0.9, and 1.0: If the *historyWeight* parameter is set to 1.0, it implies that trust score values determined by a parent node for a child node during their current association would not be considered and the check for CF status will be conducted only based on the trust scores estimated during the previous associations between the two nodes. On the other hand, a *historyWeight* value of 0.3 implies that relatively much larger importance (100% − 30% = 70% weight) will be given to the trust scores determined during the current association between the two nodes. The number of CF nodes is set to values of 20 and 40; as the total number of nodes is 100, this corresponds to operating the network with 20% and 40% of nodes as CF nodes, respectively. Until we have turned on the above-said number of nodes as CF nodes, any non-CF node in the network has an equal chance of becoming a CF node; the probability (*CFProb*) with which any non-CF node can be come a CF node is set to 0.005. Simulations are considered for fixed transmission range per node values of 25 and 35. Though *TrustScoreBuffer* is filled with the raw trust score data computed on the individual data in *BeaconWindow*, the parameters *MaxTrustBufferSize* and *MaxBeaconWindowSize* are treated independently, and the values are assigned individually without any dependency on each other.

Each data point in Figures 12.4 through 12.11 is the average of results obtained from 100 mobility profiles. For each mobility profile, we run the simulation for 1000 s. A simulation (corresponding to either the MST- or LET-based DG-trees) is run for fixed values of *BW Size* (10 and 50), *TSB Size* (10, 30, and 50), *TrustThreshold* (0.5, 0.7, and 0.9), *historyWeight* (0.3, 0.5, 0.7, 0.9, and 1.0), transmission range (25 and 35 m), maximum node velocity (v_{max} = 3 and 10 m/s), and number of CF nodes (20 and 40), resulting in a total of 2*3*3*5*2*2*2 = 720 combinations of scenarios, for which the results are shown in Figures 12.4 through 12.11. We have also collected results for *TrustThreshold* values of 0.6 and 0.8 and *TSB Size* of 70. We do not present the results obtained for these combinations of *TrustThreshold* and *TSB Size* values due to lack of space and the observation of trends similar to those combinations chosen for presentation.

12.5.2 Data Generation Model

The data measured from the network field is assumed to be the temperature of the field. We want only one representative data—either the maximum, minimum, or mean temperature of the network field. In the simulations, we assume the data is sensed at each node and broadcast to its neighbor nodes (within the transmission range) through beacons. If a node is not a CF node (i.e., a benevolent regular node, neither compromised nor faulty), then the temperature data generated by the node is in the range [*MeanData* − *STDData* … *MeanData* + *STDData*]. If a node is a CF node, then the data generated by the node is within the range [0 … 5*MeanData*].

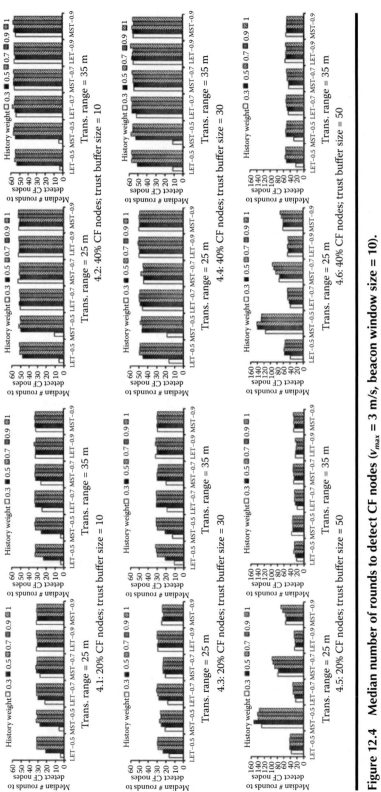

Figure 12.4   Median number of rounds to detect CF nodes (v_{max} = 3 m/s, beacon window size = 10).

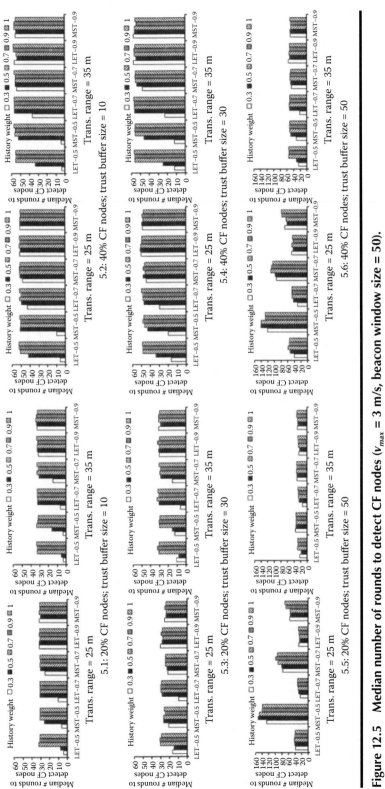

Figure 12.5   Median number of rounds to detect CF nodes (v_{max} = 3 m/s, beacon window size = 50).

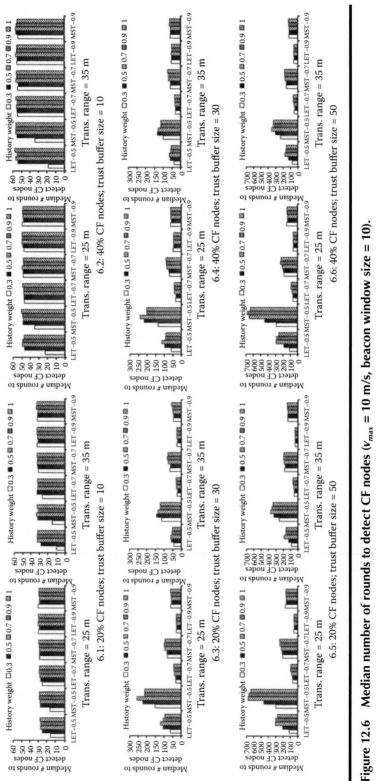

Figure 12.6 Median number of rounds to detect CF nodes (v_{max} = 10 m/s, beacon window size = 10).

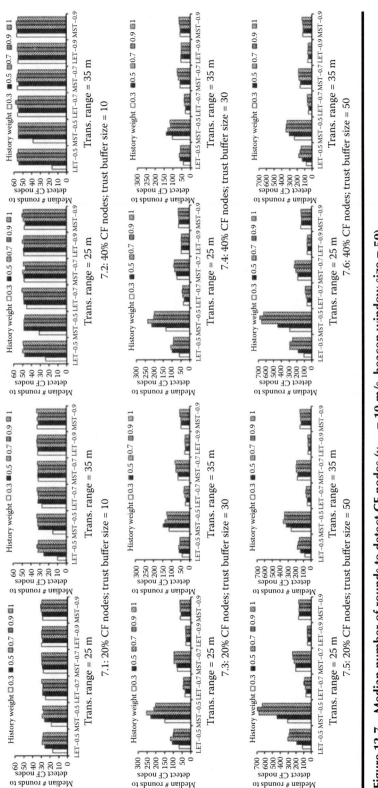

Figure 12.7   Median number of rounds to detect CF nodes (v_{max} = 10 m/s, beacon window size = 50).

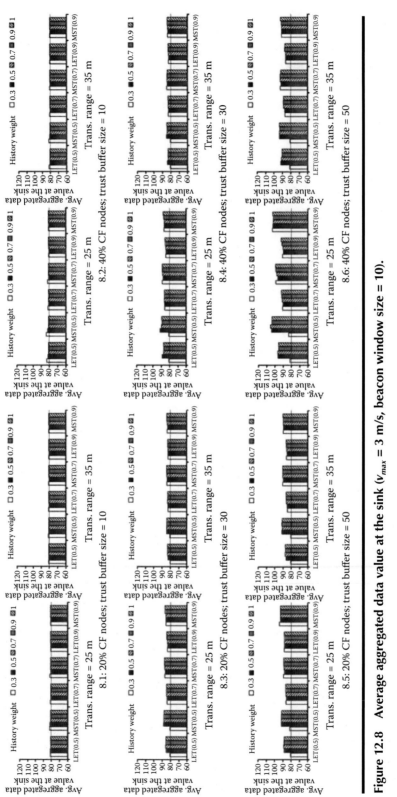

Figure 12.8   Average aggregated data value at the sink (v_{max} = 3 m/s, beacon window size = 10).

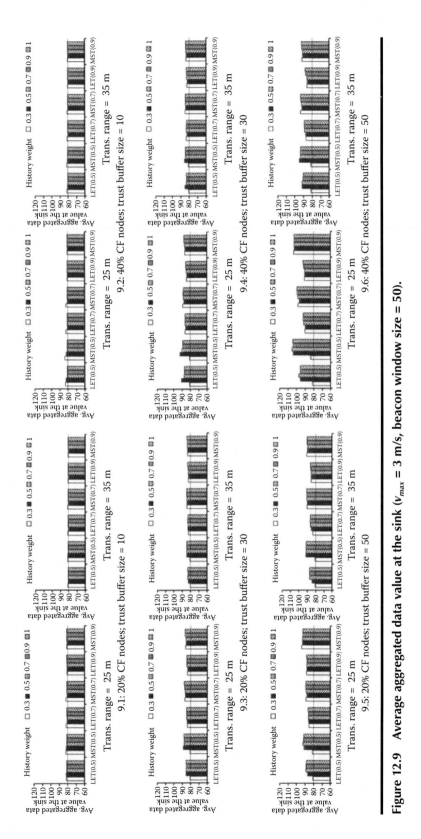

Figure 12.9   Average aggregated data value at the sink (v_{max} = 3 m/s, beacon window size = 50).

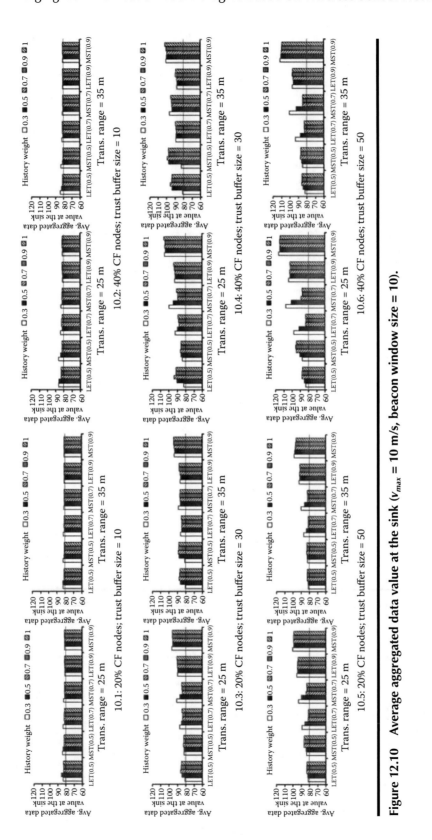

Figure 12.10 Average aggregated data value at the sink (v_{max} = 10 m/s, beacon window size = 10).

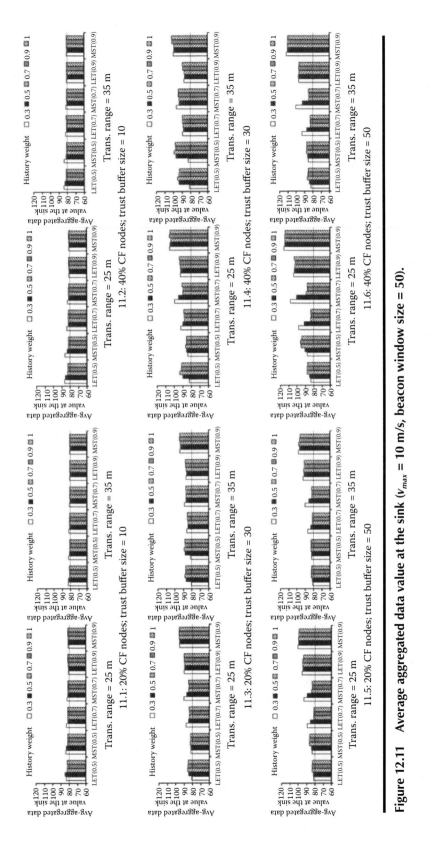

Figure 12.11 Average aggregated data value at the sink (v_{max} = 10 m/s, beacon window size = 50).

For the sake of the simulations, data at a CF-node is generated for sensing by each node according to a uniform distribution with mean 80 and standard deviation 20. Accordingly, for every data point, we generate a random number from the range $x \in [0 \ldots 1]$, and the individual data point generated is then $80 \pm 20{*}x$. Data at a CF node is generated uniform-randomly from the range $[0 \ldots 5{*}80]$.

12.5.3 Selection of Compromised or Faulty (CF) Nodes

The simulations are conducted with 100 nodes (*numNodes*). Among these 100 nodes, a fixed number of nodes (*MaxCFNodes*) are set to become CF nodes (compromised or faulty) nodes at some time instant during the simulation and will continue to remain as CF nodes. Any node could become a CF node. Until the *MaxCFNodes* number of nodes have become CF nodes, we run the *CFEnable* routine at the beginning of each round, starting from round 10 (we wait for at least 10 rounds before setting any node to become a CF node). The probability (*CFProb*) with which a node can become a CF node is an input parameter to our simulation. Until the *MaxCFNodes* number of nodes are compromised or become faulty, any node can become a CF node with a probability of *CFProb* when the *CFEnable* routine is run.

The CFEnable routine runs as follows: For every node (node ID 0 to *numNodes*—1) that is not yet compromised or faulty, we generate a random number from 0 to 1. If the random number generated is less than or equal to *CFProb*, then that node is considered to have become a CF node and is added to the list of CF nodes. Any number of nodes (0 or more) could become a CF node during a particular round.

12.5.4 Data Gathering per Round

For each mobility profile, we run the simulation for 1000 s. There are four rounds of data gathering per second, set to be at a fixed rate: the time between two successive rounds of data gathering is 0.25 s. If a DG-tree existed during the previous round, we check whether all the edges in the DG-tree exist during the current round. An edge is said to exist if the Euclidean distance between the two constituent end nodes of the edge is less than or equal to the transmission range of the nodes. If the DG-tree exists during a round, then we are ready for data aggregation. We let each sensor node generate the data, whose values depend on whether the node is a CF node. A node is assumed to broadcast the generated data to its neighborhood, and thereby, a parent node receives a beacon from each of its child neighbor nodes (i.e., the immediate downstream nodes). If the DG-tree that existed during the previous rounds no longer exists, then we determine the appropriate DG-tree (using either the MST-based or LET-based DG-tree algorithm of [19]) and then initiate data aggregation for the current round.

12.5.5 Beacon Management

The newly received beacon from a child node (of a unique node ID) is stored in the *BeaconWindow* (BW) maintained for the particular child downstream node. Before inclusion of the new beacon data, if BW size equals *MaxBeaconWindowSize*, then the earliest recorded data in the BW is purged and the newly received beacon data is appended to the BW. If the BW size is less than *MaxBeaconWindowSize*, then the newly received beacon is simply appended to the BW. The raw sensed data collected from a child neighbor node and accumulated in the BW maintained for the node is used to assess the trust value of the child node.

12.5.6 Performance Metrics

We measure the following for each simulation scenario:

1. *Median value for the number of rounds* to detect the CF nodes (for every CF node correctly identified, we record the difference between the round at which the node is identified to be a CF node and the round at which the node was set as a CF node, and determine the median value of all these differences), averaged over the 100 mobility profiles
2. *Average of the data values collected at the sink node* based on the data aggregated from the non-CF nodes and the number of non-CF nodes that participated in the data aggregation process

12.5.7 Simulation Results and Their Interpretation

The MST-based DG-trees have been observed to be more energy efficient [21], but more unstable, in the presence of node mobility [19]. The LET-spanning tree–based DG-trees are inherently more stable due to the presence of links that are predicted to have a relatively larger lifetime [19].

For a given simulation condition, in the presence of CF nodes, the LET-based DG-trees have been observed to yield more accurate values for the total aggregated data compared to those determined using the MST-based DG-trees. We also observe the LET-based DG-trees to incur a lower median value for the number of rounds to detect the CF nodes (for both 20% and 40% CF node scenarios). For a fixed v_{max}, percent of CF nodes, transmission range, and TSB size, there is no significant impact on the beacon window size (i.e., for BW size values of 10 or 50) on the median number of rounds to detect the CF nodes, as well as on the average value for the data aggregated at the sink in the presence of these CF nodes.

With regards to the impact of the *historyWeight* parameter, we observe that the SDA framework incurs the lowest value for the median number of rounds (to detect CF nodes) when operated with a *historyWeight* parameter value of 0.3. The effect of the *historyWeight* parameter is observed more prominently when operated under lower *TrustThreshold* values (especially for 0.5). This is because when we operate at lower *TrustThreshold* values, it is more likely to take a longer time (number of rounds) for the estimated average trust score to fall below the smaller *TrustThreshold* value; hence, it is critical to quickly recognize the erroneous data generated at the CF nodes and detect their presence in the network so that the aggregated data (if the CF node is an intermediate node) or individual data (if the CF node is a leaf node) received from these nodes can be discarded.

With regards to the impact of the *TrustThreshold* and *TSB Size* parameters, for low-node-mobility scenarios, we observe that the larger the *TrustThreshold*, the easier it is to identify and declare a compromised or faulty node a CF node. For smaller values of the *TrustThreshold* parameter, we incur more rounds to detect the presence of CF nodes, as more samples need to be collected when the node is indeed a CF node, especially for high values of the *TSB Size*. Note that the check for CF status is done only when the number of raw trust score values stored in the TSB is at least half of the maximum *TSB Size*. Until then, the presence of CF nodes only corrupts the values for the aggregated data. Hence, it is more sensible to operate the network at larger *TrustThreshold* values and lower *TSB Size*.

For low-node-mobility scenarios, we observe that the median number of rounds to detect CF nodes does not depend much on the values for the *TrustThreshold* and gets slightly lower for larger values of *TSB Size*. As the DG-trees are likely to exist for a longer time in low-node-mobility scenarios, the trust estimates are more accurately captured and the CF nodes are more easily identified.

There is no need to go through several rounds of data gathering and several refreshes to the TSB before identifying the CF nodes. In other words, there is no significant impact of the *TrustThreshold* and the *TSB Size* on the median number of rounds to detect the CF nodes, and the detection is done much earlier in low-node-mobility scenarios than in high-node-mobility scenarios (see analysis below). As a result, there is also less corruption in the data at low-node-mobility scenarios.

For larger values of node velocity, the DG-trees are more unstable. As a result, the parent–child association between any two nodes does not last for a longer time. Since the trust calculations are conducted only after the trust buffer has accumulated a reasonable number of trust estimate values (depending on the trust buffer size), the median number of rounds to detect the presence of CF nodes shoots up to significantly high values in the presence of node mobility, especially for the MST-DG trees (could be as large as six or seven times the values obtained for lower node mobility). The LET-DG trees are more stable, and hence incur relatively lower values for the median number of rounds to detect the CF nodes. The longer it takes to detect the presence of CF nodes, the more the corruption to the aggregated data value at the sink.

For both types of DG-trees, it is imperative that we operate at a low trust buffer size in the presence of node mobility. Given the negative influence of TSB size on the median numbers of rounds and data corruption, for a given TSB size, it is more logical to operate the network at larger values of *TrustThreshold* in high-node-mobility conditions, so that the compromised or faulty nodes can be swiftly identified when the estimated average trust score falls below the much larger *TrustThreshold* values. Also, at larger values of the maximum TSB size, we are more likely to end up with corrupt data because the estimated average trust score is begun to be computed only when the TSB size reaches at least half the maximum buffer size.

Note that there is no distinction among nodes (whether or not CF nodes) at the time of formation of the DG-trees. Only at the time of data aggregation, we take into consideration whether the data (individual data in the case of the leaf node or aggregated data in the case of the intermediate node) is received from a CF node, and accordingly, a decision to consider or discard the data is made. At larger values of the transmission range (35 m), we observe the connectivity of the network to be more likely maintained among the nodes (CF node to CF node, CF node to non-CF node, non-CF node to non-CF node), and hence we have good chances of maintaining a relatively more accurate trust buffer at every parent node for each of its immediate downstream child nodes. This translates to an earlier detection of the presence of CF nodes, even for larger TSB sizes (that require a longer wait time to detect the presence of CF nodes). On the other hand, with a lower transmission range of 25 m, operating the network at larger *TrustThreshold* as well as a larger TSB size leads to more data corruption. Thus, the median number of rounds to detect CF nodes is lower for transmission ranges of 35 m, compared to that incurred with 25 m. In the case of 25 m transmission range, we observe relatively less network connectivity in the presence of CF nodes, as well as node mobility. However, note that when energy loss due to transmissions and receptions is considered, the larger the transmission range, the larger the energy loss. Hence, we could cite this as an energy–trust trade-off.

For a larger number of CF nodes, we would naturally expect to incur a larger value for the median number of rounds to detect the CF nodes. This is the case observed with TSB size values of 10 and 30. However, for larger values of the TSB size (50), we observe them to sometimes incur a lower value for the median number of rounds, especially when operated with 40% of CF nodes (compared to 20% of CF nodes). This could be attributed to the observation that with a larger TSB size, we could effectively detect the presence of a bunch of CF nodes in a round of data aggregation, as well as discard data from the entire subtree rooted at a CF node (the subtree could in turn have one or more CF nodes). As a result, for larger TSB sizes, it is possible that the CF nodes are identified as a bunch and data received from them (either in individual or aggregated form) is discarded. However, the trade-off is that until the estimated average trust score calculations are begun, the presence of more CF nodes (especially in the case of MST-DG trees at high-node-mobility conditions, as the parent–child node associations are short-lived) could contribute to a relatively larger corruption of the data.

12.5.8 Significant Observations from the Simulation Results

We observe the stability-oriented link expiration time (LET)–driven spanning tree–based data gathering trees to be more suitable to assess the trust levels of the nodes and aggregate data using communication topology that bypasses (discards data) the compromised or faulty nodes. The median number of rounds to detect CF nodes (and likewise the corruption in the aggregated data) incurred for the MST-based DG-trees is much larger than that incurred for the LET-based DG-trees, especially in high-node-mobility scenarios.

The beacon window size used to assess the trust levels of the nodes is (by itself) not a significant parameter to influence the calculations of the estimated average weighted trust score; as long as we have a reasonable value for the beacon window size (values of 10 and 50 are used in the simulations), we will be able to effectively assess the trust levels of the nodes. In the presence of moderate to high node mobility, even operating with a BW of size 10 would be more than sufficient to effectively assess the trust level of neighbor nodes.

The trust buffer size (the buffer that stores the raw trust scores estimated for the nodes at subsequent rounds of the current association and previous associations, if any exist, for data gathering) plays a significant role. Larger trust buffer size could contribute to collectively detecting one or more CF nodes in a single round itself (an intermediate CF node and its subtree containing one or more CF nodes), leading to an overall reduced value for the median number of rounds to detect CF nodes, but the data aggregated at the sink is also more likely to be corrupted (at least until the TB size reaches half of its maximum size for the *check for CF status* to begin).

Though operating at a larger transmission range results in more energy consumption and premature node failures (in energy-constrained scenarios), we observe larger transmission ranges to be effective to maintain the connectivity of the data gathering trees (especially in the presence of a larger percentage of CF nodes and high-mobility scenarios), leading to less time to detect CF nodes, as well as aggregate relatively less corrupted data at the sink.

Though it is essential to recognize the trust assessment data collected during the previous associations of two nodes (a parent node and its child node), relatively more weight has to be given to the trust data collected during the current association in a DG-tree for effective trust assessment. In our simulations, we observe 70% weight to current association and 30% weight to history to be more appropriate.

The *TrustThreshold* value (below which a node is classified as a CF node) has to be significantly larger (0.7–0.9) to quickly detect the CF nodes, especially in highly mobile scenarios. Lower *TrustThreshold* values contribute to much higher values for the median number of rounds to detect CF nodes, especially when operated with larger TSB size. Hence, it is more sensible to operate with a larger *TrustThreshold* value and lower TSB size in dynamically changing communication topologies.

12.6 Future Work

As part of future work, we will design the structure of the messages (incorporating all the fields envisioned in the design proposed earlier) exchanged as part of the pairwise key establishment mechanism for the SDA framework and implement it in the environment of mobile sensor networks, including the energy consumption calculations. We will evaluate the energy consumption overhead associated with the exchanges of messages as part of the pairwise key establishment mechanism. We will also consider the impact of node mobility on the establishment of pairwise keys between any two nodes in the network. While node mobility could increase the number of node pairs between which secret keys are newly established, it could also increase the energy consumption overhead due to the exchange of messages to discover data gathering trees as well as establish or renew secret keys.

Note that the underlying assumption behind the construction of the DA-tree is that it is constructed in a secure fashion and is not compromised of nodes that could launch routing attacks to attract traffic (like in warmhole attacks or sinkhole attacks) or generate traffic with multiple identities (like in Sybil attacks). The proposed trust evaluation model cannot handle nodes that could launch the above routing attacks, as the model assumes that the DA-tree is securely constructed without involving nodes launching these attacks. We have illustrated the use of our recently proposed protocols to construct DA-trees in a distributed fashion. Developing a protocol to construct DA-trees that are robust to routing attacks (like warmhole, sinkhole, or Sybil attacks) in mobile sensor networks is beyond the scope of this chapter. We have not come across any such protocol so far for mobile sensor networks. As part of future work, we will develop secure data aggregation protocols that are robust to these routing attacks during the process of constructing the DA-tree itself, and evaluate the performance of the trust evaluation model in the presence of compromised nodes that launch the routing attacks.

12.7 Conclusions

The proposed secure data aggregation (SDA) framework will be the first such comprehensive framework for defense against both insider attacks (trust evaluation model) and outsider attacks (pairwise key establishment mechanism) for mobile sensor networks. The SDA framework provides the capability for WMSNs to develop autonomic and innate defense (self-discovery and assessment) capabilities to detect adversarial actions, as well as detect and be resistant to network disruption attacks and energy depletion attacks (e.g., denial-of-service and false packet injection attacks). The proposed pairwise key establishment mechanism uses the DA-tree itself as the underlying communication topology to establish pairwise keys and does not require the need for generating a key ring for each node (predeployment) and path-based key establishment (postdeployment). With node mobility, the DA-trees change with time, and we anticipate the SDA framework to establish as many pairwise secret keys as possible between the sensor nodes just based on their association in a DA-tree. Stringent mechanisms are incorporated in the design of the pairwise key establishment mechanism to ensure confidentiality, integrity, and authentication of data during aggregation. Similarly, the trust evaluation model is designed to swiftly identify the CF nodes and bypass the data aggregation process around these nodes. Simulation studies indicate that stability-based DA-trees and data gathered as part of the most recent associations between two nodes are to be preferred for effectively assessing the trust levels of the sensor nodes. With regards to the various operating parameters of the trust evaluation model, we observe the trust buffer size, trust threshold, and transmission range of the sensor nodes to play significant roles (whereas little impact is observed by varying the values of the beacon window size).

We anticipate the contributions of this chapter to lay the groundwork for developing trust scores based on secure data aggregation protocols that can construct the DA-trees in the presence of compromised or faulty nodes (e.g., the CF nodes could be given preference to be accommodated as leaf nodes of the DA-trees so that data originating from the leaf nodes, with trust scores below a threshold, could be simply dropped at the intermediate nodes without much impact on data aggregation). As stable DA-trees have been observed to enhance the accuracy of the outlier detection and trust evaluation mechanisms, we foresee the development of algorithms that give preference for nodes with stable and trust-worthy neighborhoods to be accommodated as intermediate nodes of a DA-tree, and thereby enhance the robustness of data aggregation with regard to both node mobility and node credibility. We also envision the possible use of various categories of machine learning algorithms and data mining techniques for outlier detection as part of trust evaluation frameworks for mobile sensor networks.

References

1. I. Akyildiz, W. Su, Y. Sankarasubramaniam, and E. Cayirci, Wireless Sensor Networks: A Survey, *Computer Networks*, vol. 38, no. 4, pp. 393–422, 2002. doi: 10.1016/S1389-1286(01)00302-4.
2. W. Heinzelman, A. Chandrakasan, and H. Balakarishnan, Energy-Efficient Communication Protocols for Wireless Microsensor Networks, in *Proceedings of the 33rd Hawaii International Conference on Systems Science*, Maui, HI, January 4–7, 2000, pp. 1–10. doi: 10.1109/HICSS.2000.926982.
3. S. Lindsey, C. Raghavendra, and K. M. Sivalingam, Data Gathering Algorithms in Sensor Networks Using Energy Metrics, *IEEE Transactions on Parallel and Distributed Systems*, vol. 13, no. 9, pp. 924–935, September 2002. doi: 10.1109/TPDS.2002.1036066.
4. N. Meghanathan, Grid Block Energy Based Data Gathering Algorithm for Lower Energy*Delay and Longer Lifetime in Wireless Sensor Networks, in *Proceedings of the 1st ISCA International Conference on Sensor Networks and Applications*, San Francisco, CA, November 4–6, 2009, pp. 79–84.
5. Y. Yang, X. Wang, S. Zhu, and G. Cao, SDAP: A Secure Hop-by-Hop Data Aggregation Protocol for Sensor Networks, in *Proceedings of the 7th ACM International Symposium on Mobile Ad Hoc Networking and Computing*, Florence, Italy, May 22–25, 2006, pp. 356–367. doi: 10.1145/1132905.1132944.
6. S. Ozdemir and Y. Xiao, Secure Data Aggregation in Wireless Sensor Networks: A Comprehensive Overview, *Computer Networks*, vol. 53, no. 12, pp. 2022–2037, 2009. doi:10.1016/j.comnet.2009.02.023.
7. S.-I. Huang and S. Shieh, SEA: Secure Encrypted-Data Aggregation in Mobile Wireless Sensor Networks, in *Proceedings of the International Conference on Computational Intelligence and Security*, Harbin, China, December 15–19, 2007, pp. 848–852. doi: 10.1109/CIS.2007.207.
8. A. S. K. Pathan, On the Boundaries of Trust and Security in Computing and Communications Systems, *International Journal of Trust Management in Computing and Communications*, vol. 2, no. 1, pp. 1–6, 2014. doi: 10.1504/IJTMCC.2014.063272.
9. V. Shmatikov and C. Talcott, Reputation-Based Trust Management, *Journal of Computer Security*, vol. 13, no. 1, pp. 167–190, 2005.
10. T. Camp, J. Boleng, and V. Davies, A Survey of Mobility Models for Ad Hoc Network Research, *Wireless Communications and Mobile Computing*, vol. 2, no. 5, pp. 483–502, 2002. doi: 10.1002/wcm.72.
11. L. Eschenauer and V. D. Gligor, A Key-Management Scheme for Distributed Sensor Networks, in *Proceedings of the 9th ACM Conference on Computer and Communication Security*, Washington, DC, November 18–22, 2002, pp. 41–47. doi: 10.1145/586110.586117.
12. R. Merkle, Secure Communication over Insecure Channels, *Connections of the ACM*, vol. 21, no. 4, pp. 294–299, 1978. doi: 10.1145/359460.359473.
13. H. Chan, A. Perrig, and D. Song, Random Key Predistribution Schemes for Sensor Networks, in *Proceedings of the IEEE Symposium on Security and Privacy*, Oakland, CA, May 11–14, 2003, pp. 197–213. doi: 10.1109/SECPRI.2003.1199337.
14. R. J. Anderson, H. Chan, and A. Perrig, Key Infection: Smart Trust for Smart Dust, in *Proceedings of the 12th IEEE International Conference on Network Protocols*, Berlin, Germany, October 5–8, 2004, pp. 206–215. doi: 10.1109/ICNP.2004.1348111.

15. A. K. Das, A Key Establishment Scheme for Mobile Wireless Sensor Networks Using Post-Deployment Knowledge, *International Journal of Computer Networks and Communications*, vol. 3, no. 4, pp. 57–70, 2011. doi: 10.5121/ijcnc.2011.3405.

16. A. Perrig, R. Szewczyk, J. D. Tygar, V. Wen, and D. E. Culler, SPINS: Security Protocols for Sensor Networks, *Wireless Networks*, vol. 8, no. 5, pp. 521–534, 2002. doi: 10.1023/A:1016598314198.

17. T. H. Cormen, C. E. Leiserson, R. L. Rivest, and C. Stein, *Introduction to Algorithms*, 3rd ed., MIT Press, Cambridge, MA, 2009.

18. G. Frank, Procedures for Detecting Outlying Observations in Samples, *Technometrics*, vol. 11, no. 1, pp. 1–21, 1969. doi: 10.1080/00401706.1969.10490657.

19. N. Meghanathan, Data Gathering Algorithms to Optimize Stability-Delay and Node-Network Lifetime for Wireless Mobile Sensor Networks, in *Proceedings of the Fourth International Conference on Sensor Networks and Applications*, New Orleans, LA, November 14–16, 2012, pp. 237–242.

20. C. Bettstetter, H. Hartenstein, and X. Perez-Costa, Stochastic Properties of the Random Waypoint Mobility Model, *Wireless Networks*, vol. 10, no. 5, pp. 555–567, 2004. doi: 10.1023/B:WINE .0000036458.88990.e5.

21. N. Meghanathan, A Comprehensive Review and Performance Analysis of Data Gathering Algorithms for Wireless Sensor Networks, *International Journal of Interdisciplinary Telecommunications and Networking*, vol. 4, no. 2, pp. 1–29, 2012. doi: 10.4018/jitn.2012040101.

Chapter 13

Distributed Data Gathering Algorithms for Mobile Sensor Networks

Natarajan Meghanathan

Contents

Abstract

We propose the design and development of two spanning tree–based distributed data gathering algorithms for wireless mobile sensor networks whose topology changes dynamically with time due to random movement of the sensor nodes. Our first data gathering algorithm is stability-delay oriented, and it is based on the idea of finding a maximum spanning tree on a network graph whose edge weights are predicted link expiration times (LETs). Referred to as the LET-DG tree, the data gathering tree has been observed to be more stable in the presence of node mobility, and also incurs a significantly lower delay per round of data gathering due to the shorter height of the tree with more leaf nodes. However, stability-based data gathering coupled with more leaf nodes has been observed to result in unfair use of certain nodes (the intermediate nodes spend more energy than the leaf nodes), triggering premature node failures, eventually leading to network failure (disconnection of the network of live nodes). As an alternative, we propose a minimum-distance spanning tree (MST)–based data gathering algorithm that is more energy efficient and prolongs the node and network lifetimes, as well as inflicts a lower coverage loss on the underlying network at any time instant—all of these at the cost-frequent tree reconfigurations. The MST-DG trees also incur a significantly longer delay per round, due to their larger height and fewer leaf nodes. We thus observe a complex stability-delay versus node-network lifetime-coverage loss trade-off that has hitherto not been explored in data gathering algorithms for mobile sensor networks.

13.1 Introduction

A wireless sensor network comprises several smart sensor nodes that can gather data about the surrounding environment, as well as process them before propagating to a control center called the sink, from which the end user typically operates to administer the network and access the nodes. Wireless sensor networks have been considered to give unprecedented levels of access to real-time information about the physical world, and the benefits of deploying such networks are widely seen these days. However, in almost all cases, the sensor networks are statically deployed and evaluated, wherein the mobility of the sensors, the users, and the monitored phenomenon are all totally ignored. Wireless mobile sensor networks (WMSNs) are the next logical evolutionary step for sensor networks in which mobility needs to be handled in all its forms. A motivating example could be a network of environmental monitoring sensors, mounted on vehicles, used to monitor pollution levels in a city. In this example, all the entities involved (i.e., the sensors, the users, and the sensed phenomenon) are moving. Likewise, one can conceptualize many such real-time scenarios to deploy sensor networks in which one or more of the participating entities move.

Like their static counterparts, the mobile sensor nodes are likely to be constrained with limited battery charge, memory, and processing capability, as well as operate under a limited transmission range. Two sensor nodes that are outside the transmission range of each other cannot communicate directly. The bandwidth of a WMSN is also expected to be as constrained as that of a static sensor network. Due to all of the above resource and operating constraints, it will not be a viable solution to require every sensor node to directly transmit its data to the sink over a longer distance [1]. Also, if several signals are transmitted at the same time over a longer distance, this could lead to a lot of interference and collisions. Thus, there is a need for employing energy-efficient data gathering algorithms that can effectively combine the data collected at these sensor nodes and send only the aggregated data (i.e., a representative of the entire network) to the sink.

Tree-based data gathering is considered to be the most energy efficient in terms of the number of link transmissions; however, almost all of the tree-based data gathering algorithms have been proposed for static sensor networks without taking the mobility of the sensor nodes into consideration. In the presence of node mobility, the network topology changes dynamically with time—leading to frequent tree reconfigurations. Thus, mobility brings in an extra dimension of constraint to a WMSN, and we need algorithms that can determine stable long-living data gathering trees that do not require frequent reconfigurations. To the best of our knowledge, we have not come across any work on stable data gathering trees for mobile sensor networks. The only tree-based data gathering algorithm we have come across for WMSNs is a shortest-path-based spanning tree algorithm [2], wherein each sensor node is constrained to have at most a certain number of child nodes. Based on our experiences from mobile ad hoc networks, minimum-hop shortest paths and trees in mobile network topologies are quite unstable and need to be frequently reconfigured [3,4]. We could not find any other related work on tree-based data gathering for WMSNs.

Most of the work on data gathering algorithms for WMSNs is focused on the use of clusters, wherein researchers have tried to extend the classical low-energy adaptive clustering hierarchy (LEACH) [5] algorithm for dynamically changing network topologies. Variants of LEACH that take into consideration the available energy level [6] and the mobility level [7] of the nodes to be chosen as cluster heads, the stability of the links between a regular node and its cluster head [8], and the setup of the panel of cluster heads to facilitate cluster reconfiguration in the presence of node mobility [9] have been proposed in the literature. Another category of research in WMSNs is to employ a mobile data collecting agent (e.g., [10–12]) that goes around the network in the shortest possible path toward the location from which the desired data is perceived to originate.

In this research, we propose two distributed spanning tree–based data gathering algorithms for WMSNs. One of these data gathering algorithms is based on the notion of link expiration time (LET) that is predicted according to a model used for the highly successful flow-oriented routing protocol (FORP) [13], a stable unicast routing protocol for mobile ad hoc networks. The LET-DG tree is a rooted directed spanning tree determined in a distributed fashion on a network graph comprising links whose weights are the predicted expiration time. The LET-DG tree has been observed to yield long-living stable trees that exist for a longer time. As observed in the simulation studies of this chapter, the drawback of using stable trees is that they tend to overuse certain nodes (especially the intermediate nodes of the data gathering tree) and lead to their premature failure. As sensor networks are often deployed with higher density, one or more node failures do not immediately bring the network to a halt. The live sensor nodes (the nodes that still have a positive available energy) maintain the coverage and connectivity of the underlying network for a longer time. Nevertheless, the unfairness of node usage persists with stable data gathering trees. As an alternative, we propose a second data gathering algorithm that is based on a distributed implementation of the minimum-distance spanning tree (MST) algorithm run on a network graph comprising links whose weights are the Euclidean distance between the constituent end nodes. The MST-DG trees have been observed to yield a much longer node and network lifetimes, at the cost of frequent tree reconfigurations. Another drawback observed with the MST-DG trees is that they are relatively taller (larger height), with fewer child nodes per intermediate node, as well as fewer leaf nodes, than the LET-DG trees, and as a result, we observe the MST-DG trees to incur a much larger delay per round of data gathering than the LET-DG trees.

The rest of the chapter is organized as follows: Section 13.2 presents the system model, including the models for the link expiration time and energy consumption, as well as stating the assumptions. Section 13.3 describes the proposed algorithm to determine the LET-DG trees in a distributed fashion. Section 13.4 presents a variation of the LET-DG algorithm to determine minimum-distance-based MST-DG trees. Section 13.5 presents an exhaustive simulation-based comparison of the LET-DG and MST-DG trees with respect to performance metrics such as tree lifetime, delay per round of data gathering, and the node and network lifetimes (due to disconnection), along with a distribution of the probability of node failures. Section 13.6 concludes the chapter and outlines ideas of future research. For the rest of the chapter, the terms *data aggregation* and *data gathering*, and *edge* and *link* are used interchangeably. They mean the same thing.

13.2 System Model, Energy Consumption Model, and Assumptions

The *system model* adopted for the data gathering algorithms presented in this chapter can be summarized as follows:

1. The underlying network graph considered in the construction of the communication topology used for data gathering is a unit disk graph [14] constructed assuming each sensor node has a fixed transmission range, R. There exists a link between any two nodes in a unit disk graph if and only if the physical distance between the two end nodes of the link is less than or equal to the transmission range, R.
2. The data gathering algorithms operate in several rounds, and during each round, data from the sensor nodes are collected, aggregated, and forwarded to the sink through the data gathering tree (LET-DG or MST-DG tree) rooted at a leader node.
3. The leader node of a data gathering tree remains the same as long as the tree exists and is randomly chosen by the sink every time a new tree needs to be determined.
4. **LET-DG tree:** The predicted link expiration time (LET) of a link $i - j$ between two nodes i and j, currently at (X_i, Y_i) and (X_j, Y_j), and moving with velocities v_i and v_j in directions θ_i and θ_j (with respect to the positive X-axis) is computed using the formula proposed in [5]:

$$LET(i, j) = \frac{-(ab + cd) + \sqrt{(a^2 + c^2)R^2 - (ad - bc)^2}}{a^2 + c^2} \tag{13.1}$$

   where $a = v_i^* \cos\theta_i - v_j^* \cos\theta_j$, $b = X_i - X_j$, $c = v_i^* \sin\theta_i - v_j^* \sin\theta_j$, and $d = Y_i - Y_j$.
5. **MST-DG tree:** The Euclidean distance for a link $i - j$ between two nodes i and j, currently at (X_i, Y_i) and (X_j, Y_j), is given by

$$\sqrt{(X_i - X_j)^2 + (Y_i - Y_j)^2} \tag{13.2}$$

The *energy consumption model* used is a first-order radio model [15] that has also been used in several of well-known previous works (e.g., [5,16]) in the literature. According to this model, the energy expended by a radio to run the transmitter or receiver circuitry is E_{elec} = 50 nJ/bit, and $\in_{amp}$ = 100 pJ/bit/m² for the transmitter amplifier. The radios are turned off when a node wants to avoid receiving unintended transmissions.

1. The energy lost in transmitting a k-bit message over a distance d is given by $E_{TX}(k, d) = E_{elec}^* k + \in_{amp}^* k^* d^2$.
2. The energy lost in receiving a k-bit message is given by $E_{RX}(k) = E_{elec}^* k$.

3. During a network-wide flooding of a control message (e.g., the tree establishment messages as described in Sections 13.3 and 13.4), each node is assumed to lose energy corresponding to transmission over the entire transmission range of the node and to receive the message from each of its neighbors. In networks of high density, the sum of the energy lost at a node due to reception of the broadcast message from all of its neighbors is often more than the energy lost due to transmitting the message.

The key assumptions behind the two data gathering algorithms are as follows:

1. A sensor node is able to obtain its current location, velocity, and direction of motion (with respect to the positive *X*-axis) at any point of time and also includes the same as a location update vector (LUV) in the TREE-CONSTRUCT message broadcast to its neighborhood at the time of constructing the data gathering trees (refer to Sections 13.3 and 13.4). With the inclusion of a LUV in the TREE-CONSTRUCT message, we avoid the need to periodically exchange beacons in the neighborhood.
2. For the LET-DG trees, a sensor node maintains a LET table comprising the estimates of the LET values to each of its neighbor nodes based on the latest TREE-CONSTRUCT messages received from them. For the MST-DG trees, a sensor node maintains a distance table comprising estimates of the Euclidean distance with the neighbor nodes that sent it the TREE-CONSTRUCT message.
3. Sensor nodes are assumed to be both time division multiple access (TDMA) and code division multiple access (CDMA) enabled [17]. Every upstream node broadcasts a time schedule (for data gathering) to its immediate downstream nodes; a downstream node transmits its data to the upstream node according to this schedule. Such a TDMA-based communication between every upstream node and its immediate downstream child nodes can occur in parallel, with each upstream node using a unique CDMA code.
4. We assume the size of the aggregated data packet to be the same as the size of the individual data packets sent by the sensor nodes. In other words, aggregation at any node does not result in an increase in the size of the data packets transmitted from the sensor nodes toward the sink.

13.3 Design of the Link Expiration Time–Based Data Gathering Algorithm

The LET-DG algorithm is a distributed implementation of the maximum spanning tree algorithm [18] on a weighted network graph with the edge weights modeled as the predicted link expiration time (LET) of the constituent end nodes. The objective of a maximum spanning tree algorithm is to determine a spanning tree such that the sum of the edge weights is the maximum. In this research, we aim to determine a maximum-LET spanning tree for wireless mobile sensor networks such that the sum of the LETs of the constituent links of the spanning tree is the maximum. The LET-DG tree is a rooted maximum-LET spanning tree with the root being the leader node chosen by the sink (as explained in Section 13.3.1).

13.3.1 Sink: Selection of the Leader Node

Whenever a sink node fails to receive aggregated data from the leader node of the LET-DG tree, the sink randomly chooses a new leader node from the *list of available nodes* currently perceived to exist with a positive residual energy, and sends it a TREE-INITIATE message to start construct-ing a tree rooted at the chosen leader node (LEADER). The sink includes a sequence number (a monotonically increasing value maintained at the sink) for the tree construction process in the TREE-INITIATE message, and the leader node includes it in its tree construction message (see Section 13.3.2) to avoid replay errors involving outdated links. If the leader node is alive (i.e., it has positive available energy), then it responds back with a TREE-INITIATE-ACK message acknowl-edging that it will start the flooding-based tree discovery. If the TREE-INITIATE-ACK message is not received within a certain time, the sink considers the chosen sensor node to be not alive, removes it from the *list of available nodes*, and sends the TREE-INITIATE message (with a higher sequence number, to avoid any parallel tree construction occurring in the network) to another randomly chosen sensor node from the *list of available nodes*. The above procedure is repeated until the sink successfully finds a leader node that accepts to initiate the tree construction process.

13.3.2 Initiation of the TREE-CONSTRUCT Message

When a sensor node is not aware of its position in the LET-DG tree, it sets a weight of $-\infty$ (nega-tive infinity; for simulation purposes, this will be a very small negative value) as its best-known estimated weight to join the tree. The leader node broadcasts a TREE-CONSTRUCT message containing a five-element tree configuration tuple *<sequence number, LEADER node ID, sender node ID, sender's estimated weight, upstream node ID>*, as well as a location update vector (LUV) comprising the four-element tuple *<X-coordinate, Y-coordinate, velocity, direction of motion—angle with respect to the X-axis>* to its neighbors. The sequence number is the value sent by the sink to the leader node for the specific tree construction process. If the sender node is the LEADER, it sets the upstream node ID to its own ID, while the other nodes set the upstream node ID to be the ID of the node that they perceive to be their best choice for the upstream node that can con-nect them to the tree. Initially, when a node does not know its position in the tree, its upstream node ID is NULL. The leader node sets the sender's estimated weight value to $+\infty$ in the TREE-CONSTRUCT message.

13.3.3 Propagation of the TREE-CONSTRUCT Message and Tree Establishment

When a node receives the TREE-CONSTRUCT message with a higher sequence number for the first time, it treats it as a sign of tree reconfiguration and resets its estimated weight to join the tree to $-\infty$, if it has not already done so. The receiving node then calculates the weight of the link to the sending neighbor of the message. A TREE-CONSTRUCT message is accepted at a node for a weight/tree configuration update and rebroadcast (in the neighborhood of the node) if the following conditions are met:

1. The upstream node ID is not equal to the ID of the node itself. This is to avoid any looping.
2. The estimated weight at the node is *lower than* the sender's estimated weight.
3. The estimated weight at the node is *lower than* the predicted expiration time of the link (LET, calculated according to Equation 13.1) on which the TREE-CONSTRUCT message was received.

If all the above conditions are true, then a node receiving the TREE-CONSTRUCT message accepts the message to update its position in the tree. The receiver node selects the sender node as its upstream node for joining or connecting to the tree, and sets its estimated weight in the tree as the *minimum of the sender node's estimated weight for the tree and LET of the link* through which the TREE-CONSTRUCT message was received. If its weight is updated, the receiver node sends a TREE-JOIN-CHILD message to the upstream sender node indicating the decision to connect to the tree by becoming its child node. The receiver node also decides to further broadcast the TREE-CONSTRUCT message to its neighbors by replacing the LUV of the sender node with its own LUV, the sender node ID with its own ID, the sender's estimated weight with its recently updated weight in the tree, and the upstream node ID set to the ID of the node through which it has decided to join or connect to the tree. The LEADER node ID and the sequence number fields are retained as is they are in the TREE-CONSTRUCT message.

A node follows the same procedure as explained above when it receives a TREE-CONSTRUCT message with the highest-known sequence number from any other neighbor node. In other words, a TREE-CONSTRUCT message corresponding to the latest broadcast process (decided using the sequence number) is accepted for an update and rebroadcast only if it can *increase* the estimated weight of the node to connect to the tree. The above procedure is repeated until the TREE-CONSTRUCT message reaches every sensor node. Each sensor node is guaranteed to broadcast the TREE-CONSTRUCT message at least once during this flooding process, because the initial estimated weight of a sensor node to join the tree is $-\infty$, and the leader node starts with a positive ∞ value and the LET values for the links are always positive. The objective of the LET-DG algorithm is to connect each node with the largest possible weight value in the tree—a measure of the estimated lifetime of the tree.

13.3.4 Propagation of the TREE-LINK-FAILURE Message

When an upstream sensor node finds out that a link to one of its downstream child nodes is broken due to failure to receive aggregated data packets, the upstream node initiates a TREE-LINK-FAILURE message and includes in it the sequence number that was used in the TREE-CONSTRUCT message corresponding to the most recently used flooding process. The TREE-LINK-FAILURE message is essentially reverse broadcast along the edges of the subtree leading toward the leader node, starting from the upstream node of the broken link. Similarly, the downstream node detects the link failure when it fails to receive a TDMA schedule from its upstream node for the next round of data aggregation and initiates a TREE-LINK-FAILURE message to inform of the tree failure to the nodes in the subtree rooted at it. If an intermediate node or leaf node does not receive the TREE-LINK-FAILURE message, it continues to wait for the aggregated data packets from its perceived downstream nodes or the TDMA schedule from its upstream node until it learns about the tree failure through the broadcast of a new TREE-CONSTRUCT message with a sequence number greater than that of the most recently used tree.

13.4 Design of the Minimum-Distance Spanning Tree–Based Data Gathering Algorithm

The minimum-distance spanning tree–based data gathering (MST-DG) algorithm is a distributed implementation of the minimum spanning tree algorithm [18] on a weighted network graph with the edge weights modeled as the Euclidean distance between the constituent end nodes. In this research, we aim to determine a minimum-distance spanning tree for wireless mobile sensor networks, such that the sum of the distances of the constituent links of the spanning tree is the minimum. Since a node loses more energy to transmit over a larger distance, we aim to reduce the transmission energy loss across the whole spanning tree by setting the edge weight to be physical Euclidean distance between the constituent end nodes. The MST-DG tree is a rooted minimum-distance spanning tree, with the root being the leader node chosen by the sink (as explained in Section 13.3.1). The overall procedure to construct the MST-DG tree is the same as that of the LET-DG tree, except with the differences in the criteria used for selecting the links that form part of the tree. In this section, we only highlight these differences in detail and provide a brief outline of the entire algorithm for the sake of completeness.

To initiate the process of constructing the MST-DG tree, the sink node randomly chooses a leader node (as is done in the case of the LET-DG tree; see Section 13.3.1) and sends it a unique sequence number, greater than the one used for the previous tree. The initial estimates of the weight at a node to connect to the tree are rather different for the MST-DG tree. The weight estimate at the LEADER node is 0, and the estimate at every other node is $+\infty$. The LEADER node broadcasts a TREE-CONSTRUCT message in its neighborhood; the message contains the same five-element *Tree-Configuration* tuple and a four-element *LUV*, as indicated in Section 13.3.2. A node only processes TREE-CONSTRUCT messages that are received with the largest known sequence number or higher. The criteria used at a receiving node to accept the message for a weight or tree configuration update is as detailed below.

1. The upstream node ID is not equal to the ID of the node itself (same as in Section 13.3.3).
2. The estimated weight at the node is *greater than* the sender's estimated weight.
3. The estimated weight at the node is *greater than* the predicted Euclidean distance of the link (calculated according to Equation 13.2) on which the TREE-CONSTRUCT message was received.

If all the above three conditions are met, then a receiving node decides to join or update its connection to the tree by becoming a child node of the node that sent the TREE-CONSTRUCT message, and sends to it the TREE-JOIN-CHILD message. It sets its estimated weight in the tree to be the *maximum of the sender's estimated weight and the Euclidean distance* of the link on which the TREE-CONSTRCT message was received. The receiving node also rebroadcasts the TREE-CONSTRUCT message in its neighborhood by replacing the LUV of the sender node with its own LUV, the sender node ID with its own ID, the estimated sender's weight with its own recently updated weight in the tree, and the upstream node ID with the ID of the node to which it sent the TREE-JOIN-CHILD message. Any subsequently received TREE-CONSTRUCT message is accepted at a node for an update and rebroadcast only if it can *decrease* the estimated weight of the node in the tree. The rest of the procedure in the propagation of the TREE-CONSTRUCT message is the same as explained in Section 13.3.3.

Note that every sensor node is expected to update its estimated weight in the tree at least once because the initial estimated weight at a sensor node is $+\infty$, and the value for the sender's weight in the TREE-CONSTRUCT message broadcast by the LEADER node is 0, and the Euclidean distance values are always greater than 0. The objective of the MST-DG algorithm is to connect each node with the lowest possible weight value in the tree—a measure of the energy consumption and fairness of node usage. The procedure to detect a link failure and propagate the TREE-LINK-FAILURE messages initiated by the upstream and downstream nodes of the broken link is the same as explained in Section 13.3.4.

13.5 Simulations

In this section, we present the results from simulation studies evaluating the performance of the LET-DG and MST-DG data gathering trees under diverse conditions of network density and mobility. The simulations were conducted in ns-2 (version 2.31) [19]. The medium access control (MAC) layer model is the IEEE 802.11 [20] model. The network dimensions are 100 × 100 m. The number of nodes in the network is 100, and the nodes are uniform-randomly distributed throughout the network. The sink is located at (50, 50). The transmission range per sensor node is varied from 20 to 50 m, in increments of 5 m. For brevity, we present only results obtained for transmission ranges of 25 m, 30 m (representative of moderate density, with connectivity of 96% and above), and 40 m (high density, with 100% connectivity). For coverage calculation purposes, we also use the *sensing range* of a sensor node, defined as half the transmission range of the node. Basically, a sensor node can monitor and collect data at locations within the radius of its sensing range and transmit them to nodes within the radius of its transmission range. The sensing range is generally considered to be half of the transmission range in the literature (e.g., [21]).

Simulations are conducted for two kinds of energy scenarios: In the first, a *sufficient energy scenario*, wherein each node is supplied with an abundant supply of energy (50 J per node) and there are no node failures due to exhaustion of battery charge, the simulations in these are conducted for 1000 s. The second scenario is an *energy-constrained scenario*, in which each node is supplied with a limited initial energy (2 J per node) and the simulations are conducted until the network of live sensors gets disconnected due to the failures of one or more sensor nodes. We conduct constant-bit-rate data gathering at the rate of four rounds per second (one round for every 0.25 s). The size of the data packet is 2000 bits; the size of the control messages used in the tree formation phase is assumed to be 400 bits, which is sufficiently large enough to accommodate the five-element tree configuration tuple and the four-element location update vector tuple of the TREE-CONSTRUCT message.

The node mobility model used is the well-known random waypoint mobility model [22], with the maximum node velocities being 3 m/s (for low-mobility scenarios) and 10 m/s (for high-mobility scenarios). According to this model, each node chooses a random target location to move with a velocity uniform-randomly chosen from $[0, \ldots, v_{max}]$, and after moving to the chosen destination location, the node continues to move by randomly choosing another new location and a new velocity. Each node continues to move like this, independent of the other nodes and also independent of its mobility history, until the end of the simulation. For a given value of v_{max}, we also vary the dynamicity of the network by conducting the simulations with a variable number of static nodes (out of the 100 nodes) in the network. The values for the number of static nodes used are 0 (all nodes are mobile), 20, 50, and 80.

13.5.1 Performance Metrics

We generated 100 mobility profiles of the network for a total duration of 6000 s, for every combination of the maximum node velocity and the number of static nodes. Every data point in the results presented in Figures 13.1 through 13.12 is averaged over these 200 mobility profiles.

The performance metrics measured in the simulations are

1. **Tree lifetime:** The duration for which a data gathering tree existed, averaged over the entire simulation time period.
2. **Delay per round:** Measured in terms of the number of time slots needed per round of data aggregation at the intermediate nodes, all the way to the leader node of the data gathering

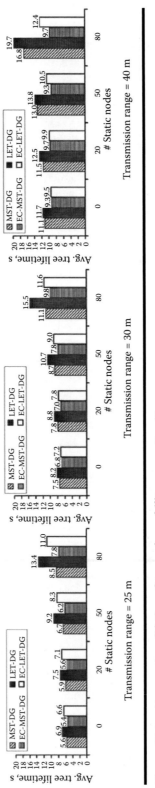

Figure 13.1 Average tree lifetime (low node mobility: v_{max} = 3 m/s).

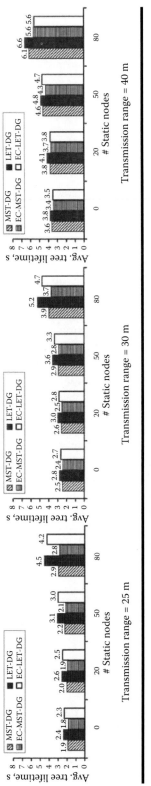

Figure 13.2 Average tree lifetime (moderate–high node mobility: $v_{max} = 10$ m/s).

tree, averaged across all the rounds of the simulation. A brief description of the algorithm used to compute the delay per round is given in Section 13.5.2, along with an illustration in Figure 13.3.

3. **Node lifetime:** Measured as the time of first node failure due to exhaustion of battery charge.

4. **Network lifetime:** Measured as the time of disconnection of the network of live sensor nodes (i.e., the sensor nodes that have positive available battery charge).

5. **Loss of coverage:** While the tree lifetime and delay per round are measured for both the sufficient energy and energy-constrained (appropriately prefixed as *EC* next to the algorithm names) scenarios, the node and network lifetimes are measured only for the energy-constrained scenarios.

   To obtain the distribution of node failure times, we counted the frequency of the number of node failures, ranging from 1 to 100, in each of the 200 mobility profile files for every combination of transmission range, maximum node velocity, and number of static nodes. The probability for *x* number of node failures (*x* from ranging from 1 to 100, as we have a total of 100 nodes in our network for all the simulations) for a given combination of the operating conditions is measured as the number of mobility profile files that reported *x* number of node failures divided by 200, which is the total number of mobility profiles used for every combination of maximum node velocity and number of static nodes. Similarly, we keep track of the time at which *x* (ranging from 1 to 100) number of node failures occurred in each of the 200 mobility profiles for a given combination of operating conditions, and the values for the time of node failures reported in Figures 13.8 and 13.9 are an average of these data collected over all the mobility profile files. We discuss the results for distribution of the time and probability of node failures along with the node and network failure times in Section 13.5.5.

6. **Fraction of coverage loss and coverage loss time:** If *f* is denoted as fraction of coverage loss (ranging from 0.01 to 1.0, measured in increments of 0.01), the coverage loss time is the time at which any *f* randomly chosen locations (*X*-, *Y*-coordinates) among 100 locations in the network are not within the sensing range of any node. Since the number of node failures increases monotonically with time and network coverage depends on the number of live nodes, our assumption in the calculations for network coverage loss is that the fraction of coverage loss increases monotonically with time. We keep track of the largest fraction of coverage loss the network has incurred so far, and at the beginning of each round, we check whether the network has incurred the next largest fraction of coverage loss, referred to as the target fraction of coverage loss. The first time instant during which we observe the network to have incurred the target coverage loss is recorded as the coverage loss time for the particular fraction of coverage loss, and from then on, we increment the target coverage loss to 0.01 and keep testing for the first occurrence of the same in the subsequent rounds.

At the beginning of each round, we check for network coverage as follows: We choose 100 random locations in the network and find out whether each of these locations is within the sensing range of at least one sensor node. We count the number of locations that are not within the sensing range of any node. If the fraction of the number of locations (actual number of locations that are not covered/total number of locations considered, which is 100) not within the sensing range of any node equals the target fraction of coverage loss, we record the time instant for that particular round of data gathering as the coverage loss time corresponding to the target fraction of coverage loss. We then increment the target fraction of coverage loss by 0.01 and repeat the above procedure to determine the coverage loss time corresponding to the new incremented value of the target fraction of coverage loss.

Each coverage loss time data point reported for particular fractions of coverage loss in Figures 13.12 and 13.13 is the average value of the coverage loss times observed when the individual data gathering tree algorithms are run over the mobility profile files corresponding to a particular condition of node dynamicity (maximum node velocity and number of static nodes) and transmission range per node. The probability for a particular fraction of coverage loss is computed as the ratio of the number of mobility profile files in which the corresponding fraction of coverage loss was observed divided by the total number of mobility profile files (we used a total of 200 mobility profile files for each operating condition).

13.5.2 Algorithm to Compute the Delay per Round of Data Gathering

The delay at a node indicates the number of time slots it takes for the node to gather data from all of its immediate child nodes. The delay of the data gathering tree is 1 plus the delay at the root node. We assign one time slot per child node to transfer data to its immediate predecessor node in the tree. We start calculating the delay at the intermediate nodes from the bottom of the data gathering tree. The delay associated with each of the leaf nodes is 0. For each intermediate node *u* at a particular level, we prepare a sorted list of the delay associated with each of its immediate child nodes. The delay associated with the intermediate node is computed through a temporary running variable, *Temp-Delay* (initialized to zero), as we explore the delay associated with each of the child nodes in the sorted list. For every child node *v* in the sorted list of the delay, *Temp-Delay* is set to the maximum of *Temp-Delay* + 1 and *Delay(v)* + 1, as we assume it takes one time slot for a child node to transfer its aggregated data to the immediate predecessor intermediate node. The delay associated with the intermediate node *u*, *Delay(u)*, is the final value of *Temp-Delay* after we run through the sorted list of the delays associated with the *Child-Nodes(u)*. The above procedure is repeated for all the intermediates nodes, from levels 1 less than the *Height* of the tree all the way to zero (i.e., the root node). Figure 13.3 illustrates the execution of the delay computation algorithm on a data gathering tree. The integer inside a circle indicates the node ID, and the integer outside a circle indicates the delay for data aggregation at the node.

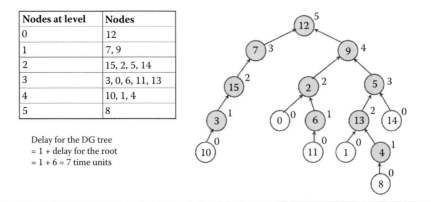

Nodes at level	Nodes
0	12
1	7, 9
2	15, 2, 5, 14
3	3, 0, 6, 11, 13
4	10, 1, 4
5	8

Delay for the DG tree
= 1 + delay for the root
= 1 + 6 = 7 time units

Figure 13.3 Example to illustrate the calculation of delay per round of data gathering.

13.5.3 Tree Lifetime

We measure the tree lifetime for both the sufficient energy scenarios (to capture the impact of node dynamicity, i.e., variations in node velocity and the number of static nodes) and the energy-constrained scenarios (to capture the impact of tree reconfigurations induced by node failures, in addition to node dynamicity). We say a tree exists topologically if the physical Euclidean distance between the end nodes of the links constituting the tree is within the transmission range of the nodes. In the energy-constrained scenarios, even though a data gathering tree may topologically exist, the tree would require reconfiguration (i.e., a new discovery through network-wide flooding of the TREE-CONSTRUCT messages) if one or more nodes in the tree fail due to exhaustion of battery charge. This has an impact on the lifetime of the data gathering trees observed in the simulations, and this is what we capture by measuring the tree lifetime under energy-constrained scenarios. Since a tree also needs to be reconfigured due to node mobility, the lifetime of the data gathering trees observed for energy-constrained scenarios is always less than or equal to that observed for sufficient energy scenarios. This statement holds true for both the LET-DG and MST-DG trees.

As the LET-DG trees are inherently more topologically stable than the MST-DG trees, we observe the difference in the absolute magnitudes of the tree lifetimes for the sufficient energy and energy-constrained scenarios to be relatively larger in the case of the LET-DG trees, especially for the moderate transmission range per node. However, at larger transmission ranges per node, the LET-DG trees sustain premature node failures due to continued use of certain intermediate nodes for stable data gathering; as a result, the MST-DG trees—with their tendency to more fairly use the nodes—show a relatively larger difference in the lifetime at sufficient energy scenarios than at energy-constrained scenarios.

For both data gathering trees, we observe that for fixed node mobility, the magnitude difference in the tree lifetime between the sufficient energy and energy-constrained scenarios increases with an increase in the transmission range per node. This can be attributed to the increased energy expenditure incurred at the nodes at larger transmission ranges, leading to certain premature node failures, as well as to a relatively longer link lifetime (i.e., the trees tend to topologically exist for a longer time at larger transmission ranges). At larger transmission ranges, the constituent end nodes of a link have more degrees of freedom to move around and still be within the transmission range of each other for a longer time. On the other hand, for a fixed transmission range per node, the difference between the sufficient and energy-constrained scenarios decreases with an increase in the maximum node velocity. This is as expected because the trees have a relatively lower topological lifetime at high node velocities. For a given maximum node velocity and transmission range per node, the difference in the magnitude for the lifetime of the data gathering trees for the sufficient energy and energy-constrained scenarios increases with an increase in the number of static nodes. This can be attributed to an increase in the topological lifetime of the trees as the number of static nodes increases.

While comparing the magnitude difference between the lifetimes of the LET-DG and MST-DG trees, we observe that the difference in magnitude decreases with an increase in the transmission range per node, as well as with an increase in network dynamicity (i.e., as more nodes are mobile). For a given maximum node velocity, the LET-DG trees incur a much longer lifetime than the MST-DG trees when operated at moderate transmission ranges per node and a larger proportion of static nodes. In the sufficient energy scenarios, when operated under moderate transmission ranges per node, the difference in the tree lifetime can be as large as 25% when all 100 nodes are mobile, and as large as 55%–60% when operated with 80 static and 20 mobile nodes. Under energy-constrained scenarios, especially at larger node mobility, as we increase the transmission range per node, the lifetime of the LET-DG trees converges to that of the MST-DG trees.

This can be attributed to the premature node failures in the LET-DG trees. For a given maximum node velocity and transmission range per node, as we increase the number of static nodes from 0 to 80 (out of a total of 100 nodes), the LET-DG trees incur about 60%–90% larger lifetime, and the MST-DG trees incur about 50%–60% larger lifetime.

For both the sufficient energy and energy-constrained scenarios, for each data gathering tree, for a fixed transmission range per node, as we increase the maximum node velocity by more than three times (i.e., from 3 to 10 m/s), we observe a more or less proportional decrease in the tree lifetime (i.e., the lifetime of the trees decreases by about one-third). For a fixed maximum node velocity, as we increase the transmission range per node from 25 to 40 m, we observe more than a proportional increase in the lifetime for both data gathering trees. This can be attributed to the significantly high network connectivity (more than a linear increase) obtained at larger transmission ranges per node.

13.5.4 Delay per Round of Data Gathering

We observe the LET-DG trees to incur significantly lower delay per round of data gathering than the MST-DG trees. The delay per round is not much affected by the dynamicity of the network and is more impacted by the topological structure of the two spanning trees. A minimum-distance-based spanning tree tends to have relatively fewer leaf nodes, and as a result, more nodes are likely to end up as intermediate nodes—leading to a much larger depth for the MST-DG trees. The MST-DG tree is also observed to be more unbalanced with respect to the distribution of the number of children per intermediate node, as well as the distribution of the leaf nodes at different levels. Not all leaf nodes are located at the bottommost level of the tree. Due to all these structural complexities, the MST-DG trees have been observed to incur a much larger delay per round of data gathering. On the other hand, the LET-DG trees have been observed to be more shallow (i.e., lower depth) with more leaf nodes, and the distribution of the number of child nodes per intermediate node is relatively more balanced. All of these factors contribute to a much lower delay per round of data gathering (Figure 13.4).

Across all the simulations, we observe the MST-DG trees to incur a 55%–75% larger delay per round of data gathering. For a given maximum node mobility, the difference in the delay per round of data gathering between the MST-DG and LET-DG trees decreases with an increase in the transmission range per node. While operating the network at larger transmission ranges per node, it is possible to obtain a slightly better distribution of the nodes across the different levels of the MST-DG tree, contributing to the reduction in the delay. For a given maximum node mobility and transmission range per node, we also observe the difference in the magnitudes of the delay per round between the MST-DG and LET-DG trees to increase with an increase in the number of static nodes. This can be attributed to the reduced chances of changes to the topological structure of the MST-DG tree in the presence of more static nodes—the unbalanced distribution of the nodes at the different levels of the tree gets to continue for a longer time—contributing to the larger delay.

We observe the energy-constrained scenarios to have only minimal impact on the delay per round of data gathering. The two data gathering trees incur only a slightly lower delay per round of data gathering (by a factor of 5%–10%) when operated in energy-constrained scenarios than when operated in sufficient energy scenarios. The reduction in the delay per round of data gathering in the presence of node failures could be attributed to the overall reduction in the number of time slots needed to gather data from around the nodes in the network. The impacts of node failures and the energy constraint on the delay per round are almost equally observed for both the LET-DG and MST-DG trees (Figure 13.5).

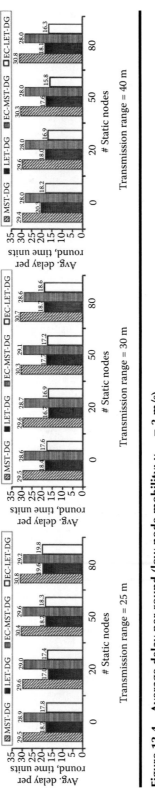

Figure 13.4 Average delay per round (low node mobility: v_{max} = 3 m/s).

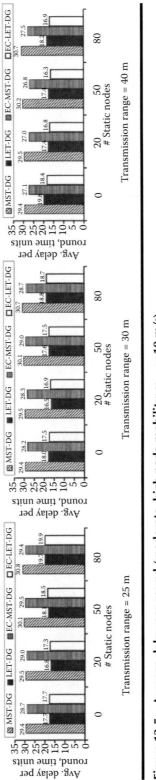

Figure 13.5 Average delay per round (moderate–high node mobility: v_{max} = 10 m/s).

13.5.5 Node Lifetime and Network Lifetime

We observe a stability–node-network lifetime trade-off between the LET-DG and MST-DG trees. While the LET-DG trees have been credited for higher stability and lower delay per round of data gathering, we observe them to be relatively unfair with respect to node usage. An intermediate node that lies on a stable tree tends to get used for a longer time and ends up spending more energy to receive data from all of its child nodes, aggregate them, and transmit to an upstream node, whereas the leaf node of a data gathering tree only spends energy to transmit its data to the upstream node. With a shallow structure and more leaf nodes, the LET-DG trees are vulnerable for premature node failures, as observed in Figures 13.6 through 13.9. However, the LET-DG trees have been observed to significantly offset the early node failures with a much better network lifetime, attributed to the lower energy spent to reconfigure the trees and the possibility of the energy-rich leaf nodes (that were lightly used before and during the first few node failures) becoming intermediate nodes in the subsequently reconfigured trees after the initial set of node failures. The impact of the latter factor could be especially observed in Figures 13.8 and 13.9, wherein we show the distribution of the node failure times and the probability of node failures.

After the initial set of node failures, attributed to the excessive use of certain nodes as part of the stable trees, we observe the LET-DG trees to have a much lower probability of node failure for much of the network's lifetime than the MST-DG trees. This could be attributed to the relatively equal expenditure of energy across all the nodes of a MST-DG tree. With fewer leaf nodes and relatively more frequent tree reconfigurations, we expect almost all of the nodes in a MST-DG tree to lose about the same amount of energy during the network lifetime. This could be confirmed by observing a much flatter curve for the probability of node failure (closer to 1) for a sufficiently larger number of node failures. The above observations can also be captured in quantitative terms through Figures 13.6 and 13.7: we observe the network lifetime incurred for the MST-DG trees to be mostly about 15%–30% more than that of the node lifetime (and at best, 70% larger when operated with 80 static nodes at v_{max} of 10 m/s and transmission range per node of 40 m), whereas the network lifetime for the LET-DG trees is mostly 50%–125% more than that of the node lifetime (and at best, can be as large as 200% more when operated with 80 static nodes at v_{max} of 10 m/s and transmission range per node of 40 m).

The impact of the difference in the node usage policies of the two data gathering trees can be observed in the difference in the magnitudes for the node lifetimes (the time of first node failure) and the network lifetimes (the time by which the network of live nodes, nodes with positive available energy, get disconnected due to failure of peer nodes) for the two data gathering trees. We observe the MST-DG trees to incur about 85%–150% larger node lifetime than the LET-DG trees for different combinations of node mobility, number of static nodes, and transmission ranges per node. On the other hand, the network lifetime sustained for the MST-DG trees is hardly 10%–15% more than that of the LET-DG trees and is only at most 25% larger.

In the case of LET-DG trees, one can observe from Figures 13.6 and 13.7 that for a given level of node dynamicity (v_{max} and the number of static nodes), the node lifetime substantially decreases (as large as 25%) with an increase in the transmission range per node, whereas the network lifetime decreases only marginally (by at most 10%) with an increase in the transmission range per node. The decrease in the node lifetime with an increase in transmission range can be attributed to the increase in the transmission energy loss and receipt of data from several downstream nodes when operated at a higher transmission range. However, when operated at a higher transmission range, LET-DG trees discover more stable routes, as well as balance the distribution of the role of the intermediate nodes and leaf nodes more evenly, resulting in a significant increase in the time

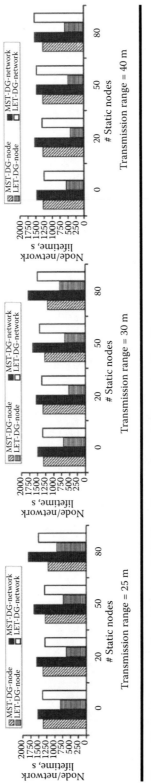

Figure 13.6   Average node and network lifetime (low node mobility: v_{max} = 3 m/s).

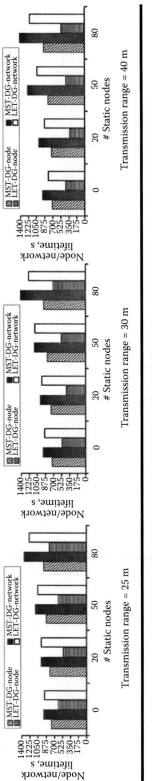

Figure 13.7   Average node and network lifetime (moderate–high node mobility: $v_{max} = 10$ m/s).

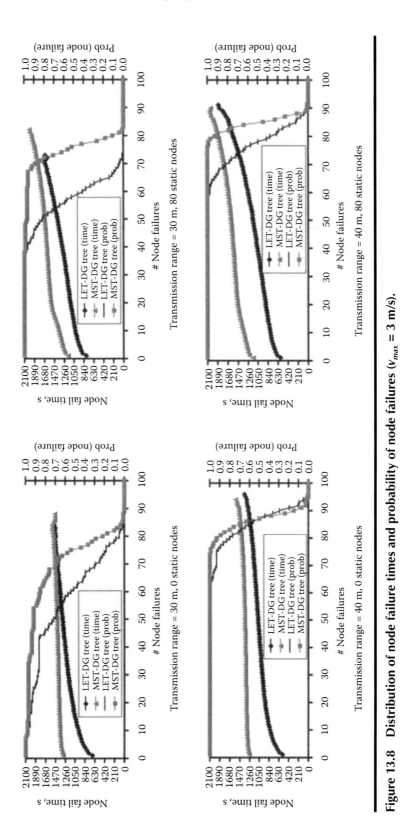

Figure 13.8 **Distribution of node failure times and probability of node failures (v_{max} = 3 m/s).**

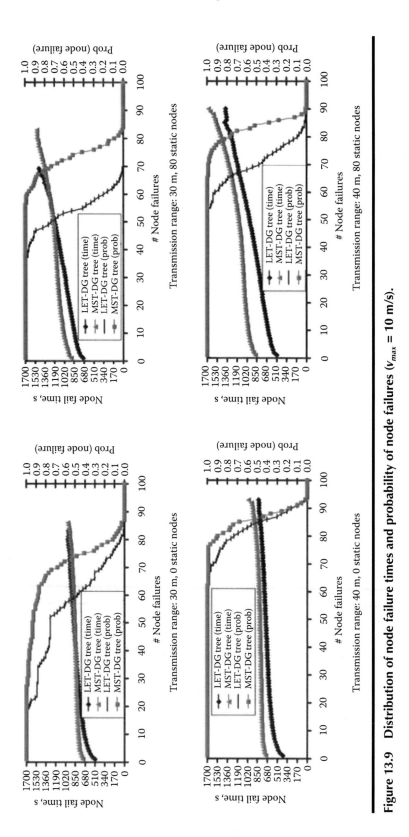

Figure 13.9 **Distribution of node failure times and probability of node failures (v_{max} = 10 m/s).**

of node failures, beyond the first node failure. In the case of MST-DG trees, we observe a very slight decrease in node lifetime (at most 5%) with an increase in transmission range per node. For the network lifetime, we observe a decrease of at most 15% at v_{max} = 3 m/s and an increase of at most 15% at v_{max} = 10 m/s. We attribute the better performance of MST-DG trees with respect to network lifetime at higher node velocities and transmission range per node to the increase in the fairness of node usage, and the possibility of the role of an intermediate node being rotated among the nodes with a regular reconfiguration of the data gathering tree at high node mobility. The energy efficiency associated with lower Euclidean distance of the links also helps to contain the transmission energy loss and aid in increasing the network lifetime.

Both the LET-DG and MST-DG trees are observed to demonstrate larger node and network lifetimes when operated in networks that have a mix of both static and mobile nodes vis-à-vis a network comprising only mobile nodes. For a given value of v_{max} and transmission range per node, both data gathering trees have been observed to sustain about 20%–25% larger node lifetime when operated in a network that is a mix of 80 static and 20 mobile nodes, compared to operating in a network of 100 mobile nodes. Under similar conditions, the network lifetimes incurred with both data gathering trees have been observed to be about 50%–70% larger. The MST-DG trees, with their vulnerability to break quickly with time in a network full of mobile nodes, are observed to be the most benefited with respect to node and network lifetimes when operated in networks that are a mix of static and mobile nodes.

13.5.6 Coverage Loss at the Network Lifetime

In this section, we compare the loss of coverage incurred with both the MST-DG and LET-DG trees with respect to a common timeline, considered the minimum of the network lifetime obtained for the two data gathering trees for every operating condition of transmission range per node, maximum node velocity, and the number of static nodes. Given the nature of results obtained for the network lifetime under different operating conditions, the minimum of the network lifetime for the two data gathering trees ended up mostly being the network lifetime observed for the LET-DG trees. For this value of network lifetime, we measured the fraction of coverage loss in the network incurred for each of the two data gathering trees, as well as the probability at which the corresponding fraction of coverage loss is to be observed.

Under the above measurement model, we observe the MST-DG trees to incur a lower fraction of coverage loss for all the operating conditions, compared to that obtained for the LET-DG trees (see Figures 13.10 and 13.11). However, the fraction of coverage loss observed for the MST-DG trees is bound to occur with a higher probability than that of the coverage loss to be incurred by using the LET-DG trees. In the case of the MST-DG trees, for a fixed v_{max} value, we observe the fraction of loss of coverage to decrease with an increase in transmission range per node from 25 to 40 m. The decrease in the loss of coverage at a higher transmission range per node of 40 m could also be attributed to an increase in the network lifetime, and is also the reason that we measure the loss of coverage at a time value (corresponding to the network lifetime of the LET-DG trees) that is lower than the network lifetime of the MST-DG trees.

On the other hand, for the LET-DG trees, for a fixed v_{max} value, we observe the fraction of loss of coverage to decrease with increase in transmission range per node from 25 to 30 m; however, with a further increase in the transmission range per node to 40 m, we observe the LET-DG trees to suffer from an increase in the fraction of loss of coverage. This could be attributed to the excessive loss of energy due to transmission at higher distances and overuse of the intermediate nodes on the long-living LET-DG trees. As a result, the network lifetime incurred for the LET-DG trees

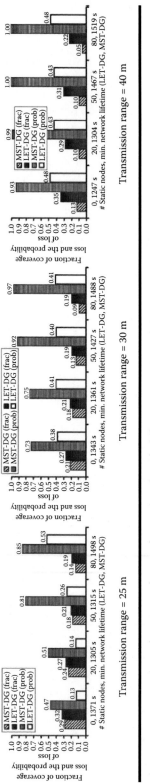

Figure 13.10 Fraction of coverage loss and associated probability (low node mobility: v_{max} = 3 m/s).

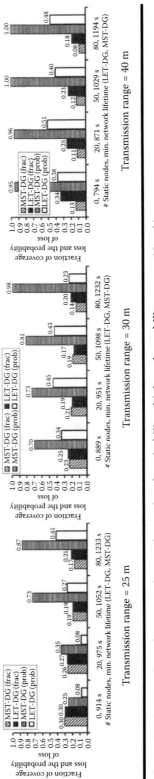

Figure 13.11 Fraction of coverage loss and associated probability (high node mobility: v_{max} = 10 m/s).

at higher transmission range per node of 40 m is lower than that incurred at 25 and 30 m, and this contributes to an increase in the loss of coverage.

In terms of the relative differences in the fraction of coverage loss, we observe the two data gathering trees to incur about the same fraction of coverage loss for both the low- and high-node-mobility conditions when operated at a low to moderate transmission range of 25 m. However, the difference in the fraction of coverage loss increases with an increase in the transmission range per node. At 40 m of transmission range per node, we observe the relative difference in the fraction of coverage loss to be as large as between 0.1 to 0.25. When operated at a transmission range of 40 m per node, with 80 static nodes and 20 mobile nodes at both low- and high-node-mobility scenarios, the fraction of coverage loss inflicted by the MST-DG tree is observed to be very low, as low as 0.06 to 0.08.

13.5.7 *Distribution of Coverage Loss*

In Figures 13.12 and 13.13, we illustrate the distribution of the time (referred to as the coverage loss time) at which particular fractions of coverage loss occur in the network (until the network lifetime) when run with the two data gathering trees. For all the operating conditions of node velocity, number of static nodes, and transmission range per node, the MST-DG tree incurs a larger coverage loss of time than that of the LET-DG trees. In other words, the time at which the network sustains a particular fraction of coverage loss (measured in increments of 0.01, from 0.01 to 1.0, or until the network of live nodes gets disconnected) is much longer for the MST-DG trees than for the LET-DG trees. The MST-DG trees have been observed to incur about 16%–26% and 6%–16% higher-coverage-loss time than that of the LET-DG trees at low-node-mobility and high-node-mobility conditions, respectively. For a fixed maximum node velocity, the difference in the absolute magnitude of the coverage loss time increases with an increase in both the transmission range per node and the number of static nodes in the network. More details on the impact of the different operating conditions on the coverage loss time incurred for the two data gathering trees is explained below.

For both the data gathering trees and a particular value of v_{max}, the time at which a particular fraction of coverage loss occurs in the network increases more with increase in the number of static nodes rather than with increase in the transmission range per node. The relatively small increase in the coverage loss time, with increase in the transmission range per node (compared to increase in the number of static nodes), can be attributed to the higher transmission energy and reception loss incurred at larger transmission ranges per node during the flooding tree discovery operations, even if the tree lifetime is expected to be larger at higher transmission ranges per node. Due to repeated failures of nodes, the data gathering trees (even though they may topologically exist) have to be frequently reconfigured through expensive flooding of the TREE-CONSTRUCT messages. On the other hand, when operated at a moderate transmission range with substantially higher number of static nodes, the loss of battery charge is relatively minimal with respect to the flooding operations, as well as the transmission and reception energy loss.

For both the data gathering trees and a particular value for the transmission range per node, the time at which a low to moderate fraction of coverage loss occurs in the network increases at about the same rate, if not better, as in the case of LET-DG trees, when we operate at 10 m/s and increase the number of static nodes from 0 to 80 vis-à-vis the scenario in which we decrease v_{max} from 10 to 3 m/s and continue to operate with all mobile nodes (0 static nodes). However, for larger values of the fraction of coverage loss and a fixed transmission range per node, the time at which coverage loss occurs can be more effectively increased by decreasing the maximum node velocity from 10 to 3 m/s, rather than increasing the number of static nodes from 0 to 80 and

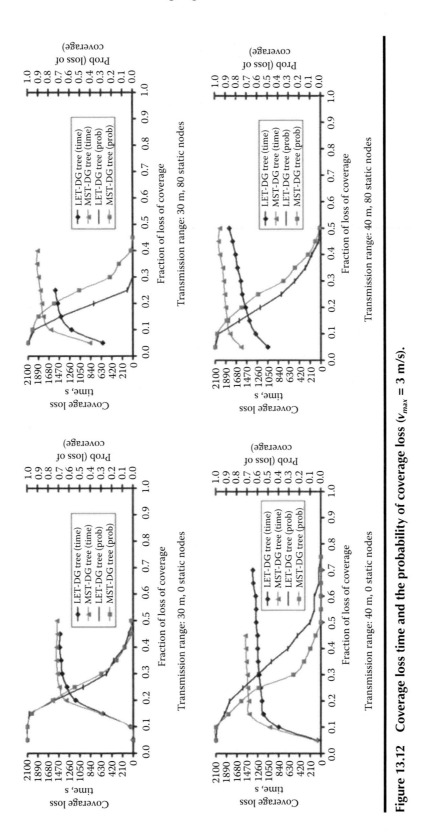

Figure 13.12   Coverage loss time and the probability of coverage loss (v_{max} = 3 m/s).

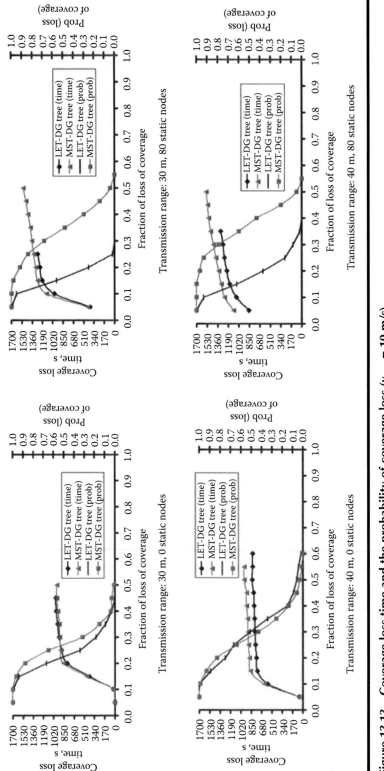

Figure 13.13 Coverage loss time and the probability of coverage loss ($v_{max} = 10$ m/s).

continuing to operate at 10 m/s. This could be mainly attributed to the potential savings in the node energy obtained from fewer tree reconfigurations at v_{max} of 3 m/s, even when operated with 0 static nodes. The 20% of mobile nodes moving at a v_{max} of 10 m/s in the network is sufficient enough to cause frequent tree reconfigurations and deprive even the static nodes of their energy by engaging them quite often in network-wide flooding.

Unlike the common timeline chosen in Section 13.5.6, in this section, for each of the two data gathering trees, we measure the time of loss of coverage until the network of live sensor nodes becomes disconnected for the individual data gathering trees. Since the MST-DG trees sustain larger network lifetimes, it is natural to expect the MST-DG trees to incur a higher probability for a particular fraction of loss of coverage to occur. That is the reason we observe the probability of a particular loss of coverage to be greater for the MST-DG trees under most of the operating conditions. However, as reported in Section 13.5.6, and Figures 13.10 and 13.11, MST-DG trees incur a lower loss of coverage (at a higher probability, though) than that observed for the LET-DG trees (at a lower probability) when the coverage loss is measured at the lower of the network lifetimes incurred for the two data gathering trees, which most of the time corresponded to the network lifetime observed for the LET-DG tree.

In the case of the LET-DG trees, for a particular value of v_{max}, the probability of a particular fraction of coverage loss increases with an increase in the transmission range per node (when the number of static nodes is fixed) and decreases with an increase in the number of static nodes (when the transmission range per node is fixed). On the other hand, for the MST-DG trees and a particular value of v_{max}, the probability of a particular fraction of coverage loss to occur marginally decreases or mostly remains the same with an increase in the transmission (when the number of static nodes is fixed) and marginally increases or mostly remains the same with an increase in the number of static nodes (when the transmission range per node is fixed). Overall, for a given value of v_{max}, the probability for a particular fraction of coverage loss to occur varies significantly for the LET-DG trees with respect to variations in the number of static nodes and transmission ranges per node, whereas no such drastic variations are observed for the MST-DG trees.

13.6 Conclusions and Future Work

We have proposed two distributed algorithms to construct stable predicted link expiration time–based data gathering (LET-DG) trees that also incur lower delay per round and energy-efficient minimum-distance spanning tree–based data gathering (MST-DG) trees that incur a larger node and network lifetimes. Performance comparison studies of the two data gathering trees under diverse simulation conditions of network mobility (the node velocity and number of static nodes vary) and density (the transmission range per node varies) indicate a trade-off between stability-delay and node-network lifetime-coverage loss. The LET-DG trees sustain for a longer time, as well as incur a significantly lower delay per round; however, due to repeated use of certain nodes as part of stable data gathering, the LET-DG trees suffer from premature node failures. Still, the network lifetime observed with the LET-DG trees is only at most 25% less than that observed with the MST-DG trees. The MST-DG trees incur a significantly longer node lifetime (time of first node failure) than that of the LET-DG trees by as large as 85%–150%; however, due to frequent tree reconfigurations and larger depth of the tree, almost all the nodes in a MST-DG tree lose about the same amount of energy. As a result, even though the first node failure occurs after a prolonged time, the subsequent node failures occur more quickly, within a short span of time, contributing to only about 15%–30% additional network lifetime beyond the time of first node failure.

The MST-DG trees have been observed to inflict a lower fraction of coverage loss on the network (at a higher probability, though) than the LET-DG trees. With the fraction of loss of coverage ranging from 0 to 1.0, measured in increments of 0.01, we observe the relative difference to be as large as between 0.1 and 0.25, observed when operated in high density (higher transmission range per node) at both low- and high-mobility scenarios. In sync with the lower fraction of loss of coverage, we also observe the MST-DG trees to sustain a larger value for the time at which the network suffers a particular fraction of loss of coverage. In other words, the coverage loss time incurred by the MST-DG trees is larger than that by the LET-DG trees for all operating conditions, and the difference ranges from 16% to 26% and 6% to 16% in the low- and high-mobility scenarios, respectively.

With respect to the impact of the operating conditions on the different performance metrics, we observe the lifetime of the LET-DG trees to be significantly larger than that of the MST-DG trees at low node mobility and moderate transmission ranges per node, and converge to that of the MST-DG trees at higher node mobility and large transmission ranges per node. As we increase the number of static nodes and reduce the number of mobile nodes, we observe the lifetime of LET-DG trees to increase at a relatively faster rate, but the node and network lifetimes observed with the use of MST-DG trees also substantially increase. The lifetime of both the LET-DG and MST-DG trees are observed to be lower at energy-constrained scenarios than those incurred at the sufficient energy scenarios. The difference in the magnitude increases with an increase in transmission range per node (for fixed node mobility) and decreases with an increase in node mobility (for a fixed transmission range per node). The delay per round of data gathering is least affected when the network is operated under energy constraints and under a mix of static and mobile nodes. The delay per round is rather more influenced by the topological structure of the two data gathering trees. The coverage loss (measured as the fraction of coverage loss and the corresponding time) inflicted by the data gathering trees on the network is influenced by three main parameters: transmission range per node, number of static nodes, and node mobility, in decreasing order of severity.

As part of future work, we plan to develop an energy-efficient version of the LET-DG tree algorithm by incorporating a variant of the expiration time of the link based on the residual energy available at the end nodes of the link and using only the minimum of the link expiration times predicted due to mobility and available energy. This way, nodes that have low energy are not likely to be selected as intermediate nodes of the LET-DG trees that tend to exist for a relatively longer time. We also plan to develop secure versions of the LET-DG and MST-DG algorithms. We will focus on securing the communication between the sink node and the leader node, as well as the tree construction process, to prevent false packet injection, spoofing, and denial-of-service attacks. We will also integrate a trust model to the data aggregation algorithms so that faulty and compromised nodes can be efficiently and effectively isolated as part of node failures and the data gathering tree can be expediently reconfigured with much loss of time.

References

1. C. Intanagonwiwat, R. Govindan, D. Estrin, J. Heidemann, and F. Silva, Directed Diffusion for Wireless Sensor Networking, *IEEE/ACM Transactions on Networking*, vol. 11, no. 1, pp. 2–16, 2003. doi: 10.1109/TNET.2002.808417.

2 M. Singh, M. Sethi, N. Lal, and S. Poonia, A Tree Based Routing Protocol for Mobile Sensor Networks (MSNs), *International Journal on Computer Science and Engineering*, vol. 2, no. 1S, pp. 55–60, 2010.

3. N. Meghanathan, Performance Comparison of Minimum Hop and Minimum Edge Based Multicast Routing under Different Mobility Models for Mobile Ad Hoc Networks, *International Journal of Wireless and Mobile Networks*, vol. 3, no. 3, pp. 1–14, 2011. doi: 10.5121/ijwmn.2011.3301.

4. N. Meghanathan and A. Farago, On the Stability of Paths, Steiner Trees and Connected Dominating Sets in Mobile Ad Hoc Networks, *Ad Hoc Networks*, vol. 6, no. 5, pp. 744–769, 2008. doi: 10.1016/j .adhoc.2007.06.005.

5. W. Heinzelman, A. Chandrakasan, and H. Balakarishnan, Energy-Efficient Communication Protocols for Wireless Microsensor Networks, in *Proceedings of the 33rd Hawaii International Conference on Systems Science*, Maui, HI, January 4–7, 2000. doi: 10.1109/HICSS.2000.926982.

6. T. Banerjee, B. Xie, J. H. Jun, and D. P. Agarwal, LIMOC: Enhancing the Lifetime of a Sensor Network with Mobile Clusterheads, in *Proceedings of the Vehicular Technology Conference Fall*, Baltimore, MD, September 30–October 3, 2007, pp. 133–137. doi: 10.1109/VETECF.2007.43.

7. G. S. Kumar, M. V. V. Paul, and K. J. Poulose, Mobility Metric Based LEACH-Mobile Protocol, in *Proceedings of the 16th International Conference on Advanced Computing and Communications*, Chennai, India, December 14–17, 2008, pp. 248–253. doi: 10.1109/ADCOM.2008.4760456.

8. S. Deng, J. Li and L. Shen, Mobility-Based Clustering Protocol for Wireless Sensor Networks with Mobile Nodes, *IET Wireless Sensor Systems*, vol. 1, no. 1, pp. 39–47, 2011. doi: 10.1049/iet-wss.2010.0084.

9. H. K. D. Sarma, A. Kar, and R. Mall, Energy Efficient and Reliable Routing for Mobile Wireless Sensor Networks, in *Proceedings of the 6th IEEE International Conference on Distributed Computing in Sensor Systems Workshops*, Santa Barbara, CA, June 21–23, 2010, pp. 1–6. doi: 10.1109/DCOSSW .2010.5593277.

10. M. Zhao and Y. Yang, Bounded Relay Hop Mobile Data Gathering in Wireless Sensor Networks, in *Proceedings of the 6th IEEE International Conference on Mobile Ad Hoc and Sensor Systems*, Macau SAR, China, October 12–15, 2009, pp. 373–382. doi: 10.1109/MOBHOC.2009.5336976.

11. G. Xing, T. Wang, W. Jia, and M. Li, Rendezvous Design Algorithms for Wireless Sensor Networks with a Mobile Base Station, in *Proceedings of the 9th ACM International Symposium on Mobile Ad Hoc Networking and Computing*, Hong Kong SAR, China, May 27–30, 2008, pp. 231–240. doi: 10.1145 /1374618.1374650.

12. W. Wu, H. B. Lim, and K.-L. Tan, Query-Driven Data Collection and Data Forwarding in Intermittently Connected Mobile Sensor Networks, in *Proceedings of the 7th International Workshop on Data Management for Sensor Networks*, Singapore, September 13, 2010, pp. 20–25. doi: 10.1145 /1858158.1858166.

13. W. Su and M. Gerla, IPv6 Flow Handoff in Ad Hoc Wireless Networks Using Mobility Prediction, in *Proceedings of the IEEE Global Telecommunications Conference*, Rio de Janeireo, Brazil, December 5–9, 1999, pp. 271–275. doi: 10.1109/GLOCOM.1999.831647.

14. F. Kuhn, T. Moscibroda, and R. Wattenhofer, Unit Disk Graph Approximation, in *Proceedings of the Workshop on the Foundations of Mobile Computing* (*DIALM-POMC*), Philadelphia, PA, October 2004, pp. 17–23. doi: 10.1145/1022630.1022634.

15. T. S. Rappaport, *Wireless Communications: Principles and Practice*, 2nd ed., Prentice Hall, Piscataway, NJ, 2002.

16. S. Lindsey, C. Raghavendra, and K. M. Sivalingam, Data Gathering Algorithms in Sensor Networks Using Energy Metrics, *IEEE Transactions on Parallel and Distributed Systems*, vol. 13, no. 9, pp. 924–935, 2002. doi: 10.1109/TPDS.2002.1036066.

17. A. J. Viterbi, *CDMA: Principles of Spread Spectrum Communication*, 1st ed., Prentice Hall, Piscataway, NJ, pp. 271–275.

18. T. H. Cormen, C. E. Leiserson, R. L. Rivest, and C. Stein, *Introduction to Algorithms*, 3rd ed., MIT Press, Cambridge, MA, 2009.

19. K. Fall and K. Varadhan, ns-2 Notes and Documentation, VINT Project at LBL, Xerox PARC, UCB, and USC/ISI. http://www.isi.edu/nsnam/ns/ (accessed March 17, 2015).

20. G. Bianchi, Performance Analysis of the IEEE 802.11 Distributed Coordinated Function, *IEEE Journal on Selected Areas in Communications*, vol. 18, no. 3, pp. 535–547, 2000. doi: 10.1109/49.840210.

21. H. Zhang and J. C. Hou, Maintaining Sensing Coverage and Connectivity in Large Sensor Networks, *Wireless Ad hoc and Sensor Networks: An International Journal*, vol. 1, no. 1–2, pp. 89–123, January 2005.

22. C. Bettstetter, H. Hartenstein, and X. Perez-Costa, Stochastic Properties of the Random-Way Point Mobility Model, *Wireless Networks*, vol. 10, no. 5, pp. 555–567, 2004. doi: 10.1023 /B:WINE.0000036458.88990.e5.

Chapter 14

Sensor Proxy Mobile IPv6: A Novel Scheme for Mobility-Supported IP-WSN

Md. Motaharul Islam

Contents

Abstract

The Internet Protocol–based wireless sensor network (IP-WSN) is gaining importance for its broad range of applications in healthcare, home automation, environmental monitoring, industrial control, vehicle telematics, and agricultural monitoring. In all of these applications, mobility in sensor networks, with special attention to energy efficiency, is a major issue to be addressed. A host-based mobility management protocol is not suitable for IP-WSN because of its energy inefficiency. So, a network-based mobility management protocol can be an alternative for the mobility-supported IP-WSN. In this chapter, we propose a network-based mobility-supported IP-WSN protocol called Sensor Proxy Mobile IPv6 (SPMIPv6). We present its architecture and message formats and also evaluate its performance considering message flows.

14.1 Introduction

Recent advancements in microelectromechanical and wireless communication systems have enabled the development of low-cost, low-power, multifunctional sensor nodes that are small in size and can communicate at short distances [1,2]. A sensor network is a special type of communication network that is composed of a large number of tiny sensor nodes that are densely deployed either inside the phenomena or very close to it [1]. Sensors are generally equipped with data processing and communication capabilities. The sensing circuitry senses the environment surrounding the sensor and transforms it into an electric signal. The sensor sends such a signal, usually via radio transmitter, to the sink node either directly or through other sensor nodes.

In the past, applications of sensor networks were thought to be very specific. The communication protocol of sensor networks was very simple and straightforward. Even some of researchers were against the use of the internetworking concept for wireless sensor networks (WSNs) for reasons such as resource constraints for layered architecture, the configuration problem of a large number of devices, and the essence of sensor nodes' distinct identity. But with the advent of the Internet of Things and the federated Internet Protocol–based wireless sensor network (IP-WSN), this demand is becoming blurred. The huge number of IPv6 addresses, the necessity of end-to-end communication, and the advancement of microelectronics has changed the concept of the research community. Now, a tiny sensor node can hold the compatible Transmission Control Protocol (TCP)/IP protocol stack [3]. So, we can now think of using the concept of the internetworking protocol in the IP-WSN [4]. We can easily think of providing an IPv6 address to an individual sensor node since it provides around 6×10^{23} addresses per square meter of the earth's space.

The IPv6 over Low-Power Wireless Personal Area Network (6LoWPAN) Working Group of the Internet Engineering Task Force (IETF) defines the manner in which IPv6 communication is to be carried out over the IEEE 802.15.4 interface [5,6]. Although 6LoWPAN helps to make the wide implementation of IP-WSN a reality and its end-to-end communication to the external world feasible, excessive signaling costs for sensor nodes because of too much tunneling through the air make it difficult. Excessive signaling costs therefore become a barrier for the application of IP-WSN, especially in the case of the mobility scenario of individual sensor nodes or the group of nodes in different areas, such as in a patient's body sensor network or in industrial automation [7]. Nowadays, most of the communication protocols are host based, which is practically infeasible for IP-WSN since an individual node needs to participate in mobility-related signaling. The PMIPv6, a network-based protocol, provides mobility support to any IPv6 host within a restricted and topologically localized portion of the network, without requiring the host to participate in any mobility-related signaling.

6LoWPAN-based IP-WSN may use the sensor network–compatible PMIPv6 to introduce and enhance the mobility scenario in a localized domain. With this thinking in mind, in this work, we focus on the mobility in IP-WSN with the attention of an energy-efficient network-based communication protocol for IP-WSN. In this regard, we have proposed the SPMIPv6 for IP-WSN. In our proposed scheme, we use a sensor network–based localized mobility anchor, mobile access gateway, and multiple smart and fully functional sensor nodes that are IPv6 header stack enabled. These IP devices, including the sensor nodes, make it feasible to implement the SPMIPv6 over IP-WSN.

There are different application areas where IP-WSN can be used, such as industrial control, structural monitoring, healthcare, smart homes, vehicle telematics, and agricultural monitoring [8]. In these scenarios, node-to-node communication is very important since these are collaborative works. In these cases, IP-WSN-based on a mesh approach can enhance the communication scenario significantly, where individual sensor nodes can act as a router or fully functional device; this is discussed in the route-over-routing issue in 6LoWPAN [8,9]. Moreover, SPMIPv6-based IP-WSN facilitates node-to-node seamless communication in different cases of the individual sensor node and sensor network mobility scenarios.

In this chapter, we discuss the following issues:

■ IP-WSN architecture based on PMIPv6, named SPMIPv6, proposed for energy-efficient mobility of an individual sensor node or a group of sensor nodes
■ Applicability of mobility issues of individual or groups of IP sensor nodes in a patient care scenario
■ Functional architecture, respective message formats, and sequence diagram

The rest of the chapter is organized as follows: Section 14.2 reviews background related to PMIPv6 and 6LoWPAN. The proposed SPMIPv6 protocol architecture, along with its mobility scenarios, sequence diagram, and message formats, is depicted in Section 14.3. Section 14.4 discusses WSN versus IP-WSN, Section 14.5 highlights IP-WSN trends toward the Internet of Things (IoT), Section 14.6 mentions a few future research directions, and Section 14.7 introduces some upcoming research challenges. Finally, Section 14.8 concludes the chapter.

14.2 Background

14.2.1 Overview of PMIPv6

The foundation of PMIPv6 is based on MIPv6 in the sense that it extends MIPv6 signaling and reuses many concepts, such as the home agent (HA) functionality [10,11]. However, PMIPv6 is designed to provide network-based mobility management support to a mobile node (MN) in a topologically localized domain [12]. Therefore, an MN is free from participation in any mobility-related signaling, and the proxy mobility agent in the serving network performs mobility-related signaling on behalf of the MN. Once an MN enters its PMIPv6 domain and performs access authentication, the serving network ensures that the MN is always on its home network and can obtain its home address on any access network. The serving network assigns a unique home network prefix to each MN, and conceptually, this prefix always follows the MN wherever it moves within a PMIPv6 domain. From the perspective of the MN, the entire PMIPv6 domain appears as its home network. Accordingly, it is needless to configure the care of address at the MN. The new functional entities of PMIPv6 are the mobile access gateway (MAG) and local mobility anchor (LMA) [13]. The MAG typically runs on the access router (AR). The main role of the MAG is to detect the MN's movements and initiate mobility-related signaling with the LMA on behalf of the MN. In addition, the MAG establishes a tunnel with the LMA for enabling the MN to use an address from its home network prefix and emulates the MN's home network on the access network for each MN. On the other hand, the LMA is similar to the HA in MIPv6. However, it has additional capabilities required to support PMIPv6. The main role of the LMA is to maintain reachability to the MN's address while it moves around within a PMIPv6 domain, and the LMA includes a binding cache entry for each currently registered MN. The binding cache entry maintained at the LMA is more extended than that of the HA in MIPv6, with some additional fields, such as the MN identifier, the MN's home network prefix, a flag indicating a proxy registration, and the interface identifier of the bidirectional tunnel between the LMA and MAG. Such information associates an MN with its serving MAG, and enables the relationship between the MAG and LMA to be maintained.

14.2.2 6LoWPAN

The 6LoWPAN [5] Working Group of the IETF has defined an adaptation layer for sending IPv6 packets over IEEE 802.15.4. The goal of 6LoWPAN is to reduce the sizes of IPv6 packets to make them fit in 127-byte IEEE 802.15.4 frames. The 6LoWPAN proposal consists of a header compression scheme, a fragmentation scheme, and a method for framing IPv6 link local addresses into IEEE 802.15.4 networks [6]. The proposal also specifies enhanced scalabilities and mobility of sensor networks. The challenge to 6LoWPAN lies in the sizable differences between an IPv6 network and an IEEE 802.15.4 network. The IPv6 network defines a maximum transmission unit as 1280 bytes, whereas the IEEE 802.15.4 frame size is 127 octets. Therefore, the adaptation layer between the IP layer and the medium access control (MAC) layer must transport IPv6 packets over IEEE 802.15.4 links. The adaptation layer is responsible for fragmentation, reassembly, header compression and decompression, mesh routing, and addressing for packet delivery under the mesh topology. The 6LoWPAN protocol supports a scheme to compress the IPv6 header from 40 bytes to 2 bytes [7,14].

14.2.3 Related Work

In comparison to conventional WSN, IP-WSN provides flexibility to access and manage individual sensor nodes locally or from a remote site through the Internet. This specific issue provides great significance to different mission-specific applications. Our chapter focuses on the energy-efficient mobility of both the IP sensor node and the sensor network, which is based on PMIPv6 and 6LoWPAN. There are a few approaches proposed in the literature regarding the IP-WSN and its mobility issue. RFC 5213 gives a complete description of PMIPv6 [11]. RFC 4944 depicts the overall aspect of 6LoWPAN, the packet format provided by IETF to ensure IPv6 communication over low-power wireless personal area networks. In [9], the authors represent the overall scenario for extending IP to LoWPAN, such as the 6LoWPAN adaptation layer, header format encapsulation, and compression, and finally focus on addressing, routing, autoconfiguration, and neighbor discovery issues. In [14], the authors propose a network mobility protocol to support 6LoWPAN, but they do not consider the 6LoWPAN node mobility. The respective authors only consider network mobility that is performed with the help of a mobile router; they do not consider a unique address for each sensor node. In [15], the authors propose a dual-addressing scheme for 6LoWPAN-based WSN. It combines a global unicast address to cope with association link changes and node mobility, and it links local addresses to reduce the overhead. However, in our proposed scheme, we consider only a global unicast address. Instead of using a global unicast address, our scheme is energy efficient since it uses SPMIPv6 for the communication protocol. In [8], the authors describe some real-life design and application spaces for 6LoWPANs, which shows the importance 6LoWPANs will have in the near future.

14.3 SPMIPv6 Protocol

14.3.1 SPMIPv6 Protocol Architecture

To introduce energy-efficient mobility, we propose Sensor Proxy Mobile IPv6 (SPMIPv6) for IP-WSN. It is a localized mobility management protocol based on PMIPv6. The SPMIPv6 architecture consists of a sensor network–based localized mobility anchor (SLMA), a sensor network–based mobile access gateway (SMAG), and many fully functional IPv6 header stack-enabled IP sensor nodes. We discuss the individual module of the proposed architecture with the help of Figure 14.1.

14.3.1.1 SLMA

The SLMA acts as a topological anchor point for the entire SMAGs and IP sensor nodes. The main role of the SLMA is to maintain accessibility to the sensor node while the node moves within or outside of the SPMIPv6 domain. The SLMA includes a binding cache entry for each sensor node, an encapsulation and de-capsulation section, and a SMAG information table. The binding cache entry at the SLMA is used for holding the information of the mobile sensor node. It includes different information, such as a sensor node address, the sensor node's home network prefix, and a flag bit indicating the sensor proxy registration.

The SLMA has sufficient memory to hold all of these necessary records. It has sufficient power supply and processing capability. It also acts as the interfacing device between the SPMIPv6

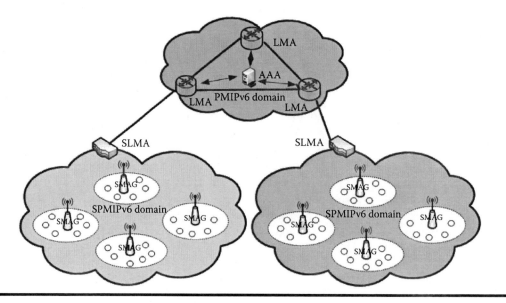

Figure 14.1 Sensor Proxy Mobile IPv6 architecture.

domain and PMIPv6 domain. As an interfacing entity, it facilitates the communication of IP-WSN with the Internet world. In this scheme, an authentication, authorization, and accounting (AAA) service has been integrated within SLMA. We name this integrated service sensor network–based authentication, authorization, and accounting (SAAA) scheme. The SAAA scheme helps the SMAG and sensor node to obtain secured mobility in the SPMIPv6 domain. It also facilitates authentication services for each fully functional sensor node.

14.3.1.2 SMAG

The SMAG acts like a sink node in the traditional sensor network. In SPMIPv6, it acts like an access gateway router, with the main function of detecting sensor node movement and initiating mobility-related signaling with the sensor node's SLMA on behalf of the sensor node. The SMAG can move with its member sensor node as a small IP-WSN domain, similar to the body sensor network of a patient. It consists of different functional modules, such as routing, neighbor discovery, sensor information table, adaptation module, and interfacing module to the sensor node and SLMA. The routing module performs efficient data transmission among individual sensor nodes and facilitates the end-to-end communication. The neighbor discovery module performs neighbor discovery and duplicate address detection functionality. The adaptation module performs the task of transmitting IPv6 packets over the IEEE 802.15.4 link, as mentioned in the 6LoWPAN adaptation layer. The sensor information table provides up-to-date information of the sensor nodes to the SLMA. It works in close association with the binding cache entry of the SLMA, and the two interfacing modules communicate with the SLMA and sensor nodes.

14.3.1.3 IP Sensor Node

The SPMIPv6 domain consists of numerous sensor nodes based on the IPv6 address. We consider the domain a federated IP sensor domain. There are two types of sensor nodes. The first type contains the tiny TCP/IP communication protocol stack with an adaptation layer and IEEE 802.15.4 interface. It can forward information to other nodes of a similar type, as well as sense information from the environment. Actually, this type of sensor node acts as a mini–sensor router. The second type of sensor node has protocol stack and environment sensing capability, but can forward the sensed information to a nearby mini–sensor router node. However, both types are able to perform end-to-end communication.

14.3.2 Operational Architecture of SPMIPv6

Figure 14.2 shows a detailed operational architecture of the SPMIPv6, which includes SLMA, SMAG, and IP sensor nodes. It also shows how these entities communicate with each other by different types of interfaces. The SMAG needs two or more interfaces for communication with different access networks, such as SPMIPv6-based IP-WSN and an external PMIPv6 network.

It includes network layer, adaptation layer, and physical layer functionality. The network layer provides different functionalities, such as addressing, routing, neighbor discovery, and the data structure for holding IP sensor node information. The most important layer is adaptation, which ensures mesh routing, compression and decompression, and fragmentation and reassembly. The physical layer provides access to different physical interfaces. SLMA holds network-related information, such as binding cache entry, encapsulation, and de-capsulation. Binding cache entry provides the data structure to hold different information, such as a new SPMIPv6 flag, a link local address for each interface, a home prefix, a bidirectional tunnel interface identifier, access technology, and a time stamp. Finally, the IP sensor node works as a very smart, tiny 6LoWPAN-based node. It holds a tiny TCP/IP protocol stack. Moreover, it holds the basic functionality of the adaptation layer. All the sensor nodes consist of an IPv6 address for local and global communications.

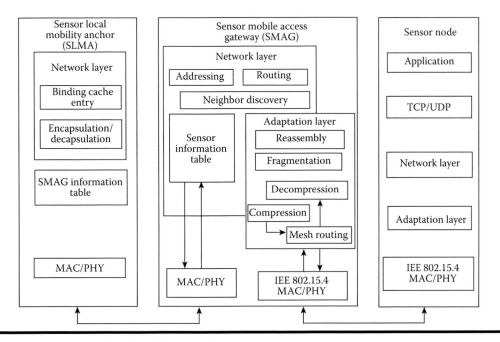

Figure 14.2   Operational architecture of SPMIPv6.

14.3.3 Application Scenarios of IP-WSN Based on SPMIPv6

This section briefly covers different SPMIPv6-based IP-WSN application scenarios [8]. A complete description is out of the scope of this chapter.

14.3.3.1 Industrial Monitoring

SPMIPv6-based IP-WSN applications for industrial monitoring can be related to a broad range of methods to increase the productivity, energy efficiency, and safety of industrial operations in engineering facilities and manufacturing plants. Many industries currently use time-consuming and expensive manual monitoring to predict failures and schedule maintenance or replacements, in order to avoid costly manufacturing downtime. SPMIPv6-based IP-WSN can be inexpensively installed and provide more frequent and reliable data. The deployment of SPMIPv6-based IP-WSNs can reduce equipment downtime and eliminate manual equipment monitoring that is costly to carry out. Additionally, data analysis functionality can be placed into the network, eliminating the need for manual data transfer and analysis.

14.3.3.2 Structural Monitoring

Intelligent monitoring in facility management can make safety checks and periodic monitoring of the architecture status highly efficient. Powered nodes can be included in the design phase of a construction, or battery-equipped nodes can be added afterward. All nodes are static and manually deployed. Some data, such as normal room temperature, is not critical for security protection, but event-driven emergency data must be handled in very critical manner.

14.3.3.3 Healthcare

SPMIPv6-based IP-WSNs are envisioned to be heavily used in healthcare environments. Although hospital scenarios can be handled differently, IP-WSN provides big potential to ease the development of new services by getting rid of cumbersome wires and simplifying patient care in hospitals and for home care. The worldwide population of people age 65 and older is expected to more than double by 2025, from 357 million to 761 million [16]. The speed with which this age-structural change is taking place implies an urgent need for solutions that will relieve the mounting pressure on our healthcare system, as well as support a better quality of life and quality of care for our aged.

14.3.3.4 Connected Home

The connected home or smart home is no doubt an area where SPMIPv6-based IP-WSN can be used to support an increasing number of services: home safety and security, home automation and control, personal healthcare, smart appliances, and home entertainment systems. In home environments, SPMIPv6-based IP-WSN networks typically comprise a few dozen nodes, and probably in the near future, they will comprise a few hundred nodes of various natures: IP sensors, actuators, and connected objects.

14.3.3.5 Vehicle Telematics

SPMIPv6-based IP-WSNs play an important role in intelligent transportation systems. Incorporated in roads, vehicles, and traffic signals, they contribute to the improvement of safety of transporting systems. Through traffic or air-quality monitoring, they increase the possibilities in terms of traffic flow optimization and help reduce road congestion.

14.3.3.6 Agricultural Monitoring

Accurate temporal and spatial monitoring can significantly increase agricultural productivity. Due to natural limitations, such as a farmer's inability to check his crop at all times of the day or inadequate measurement tools, luck often plays too large of a role in the success of harvests. With an SPMIPv6-based IP-WSN, indicators such as temperature, humidity, and soil condition can be automatically monitored without labor-intensive field measurements. For example, SPMIPv6-based IP-WSNs could provide precise information about crops in real time, enabling businesses to reduce water, energy, and pesticide usage and enhance environment protection. The sensing data can be used to find optimal environments for the plants. In addition, the data on the planting condition can be saved by sensor tags, which can be used in supply chain management.

14.3.4 Mobility Scenario of IP-WSN Based on SPMIPv6

To represent the mobility scenario, we consider a state-of-the-art technology-based patient care unit in a specialized hospital. Figure 14.3 represents such a scenario where a patient can get both special care in the hospital and personal home care from a remote specialized hospital. In this scenario, we try to depict different types of mobility in the SPMIPv6 environment [17]. Sensor nodes

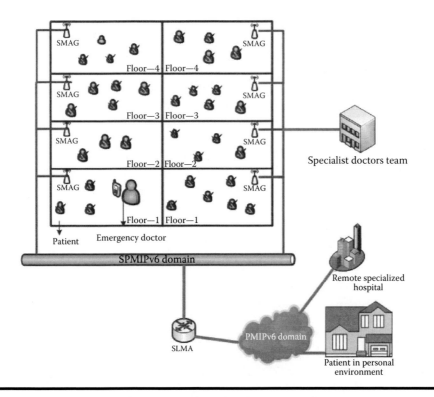

Figure 14.3   Mobility scenario in SPMIPv6-based patient care unit.

are deployed on the body of the patients, as well as all over the environment. The patient care unit ensures real-time care and observation of the patients from a central specialized doctor's forum. In case of emergency, patients can move from one place to another with their complete setup, so that seamless connectivity with the sophisticated medical equipment remains established and doctors can monitor the patient online from the central doctor's research group. All the floors are considered a single SPMIPv6 domain. Inside the SPMIPv6 domain, there are six floors considered individual SMAG domains.

The mobility issues considered in this scenario are

Case I: Movement of nodes within the same SMAG domain of the SPMIPv6 domain
Case II: Movement of nodes between different SMAGs of the same SPMIPv6 domain
Case III: Movement of nodes between different SMAGs of different SPMIPv6 domains
Case IV: Movement of a SMAG-based personal area network (PAN) within the same SPMIPv6 domain
Case V: Movement of a SMAG-based PAN between different SPMIPv6 domains
Case VI: Patient monitoring in personal home environment

These scenarios are explained below:

Case I: In this case, the mobility of the nodes will be handled by the appropriate SMAG, without the involvement of the SLMA. This, the simplest mobility scenario, arises frequently in hospital management: a patient can move within the PAN of a single branch of the hospital for different purposes, such as exercise and fresh air.

Case II: In this case, mobility will be handled by the appropriate SMAG with minimal initiative from the SLMA. The initial coordination will be performed by the SLMA alone; then the SMAG will oversee the remaining procedures. In our hospital management model, a patient can move from one PAN to another PAN in the same branch of the hospital.

Case III: In this case, mobility is interdomain, using the public PMIPv6 domain. The LMA, AAA, and SLMA will coordinate with one other. In our hospital management model, a patient can move on an emergency basis from one PAN of a hospital branch to a PAN of another branch of the same hospital.

Case IV: In this case, mobility is based on the network mobility (NEMO) protocol [14], confined to the same domain. Only the SLMA and corresponding SMAGs will be involved. In our hospital management model, a patient with the whole setup can move from one PAN to another PAN.

Case V: This case is also based on the NEMO protocol, but is much different from Case IV. In our hospital management model, a patient can move on an emergency basis with his or her whole setup from one branch of a hospital to the more specialized branch of the same hospital.

Case VI: Due to the increasing number of the aging demographic group, we consider this case so that a patient can be monitored continuously from the patient's personal environment, as discussed in our recent paper [16].

Although some of the mobility scenarios may seem complicated in cases of IP-based wireless sensor networks, they are still considered for future research issues. Because of the essence of a huge deployment of IP sensor networks in healthcare, departmental stores, and industrial automation, this mobility scenario will eventually come true in the future.

14.3.5 Message Flow in SPMIPv6

In the SPMIPv6 architecture, authentication features are integrated in SLMA instead of these two functions being performed separately. The proxy binding update message and SAAA query message are combined together and named binding update and authentication query. Also, the SAAA reply and proxy binding acknowledge message are merged together and named binding acknowledge and authentication reply. For this reason, the number of communications is greatly reduced in SPMIPv6 compared to in conventional PMIPv6. The reduced communication scenario has been drawn and compared with the conventional PMIPv6 sequence diagram. The steps of the sequence diagram in both PMIPv6 and SPMIPv6 are discussed below.

Figure 14.4a depicts the sequence diagram of the message flow with respect to PMIPv6. For simplicitym in Figure 14.4a, we combine more than one communication within a single step.

Step 1: When a sensor node first attaches to a SMAG domain, the access authentication procedure is performed using the sensor node address.

Step 2: After successful access authentication, the SMAG obtains the sensor node's profile from the AAA service policy store. This profile contains the sensor node's address, SLMA address, supported address configuration mode, and other associated information.

Step 3: The SMAG sends a proxy binding update (PBU) message, including the MN address, to the sensor node's SLMA on behalf of the sensor node.

Step 4: Once the SLMA receives the PBU message, it checks the policy store to ensure that the sender is authorized to send the PBU message. If the sender is a trusted SMAG, the SLMA accepts the PBU message.

Step 5: The SLMA sends a proxy binding acknowledgment (PBA) message, including the MN's home network prefix option, and establishes a route for the sensor node's home network prefix over the tunnel to the SMAG.

Figure 14.4b shows the sequence diagram of the overall message flow with respect to SPMIPv6. The number of communications is noticeably reduced in SPMIPv6 in comparison to in the conventional PMIPv6 shown in Figure 14.4a.

Step 1: When the location of an IP sensor node is changed, it sends a solicitation message for the nearest SMAG discovery mechanism.

Step 2: SMAG sends a binding update and authentication query to SLMA integrated with SAAA.

Step 3: In response to the binding update and authentication query message, SLMA integrated with SAAA sends back a binding acknowledge and authentication reply message.

Step 4: Finally, SMAG sends the advertisement message to the respective IP sensor node, and thus the IP sensor node is connected to the nearest SMAG.

Step 5: Now, the sensor node is able to communicate with the corresponding node based on SPMIPv6. Thus, data can be transmitted from the IP sensor node to the correspondence node, and vice versa.

14.3.6 Proxy Binding Message Format of SPMIPv6

In the proposed proxy binding update (PBU) and proxy binding acknowledgment (PBA) messages in Figure 14.5, we have specified a flag bit S. If this S flag is set, it indicates SMIPv6 operations.

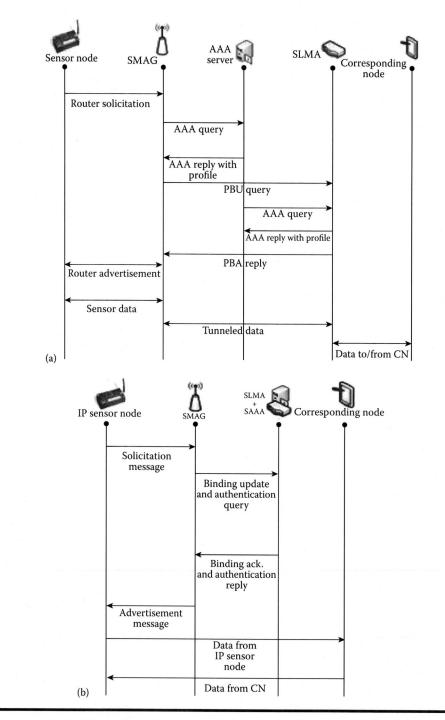

Figure 14.4 (a) Sequence diagram in PMIPv6. (b) Sequence diagram in SPMIPv6.

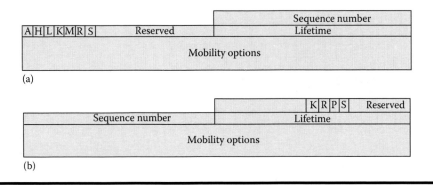

(a)

(b)

Figure 14.5   (a) SPMIPv6 PBU message format. (b) SPMIPv6 PBA message format.

If the S bit is not set, this indicates non-SPMIPv6 operations. The meanings and descriptions of the other flags are described in different request for comments (RFCs) [11,18–20]. The mobility options header plays an important role for the different types of mobility scenarios. Some of the mobility header options are handoff indicator options, access technology-type options, link-layer identifier options, link local address options, and time-stamp options.

14.3.7 Gateway Router Solicitation and Advertisement Message Format of SPMIPv6

Figure 14.6 depicts the router solicitation (RS) and router advertisement (RA) message formats of SPMIPv6. Figure 14.6a shows the RS message that is sent by the IP-WSN sensor node. The IEEE 802.15.4 MAC header's source and destination addresses are used as the MAC address of the source and destination sensor nodes, respectively. Dispatch of the IP-WSN addressing header indicates the compressed IPv6 header. The header compression section compresses different fields of the IP-WSN addressing header. The router solicitation message's RS options enable the IP-WSN gateway to obtain the sensor node's MAC address, link local address, and sensor node's ID. The RA message format has two options, such as home network prefix (HNP) and address options. The HNP options are used to emulate the IP-WSN sensor node's home network.

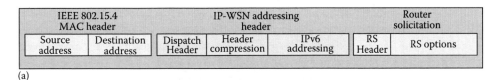

(a)

(b)

Figure 14.6   (a) SPMIPv6 router solicitation (RS) message format. (b) SPMIPv6 router advertisement (RA) message format.

14.4 WSN versus IP-WSN

A sensor network is a special type of communication network that is composed of a large number of tiny sensor nodes that are densely deployed either inside the phenomenon or very close to it. Sensors are generally equipped with data processing and communication capabilities. The sensing circuitry senses the environment surrounding the sensor and transforms it into an electric signal. The sensor sends such a signal, usually via radio transmitter, to the sink node either directly or through other sensor nodes. IP-WSN is a special type of sensor network in which communication among sensor nodes is based on the Internet Protocol. In IP-WSN, an individual node is identified by IP address instead of the specific ID used in traditional WSNs.

The IP-WSN consists of numerous sensor nodes based on IPv6 addresses. There are two types of sensor nodes. The first type contains the tiny TCP/IP communication protocol stack with an adaptation layer and IEEE 802.15.4 interface. This type can forward information to another node of a similar type, as well as sense information from the environment. Actually, this type of sensor node acts as a mini–sensor router and is considered a fully functional device. The second type of sensor node has the protocol stack and environment sensing capability, but can only forward the sensed information to nearby mini–sensor router nodes. These types of sensor nodes are considered reduced-function devices. Nevertheless, both types are able to perform end-to-end communication. IP-WSN nodes known as reduced-function devices (RFDs) will be identified within a particular wireless personal area network domain by a 16-bit short address. The sensor mobile access gateway (SMAG) known as a full functional device (FFD) will hold IEEE EUI-64, and finally, a border router (BR) will hold 128 bits of global unicast addressing. The RFD is assigned a 16-bit short address, which is unique within a WPAN or SMAG domain, and remains fixed irrespective of its location within the WPAN. All three levels of addresses are created hierarchically. Sixteen-bit short addresses are assigned to a RFD at the time of deployment. A EUI-64 identifier is formed based on the EUI-64 bit identifier.

14.5 IP-WSN Trends toward Internet of Things

IoT has been defined from various different perspectives, and hence numerous definitions for IoT exist in the literature. The reason for the apparent fuzziness of the definition stems from the fact that it is syntactically composed of two terms: *Internet* and *Things*. The first one pushes toward a network-oriented vision of IoT, while the second tends to move the focus to generic objects to be integrated into a common framework [21]. However, the terms *Internet* and *Things*, when put together, assume a meaning that introduces a disruptive level of innovation into the Information and Communication Technology (ICT) world. In fact, IoT semantically means a worldwide network of interconnected objects, uniquely addressable, based on standard communication protocols [22]. This implies that a huge number of possibly heterogeneous objects are involved in the process. In IoT, the unique identification of objects and the representation and storage of exchanged information are the most challenging issues. This brings the third perspective of IoT: the semantic perspective. The main concepts, technologies, and standards are highlighted and classified with reference to the three visions of IoT: Things oriented, Internet oriented, and Semantic oriented. The IoT paradigm will lead to the convergence of the three visions accordingly. A Things-oriented vision includes radio frequency identifiers (RFIDs), smart items, and sensors and actuators. The Internet-oriented vision encompasses IPSO (IP for smart objects), web of things, connectivity, and communicating things. The semantic-oriented vision consists of reasoning over data, semantic technologies, and smart semantic middleware. While the perspective of things focuses on integrating generic objects into a common framework, the perspective of Internet pushes toward a network-oriented definition. According to the IPSO Alliance [23], a forum formed in 2008, the IP stack is a lightweight protocol that already connects a large number of communicating devices and runs on battery-operated devices. This guarantees that IP has all the qualities to make IoT a reality. It is likely that through an intelligent adaptation of IP, by incorporating the IEEE 802.15.4 protocol into the IP architecture, and by adoption of 6LoWPAN, a large-scale deployment of IoT can become a reality. Thus, IP-WSN has a close relationship with the deployment of IoT.

14.6 Future Research Directions

The topic presented in this chapter has opened up many avenues for future works. For the next step, further investigation may be carried out to enhance the shortcomings of the IP-WSN framework we proposed. A wireless node for the Internet of Things is a smart IP-based sensor, developed on a traditional wireless sensor network (WSN). These smart devices are going to revolutionize the cyber physical space. There are a lot of challenges and opportunities remaining in this visionary field that need to be dealt with with care. Out of the many challenges, we focus on a few aspects, such as architectural design, efficient addressing, energy consumption, and state-of-the-art applications. In summary, this chapter provides comprehensive and effective solutions for some important challenging issues of an integrated framework of SPMIPv6-based IP-WSN. In the future, we have a keen interest to work on a large-scale federated IP-WSN or the cloud of Internet of Things. We look forward to the future for a smart IoT-assisted connected world.

Despite the benefits gained by introducing SPMIPv6 to IP-WSN, it has some limitations also. The SLMA in SPMIPv6 maintains binding cache entry (BCE) for the nodes to keep its binding information, and the BCE should be updated at each sensor node movement. In addition, all the mobility-related signaling messages and data packets should be passed through and processed by the SLMA. This extensive access will significantly increase the load of the SLMA, eventually leading to a bottleneck. The SMAGs in SPMIPv6 are responsible for performing mobility-related signaling with SLMA on behalf of the sensor nodes. However, SMAGs can also be overloaded when a large number of sensor nodes are attached to them. No doubt, it is a complicated task to address these issues. Further investigation may be carried out in this regard.

14.7 Upcoming Challenges

To incorporate all the features, there are a lot of challenges to overcome. In the next section, we discuss the major challenges and problem statement.

14.7.1 Architectural

Architecture plays a key role in designing an IP-WSN framework. Suitable architecture, along with appropriate protocol, is a demand of the current age for an integrated framework of smart IP-WSN. In this regard, addressing and routing are also important and require special care. There are many protocols in the existing literature that focus on the energy perspective, but none of them highlight the energy efficiency by taking architectural issues into consideration. Suitable protocol architecture can make today's smart IP-WSN, based on an IP-based wireless sensor network

(IP-WSN), more energy aware. In this regard, we emphasized the architecture of the large-scale IP-WSN, which is the basic building block of smart IoT.

14.7.2 Energy Efficiency

Energy is the most crucial resource for the battery-powered smart IoT, but how to mitigate the energy consumption is a challenging question [24–27]. In this regard, researchers have argued different points of view. In this chapter, we focused on the SPMIPv6 protocol to reduce the energy consumption in the envisioned mobility scenarios. Better performance might be proved in two ways: shorter addressing based on a hierarchical approach and less data transmission by the end sensor devices, because of using SPMIPv6 architecture.

14.7.3 Resource Utilization

It has been observed that most of the sensor nodes remain idle for most of their lifetime, resulting in underutilization of resources. There is a lot of ongoing research to utilize WSN resources in an efficient way. Virtualization of sensor network (VSN) is one of the novel approaches to utilize the physical infrastructure of WSN. VSN can be simply defined as the virtual version of WSN over the physical sensor infrastructure. By allowing sensor nodes to coexist on a shared physical substrate, VSN may provide flexibility, cost-effectiveness, and manageability.

14.7.4 Other Challenges

There are other miscellaneous challenges to integrating the network of smart IP-WSN. The current storage infrastructure and file systems that back up and form the backbone are outdated for a huge number of smart IP-WSNs. The current infrastructure also cannot easily handle the critical and highly secure information flows. It is also necessary to focus on the exponential growth of the growing demand of real-time, reliable data processing of the IP-WSN.

14.8 Conclusions

In this chapter we presented a Sensor Proxy Mobile IPv6 protocol that enhances the mobility issue in IP-WSN. Mobility in sensor networks is the most challenging issue that must be addressed, with special attention to energy efficiency. Since SPMIPv6 is a sensor network–based localized mobility management protocol, it meets the demand of energy efficiency in terms of reducing signaling costs and mobility costs. Here, we also presented its architecture, and messages formats and evaluated its performance by analyzing signaling costs and mobility costs in terms of message flows. We focused on 6LoWPAN-based IP-WSN of the same vendor and protocol stacks. In the future, the sensor networks consisting of multivendor and heterogeneous protocol stacks may be an interesting research area.

References

1. Akyildiz, I. F., Su, W., Sankarasubramaniam, Y., Cayirci, E. A survey on sensor networks. *IEEE Communications Magazine* 2002, 40, 102–114.
2. Kemal, A., Younis, M. A survey on routing protocols for wireless sensor networks. *Ad Hoc Networks* 2005, 3, 325–349.

3. Luo, X., Zheng, K., Pan, Y., Wu, Z. A TCP/IP implementation for wireless sensor networks. In *Proceedings of IEEE International Conference on Systems, Man, and Cybernetics*, Hague, Netherlands, October 10–13, 2004, pp. 6081–6086.

4. Rodrigues, J. J. P. C., Neves, P. A. C. S. A survey on IP-based wireless sensor network solutions. *International Journal of Communication Systems* 2010, 963–981. doi: 10.1002/dac.1099.

5. Kushalnagar, N., Montenegro, G., Schumacher, C. IPv6 over low-power wireless personal area networks (6LoWPANs): Overview, assumptions, problem statement, and goals, RFC 4919. Internet Engineering Task Force, Fremont, CA, 2007.

6. Montenegro, G., Kushalnagar, N., Hui, J. W., Culler, D. E. Transmission of IPv6 packets over IEEE 802.15.4 networks, RFC 4944. Internet Engineering Task Force, Fremont, CA, 2007.

7. Zach, S., Bormann, C. *6LoWPAN: The Wireless Embedded Internet.* John Wiley & Sons, Hoboken, NJ, 2009.

8. Kim, E., Kaspar, D., Chevrollier, N., Vasseur, J. P. Design and application spaces for 6LoWPANs, draft-ietf-6lowpan-usecases-03. Internet Engineering Task Force, Fremont, CA, July 2009.

9. Hui, J. W., Culler, D. E. Extending IP to low-power, wireless personal area networks, *IEEE Internet Computing* 2008, 12, 4, 37–45.

10. Kong, K.-S., Lee, W., Han, Y.-H., Shin, M.-K., You, H. R. Mobility management for all-IP mobile networks: Mobile IPv6 vs. proxy mobile IPv6. *IEEE Wireless Communications* 2008, 15, 36–45.

11. Gundavelli, S., Leung, K., Devarapalli, V., Chowdhury, K., Patil, B. Proxy Mobile IPv6, RFC 5213. Internet Engineering Task Force, Fremont, CA, 2008.

12. Kim, M. S., Lee, S. K. A novel load balancing scheme for PMIPv6 based networks. *International Journal of Electronics and Communications* 2009, 64, 6, 579–583.

13. Chalmers, R. C., Almeroth, K. C. A mobility gateway for small-device networks. In *Proceedings of Second IEEE Annual Conference on Pervasive Computing and Communications*, Washington, DC, June 2004.

14. Kim, J. H., Hong, C. S., Shon, T. A lightweight NEMO protocol to support 6LoWPAN. *ETRI Journal* 2008, 30, 685–695.

15. Yang, S. et al. Dual addressing scheme in IPv6 over IEEE802.15.4 wireless sensor network. *ETRI Journal* 2008, 30, 5, 674–684.

16. Hwang, S. M. et al. Multi-modal sensing smart spaces embedded with WSN based image camera. In *Proceedings of the 3rd International Conference on Pervasive Technologies Related to Assistive Environments*, Samos, Greece, 2010.

17. Bag, G., Hamid, M., Saif Shams, S. M., Kim, H. K., Yoo, S. W. Inter-PAN mobility support for 6LoWPAN. In *Proceedings of IEEE Computer Society Third International Conference on Convergence and Hybrid Technology*, Busan, South Korea, November 2008.

18. Johnson, D., Perkins, C. et al. Mobility support in IPv6, RFC 3775. Internet Engineering Task Force, Fremont, CA, 2004.

19. Devarapalli, V., Wakikawa, R., Petrescu, A., Thubert, P. Network mobility (NEMO) basic support protocol, RFC 3963. Internet Engineering Task Force, Fremont, CA, 2005.

20. Soliman, H., Castelluccia, C., Malki, K. El, Bellier, L. Hierarchical Mobile IPv6 Mobility Management (HMIPv6), RFC 4140. Internet Engineering Task Force, Fremont, CA, 2005.

21. Atzori, L., Iera, A., Morabito, G. The Internet of Things: A survey. *Computer Networks* 2010, 54, 15, 2787–2805.

22. Huang, Y., Hsieh, M. Y., Chao, H. C., Hung, S., Park, J. Pervasive, secure access to a hierarchical sensor-based healthcare monitoring architecture in wireless heterogeneous networks. *IEEE Journal on Selected Areas in Communications* 2009, 27, 4, 400–411.

23. Ma, H.-D. Internet of Things: Objectives and scientific challenges. *Journal of Computer Science and Technology* 2011, 26, 6, 919–924.

24. Heinzelman, W. R., Chandrakasan, A., Balakrishnan, H. Energy-efficient communication protocol for wireless micro sensor networks. In *IEEE Proceedings of International Conference on System Sciences*, Maui, HI, January 2000.

25. Pathan, A. S. K., Hong, C. S. SERP: Secure energy-efficient routing protocol for densely deployed wireless sensor networks. *Annals of Telecommunication* 2008, 63, 9–10, 529–541.

26. Razzaque, M. A., Hong, C. S. Analysis of energy-tax for multipath routing in wireless sensor networks. *Annals of Telecommunication* 2009, 65, 1–2, 117–127.

27. Dhanajay, S., Lee, H. J., Chung, W. Y. An energy consumption technique for global healthcare monitoring applications. In *Proceedings of International Conference on Information Sciences*, Seoul, Korea, November 24–26, 2009.

Index

Page numbers followed by f and t indicate figures and tables, respectively.